DESIGNING TECHNICAL REPORTS

SECOND EDITION

DESIGNING TECHNICAL REPORTS

Writing for Audiences in Organizations

J.C. Mathes   Dwight W. Stevenson

UNIVERSITY OF MICHIGAN

Macmillan Publishing Company
NEW YORK

Collier Macmillan Canada
TORONTO

Editor: Barbara A. Heinssen
Production Supervisor: Linda Greenberg
Production Manager: Sandra Moore
Text and Cover Designer: Natasha Sylvester
Cover Photograph: Jeff Zaruba, *The Stock Market*
Illustrations: Graphic Typesetting Service, Inc.
This book was set in Melior by Graphic Typesetting Service, Inc.,
and printed and bound by Halliday Lithograph.
The cover was printed by The Lehigh Press, Inc.

Acknowledgments appear on pages 493–495,
which serve as an extension of the copyright page.

Printed in the United States of America

Macmillan Publishing Company
866 Third Avenue, New York, New York 10022

Collier Macmillan Canada, Inc.
1200 Eglinton Avenue, E., Suite 200
Don Mills, Ontario, M2C 3NI

Library of Congress Cataloging-in-Publication Data
Mathes, J.C. (John C.), 1931–
Designing technical reports; writing for audiences in
organizations / J.C. Mathes and Dwight W. Stevenson, —2nd ed.
p. cm.
Includes index.
ISBN 0-02-377095-3 (ppk.)
1. Technical writing. I. Stevenson, Dwight W., 1933–
II. Title.
T11.M36 1991 90-23869
808'.0666—dc20 CIP

Printing: 1 2 3 4 5 6 7 Year: 1 2 3 4 5 6 7

As husbands, we dedicate this book to Rosemary and Mary Ann.

As fathers, we dedicate it to John, Ann, and Melissa.

As children, we dedicate it to John C. and Cletus Mathes and Dwight E. and DeLoris Stevenson.

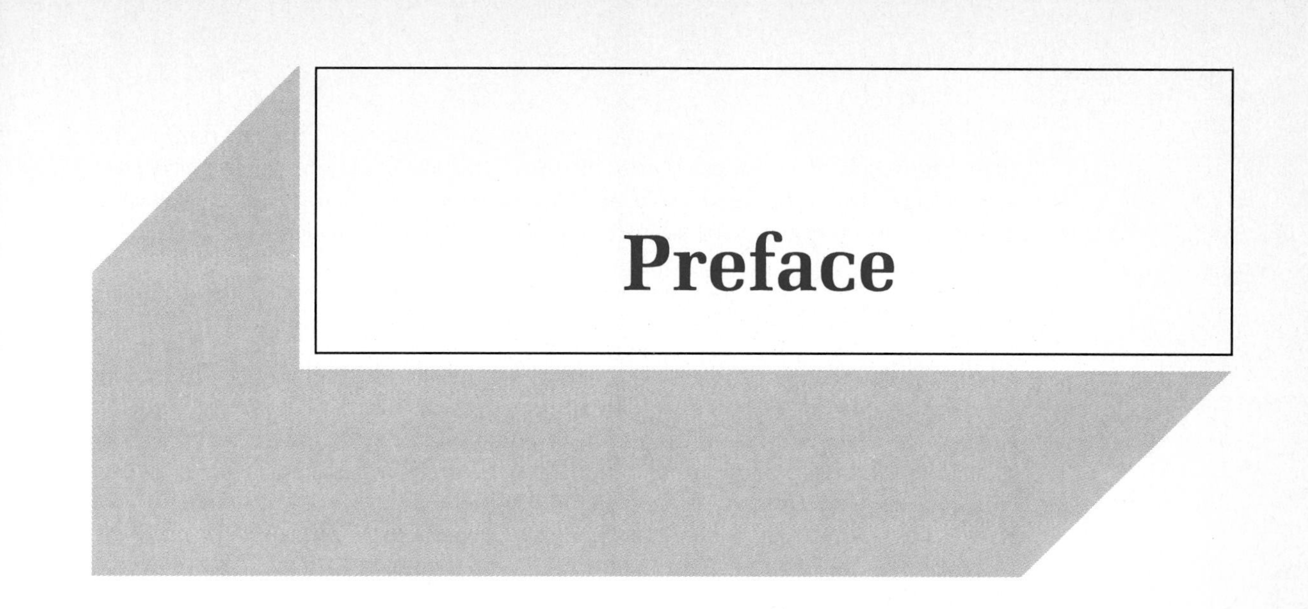

Preface

Designing Technical Reports, Second Edition, is intended for students headed for roles in organizations within industry, business, and government—and for professionals in those roles already. The purpose of the book is to train these students and professionals in the written communication required of practicing professionals and managers on the job.

Our Approach

The distinctive feature of our approach is that we present a systematic procedure for designing technical and professional reports. This procedure will enable students as well as professionals to analyze the audiences for reports, to state the purposes for reports, to select and arrange report materials, to design and integrate graphic and tabular materials, and to prepare and edit a report text. We treat writing as a design task, one which requires predesign analysis and decisions and then moves in a deliberate way through a logical sequence of design decisions. In a comprehensive sense, the purpose of this book is to improve report writers' mastery of the entire process of report writing from the prewriting stages through the final editing and text-production stages.

Features New to This Edition

For readers familiar with the first edition of this book, our basic process approach and our design philosophy in this second edition will be expected and—we hope—welcome. Since the first edition appeared, we have had many new opportunities to work with professionals in a wide variety of organizations, disciplines, and countries. We also have benefited from the efforts of numerous researchers, teachers, and students and from the work of colleagues and former graduate students. In the process, we learned that we needed to address several new topics in the second edition, and that we needed to substantially modify and supplement our treatments of some of the topics from the first edition.

Accordingly, there are three new and important features in this edition of the book.

1. Multidisciplinary approach First, we have shifted away from our steadfast focus on engineering. In the second edition, we address a wider range of subject matters and illustrate successful performance with varied examples drawn from business, banking, health care, social service, criminal justice, and insurance as well as engineering.

We have done this because we believe that the common denominator for professionals is not their technical specialization; rather, it is their responsibility to communicate to diverse audiences and for action-oriented purposes within corporate and administrative settings. It makes little difference that the specific subject matters of professionals differ. Professionals are specialists in different subject matters; they work in a variety of disciplines, each with its own conventions, and they write about different topics. But they are alike in that they all live within organizational contexts, they all must deal with diverse readers, and the major output of their professional activity is paper that is intended to produce results. Engineers, business people, health care professionals, social workers, judges, bankers, insurance agents, and those in the military at first glance are people with distinctively different needs. Yet beneath the surface-level differences among their writing, the realities of organizational behavior, of reading behavior, and of composing behavior give these professionals much in common. This book focuses on these common features.

To make these points of common need and experience evident, we have used numerous examples drawn from the real work of business, industry, and government; also, we have prepared an extended case study based upon the Three Mile Island nuclear power plant accident to provide a real-world context in which to explain the principles of organizational communication.

2. Emphasis on short, informal reports We have shifted the emphasis of this edition to the needs of writers of short, informal reports—the "bread and butter" of a professional's writing—deemphasizing the writing of long, formal reports that characterized the earlier edition of the book. In the first edition we included an appendix consisting mostly of formal engineering reports; in the second edition, we integrate numerous short, informal reports (drawn from a variety of disciplines) as examples throughout the text.

3. New topics We have expanded the book's coverage considerably, from ten chapters to sixteen. We have completely rewritten all of the chapters drawn from the first edition of the book; we have also introduced new chapters to treat topics that we ignored in the earlier edition. Among the new chapters: *"Writing and Editing Paragraphs"* (Chapter 9) applies research-based principles of technical writing and composition to paragraph-level structures; *"Group Writing"* (Chapter 13) emphasizes the importance of collaborative writing; *"Writing in the Multinational Context"* (Chapter 14) illustrates the cultural relativity of appropriate document design; *"Accepting Ethical and Legal Responsibilities"* (Chapter 15) includes advice on writing successfully in the demanding legal and ethical contexts within which professionals must work; *"Using Essential Visuals in Reports"* (Chapter 11) emphasizes a functional and process approach to the design and use of graphic elements in professional documents; *"Using Electronic Communication Tools"* (Chapter 16) shows how the computer is redefining communication processes and products.

We believe that we preserved the spirit of the earlier edition while revising and modifying the book in ways that will help students as they cope with the task of becoming professionals.

Acknowledgments

In closing, let us pay special thanks to these colleagues and former colleagues for all of their help: Professors Leslie Olsen, Tom Huckin, Jim Zappen, Tom Sawyer, David Kieras, Lisa Barton, Ben Barton, Maurita Holland, Peter Klaver, and Howard Klee. For teaching their professors more than their professors taught them, we would like to thank several of our former graduate students whose research has contributed to this revision and continues to contribute to the profession: Barbara Mirel, John Brockmann, Alice Morehead, Terry Skelton, and Lee Brasseur. We particularly want to thank our manuscript reviewers, whose many detailed comments and suggestions we tried to respond to: R. J. Adams, Purdue University; David H. Covington, North Carolina State University; Ruth E. Falor, The Ohio State University; Dean G. Hall, Kansas State University; Dixie Elise Hickman, University of Southern Mississippi; M. Jimmie Killingsworth, Memphis State University; Carolyn Miller, North Carolina State University; Sheryl Pearson, University of Michigan—Dearborn; Helen Quinn, University of Wisconsin—Stout; Thomas M. Rivers, University of Southern Indiana; Michael Rose, Lane Community College; Helen Schwartz, Oakland University; Jack Selzer, Pennsylvania State University; Barbara A. Smith, Alderson-Broaddus College; Susan Wells, Temple University; and James P. Zappen, Rensselaer Polytechnic Institute. Thanks also to our editor at Macmillan, Barbara Heinssen, and others of the Macmillan staff, especially Peter Knapp and Linda Greenberg. Finally, to the more than four hundred teachers of technical and business communication who participated in our conferences at The University of Michigan, we are indebted for your suggestions and help; to all of you, our thanks. To all of our students, our best wishes.

J. C. Mathes and Dwight W. Stevenson

Contents

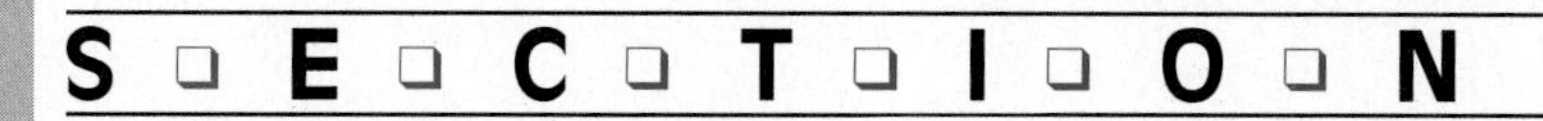

Determining the Communication Context

CHAPTER 1

Writing for Audiences in Organizations

This book focuses on a specific professional task: writing for audiences in organizations. For most professionals in organizations, writing is a primary activity of central importance in the success of those organizations and of their own careers within them. A social worker in a hospital, a business manager in an office, a chemist in a research laboratory, an engineer in a farm implement manufacturing company, a foreign trade expert in a government agency, and a parole officer in a court system all have numerous and very different professional activities to perform. Yet, most of these activities result in written communication which is sent to others in the organization and sometimes to readers in other organizations as well. For these professionals—and for you—writing is the output, the product by which professional work becomes available to others and can therefore be turned into decisions and actions. Writing is also a concrete thing by which all professionals become known within their organizations and beyond. Your name is on it and your voice is behind it. For these reasons, writing is an essential skill for all professionals.

Our purpose, therefore, is to explain how you as a professional can communicate effectively within your organization and with people in other organizations. We concentrate on the everyday writing by professionals: reports, memoranda, and letters that all of us must write regularly in order to do our jobs. This is writing by professionals in their organizational roles to others in their organizational roles. It is writing that has an organizational function—"instrumental writing," we call it. Writing to get something done. Your effectiveness as a professional is determined, in part, by how successfully your writing serves this function.

To design your written communication, you need to put yourself into an organizational context. Then you need to understand the organizational function of written communication and to appreciate the problem of written communication. In addition, because of the interrelationships among organizations and other social institutions, you need to understand the legal and social contexts of written communication.

In the following discussion, we explore all of these concepts. As you will see, all of these considerations will provide the foundation for the report design process which we explain later.

❑ 1.1 ❑ Writing in Organizations

In order to communicate effectively, first you must understand the organizational context of writing. This context has been analyzed and defined in organizational and management research literature, which provides a framework for us to introduce the functions of written communication and the constraints that any organization imposes on written communication. As a future professional in an organization, you need to know about these functions and constraints to communicate effectively.

In an organization, people and groups of people must coordinate their activities in order to achieve specific organizational goals.[1] A petrochemical corporation, for example, is an organization designed to produce gasoline and other petrochemical products. A hospital is an organization designed to heal patients. A public utilities commission is a state agency designed to regulate companies which produce electricity. A public interest group is an organization designed to influence other organizations, usually government organizations and public agencies. All of these organizations, although quite distinct in what they seek to accomplish, achieve their goals when a number of people with different responsibilities interact so that their collective activities produce the desired results.

On the job, then, you behave quite differently from the way you behaved in college. As a professional, your role consists of processing information which comes to you from the organization, of responding by making judgments and performing other professional activities, and of transmitting information into the organization so that organizational goals can be achieved. Your communication activities thus have a function. They exist to achieve organizational goals.

In college, however, in a sense you were the goal. The college as an organization is designed to educate you as a future professional. You, an educated graduate, are the product or output of the college. Your communication activities—your term papers and laboratory reports—had no organizational function other than to cause you to learn through the act of writing or to demonstrate that you had mastered some subject matter or skill. In other words, your writing was focused upon you—what *you* were learning and what *you* had achieved. It was not focused upon the activities of the organization itself. In instrumental terms, it had no "real" communication purpose. No one did anything with your writing except to evaluate you. Good or bad, your writing had no effect on the college as an organization. Indeed, the organization continued to function unchanged no matter what you wrote or how well or badly you did it. (Of course, if enough students' writing were either hugely successful or hugely unsuccessful, a college might be forced to change its procedures, or at least it would have to adapt to changing enrollments.) The point is that, as a student, you didn't write to coordinate organizational activities; you wrote to learn and to impress a professor with what you knew and thought.

In organizations, reports are written by professionals with specific responsibilities and roles. In an organization, then, you interact and communicate in terms of your role, not as an independent individual.[2] Your education has prepared you to apply analytical skills and formulate judgments that acquire meaning only when they are transformed into information for the organization. The organization itself responds only to this output. Your output is what causes machinery to be tooled, health care to be delivered, loans to be granted, sales to increase, and people to live better. You cause things to happen. And you communicate in order to do so.

Communication in organizations, therefore, coordinates the activities of various persons to achieve organizational goals. To a significant extent, this depends on written communication as we discuss it in this book. The concept of planned coordination, therefore, underlies the fundamental principles of report design.

As you can see, writing in organizations is an important professional activity, and it is very important for you to know specifically how it functions.

❑ 1.2 ❑
The Function of Written Communication

In one way or another, all written communication in organizations is designed to fulfill some organizational purpose and to contribute to the achievement of some organizational goal. The organizational activities that written communication supports are numerous and diverse, varying according to the needs of different organizational units and departments. To analyze the specific functions of written communication, therefore, the activities of these units need to be examined. You can design a report in many ways; the way you choose to design it will depend on your role within your department as well as on the roles of the other persons who will receive your report. Organizational activities have been classified in terms of five basic types of systems or units:[3]

1. Production units concerned with the work that gets done.
2. Supportive units concerned with procurement, disposal, and institutional relations.
3. Maintenance units for tying people into their functional roles.
4. Adaptive units concerned with organizational change.
5. Managerial units for directing, adjudicating, and controlling the many units and activities of the organization.

In other words, no matter what the specific nature and purpose of a particular organization—a hospital, a bank, or a steel company—the organization's activities can be described as related to one of these five types, and the organization itself is structured as some combination of these five types of units.

If you are in an industrial research and development department, you transform input from other units, such as marketing, to produce output for other units, such as engineering. From the communication perspective, your output consists of reports that provide input to persons in other units.

In a school system, the production unit consists of the teaching activities in the classroom, which "transform" the "input" of students to the "output" of graduates. In an automobile company, supportive units include purchasing agents who deal with vendors in other organizations to obtain parts. A maintenance unit includes the personnel director who hires the professionals who go to work in all departments of the corporation. Various adaptive units help the corporation cope with anticipated changes in the marketplace: the market research, the research and development, and the long-range planning groups, which provide input and output to each other as well as to management. The activities of all of these various units have to be planned, coordinated, and managed. In a university some administrator must decide how many students to admit to each college and school, and another administrator must allocate the budgets to those units.

When it comes to writing reports, therefore, you should not think of routine reports for the organization as a whole. Rather, you must think of reports according to how they function within and across these units. The production department in a shipyard, for example, requires many different types of reports in order for ships to be designed and constructed: engineering change requests, problem identification reports, requests for engineering services, production

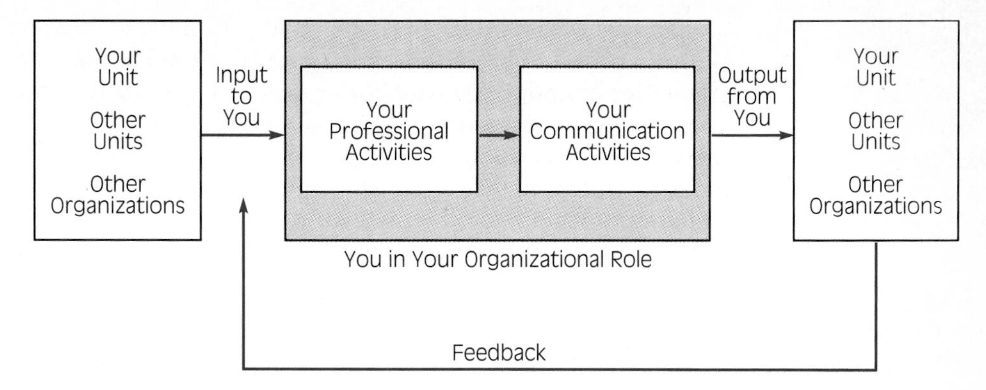

FIGURE 1.1 A Professional in an Organization Transforms Input to Output.

change notices, and production status reports. Ship construction depends on various support units and their reports: letters to vendors of parts, memoranda to the owners of the ships being constructed and their naval architects, reports to the Navy, and letters to regulatory agencies such as the American Bureau of Shipping.[4] Adaptive units depend on recommendations to management and directives from management. Maintenance units in all organizations depend on the ubiquitous progress report, and managerial units depend on standard operating procedures.

As a professional, you need to view your communication activities as your own "products" or "output" (Figure 1.1). That is, in your organizational role you receive input, usually as some form of communication, perform some professional tasks, and then communicate to provide the output as a result of those tasks. Your professional and communication activities essentially "transform" input to output for your unit in the organization.

Your communication activities, furthermore, are quite complex. Within your department you must interact and communicate with others in order to do your job. You interact intensively with others in your group, including your supervisor. Also, you interact with other groups in your department as well as with other individual group members and supervisors. You occasionally interact directly with your manager. These interactions, of course, are often more than two-way interactions; you often interact with many persons, using the same communication. Within your department, these interactions usually take the form of oral and interpersonal communication rather than written communication. When written, it is often in the form of brief memos and handwritten notes.

Occasionally, however, even within the department, you must prepare written reports. For example, you might write up minutes of a meeting or a trip so that others in your group as well as those in other groups get necessary information. Or, the maintenance function within the department might require you to write progress reports for your supervisor and group leader or manager.

Most of the written reports you receive and write, however, cross departmental and unit boundaries. Every organization consists of departments and other types of units whose activities need to be coordinated to achieve certain tasks or goals. A change in government regulations can require research activities that will lead to design alterations requiring modification of production processes and new marketing strategies. The same change also might require operating procedures to be revised. If you are in a design group, you have to interact with manufacturing departments in order to achieve production goals. In a hospital, if you are in a test laboratory, you have to interact with the medical care units in order for them to serve patients.

Effective interaction among departments usually requires effective written communication; oral communication by itself is insufficient. Departments are separated from each other geographically and, more importantly, they are separated by organizational functions. Written communication is an important means by which their activities are integrated in order to achieve primary organizational goals such as producing a product, training people, or administering the law.

Because you will have diverse written communication tasks as a professional, we stress the function of the reports you write rather than the type of report you write. You design your reports in terms of function rather than in terms of content. Your challenge is to determine the specific function of a report you are writing, and then to design it accordingly.

The functional approach to report writing will enable you to write reports in different communication situations in most kinds of organizations. Engineers, for example, don't all work for manufacturing industries. They work for city governments, hospitals, accounting firms, banks, and even companies which design games. To prepare an engineer for a career, then, it is far more useful to approach written communication in terms of function than in terms of a type of report, such as a test report. Using this functional approach to written communication, we have taught scientists, engineers, environmentalists, agronomists, and economists how to prepare effective reports. We have taught firemen, policemen, and parole officers. We have even taught judges how to write their legal opinions. The common denominator is that all of these people are professionals in organizations. They don't write about the same things. They don't even write the same types of reports. But they all write in environments which are organizationally and functionally similar.

❑ 1.3 ❑ The Problem of Written Communication

The problem of written communication is that in order for you to accomplish your communication tasks you often must interrupt the ongoing processes of the organization. An organization is not a factory in which people function on an "assembly line" of decision and action. They cannot be automated so that they merely process information though formally established lines of communication. Instead, as a professional you must interrupt their routines and the routine of the organization in order to make things happen.

This problem of written communication is introduced by Herbert Simon in his book *Administrative Behavior*. He says:

> There is nothing to guarantee that advice produced at one point in an organization will have any effect at another point in the organization unless the lines of communication are adequate to its transmission, and unless it is transmitted in such form as to be persuasive.[5]

If you think about it, you can see that Simon has defined the dilemma which you will face with almost every significant piece of writing you will produce within an organization. That is, you will ordinarily present information and advice with much of what you write. (Conclusions and recommendations are normal parts of most reports and memoranda.) You will present this information and advice for yourself and for your department to persons in other departments, frequently in other units of the organization. However, as Simon indicates, there is nothing to guarantee your success because two factors are necessary for your advice to have any effect on the functioning of the organization: one, the lines of transmission must be adequate to carry the advice to the people who need it; two, the communication itself must be persuasive.

Your first problem is to establish adequate lines of transmission for your report. In terms of formal lines of authority, however, the established organizational lines of transmission are often quite inappropriate and inadequate. The formal lines of communication in your organization represent links of authority to support ongoing organizational processes. These links are outlined on the organization chart, where each division is broken down into units and departments and in turn combined with other divisions in a hierarchical structure with the Chief Executive Officer (CEO) at the top. Most of the reports you write, however, don't follow these formal lines.

Your report does not follow formally established lines of communication because it has a unique purpose. It involves specific departments and specific people, many of whom are in a horizontal or equal relationship with each other on the formal organization chart. This requires its lines of communication to cut across the formal structure of the organization, which are vertical on the organization chart to represent the chain of command. Only the most routine status and administrative reports might have lines of communication generally corresponding to lines on the organization chart. But these types of reports are seldom of great concern to you as a professional. It is all the rest of your writing that will really affect both the organization and you.

In a test facility of one automotive corporation where we have analyzed the report writing situations, a typical report addresses various engineering design departments—engineers are designing automobiles, and the test facility tests the designs. The working relationships therefore are horizontal—between design engineers and test engineers, including their supervisors and managers, in two different departments of the production subsystem in two separate locations 70 miles apart. Any specific test report has additional audiences—in service, quality control, manufacturing, for example—depending on its subject. A report on standardizing "front end sheet metal" involves engineers in manufacturing processes, quality control, and production in addition to those in engineering. Thus, the lines of communication appropriate for these test reports go in several directions, most of them horizontal.

By protocol, however, the test reports follow the vertical lines of the formal organizational structure rather than the appropriate lines of communication. A test report goes from an engineer to his or her supervisor, then to his or her manager, then to the chief engineer of testing and development, then to the director—all persons within the test facility—then to the director of truck engineering, the chief engineer, a manager, a supervisor, and finally a design engineer—all persons within engineering—as well as on to several managers, supervisors, and engineers in manufacturing processes, quality control, and production. The result could be that the "advice" produced in the test facility might not reach all persons who need it.

The formal lines of authority are inconsistent with the appropriate lines of communication. You have to make sure that you establish adequate lines of communication for your test report. You have to make sure that it gets to the right people so that the appropriate decisions are made and actions taken. You can't rely on the formal lines of authority and communication because they don't guarantee that your "advice" will have its intended effect, as Simon points out.

Your second problem, as Simon observes, is to make your report persuasive. Routine transfer of information usually is inappropriate and is almost impossible for most report writing situations. Until actual organizational processes change, no communication has occurred. Routine transfer of information does not always lead to action or to the appropriate action, especially when the information is transmitted between departments in different units of the organization. Even between departments which are closely related, routine transfer of information does not always lead to action. Each department, such as production engineering or quality control, has its own goals and, therefore, priorities. These serve to filter out information that isn't persuasive.

The mortality rate for information that isn't persuasive, therefore, is exceedingly high. If you don't get someone's attention and persuade them to make a decision or to act, you might as well not have written your report. You've wasted your time and theirs. To effect change usually requires persuasiveness. You have to write your test report so that the design engineer is convinced that a design change to standardize front end sheet metal is necessary and so that management agrees to change production processes in two different plants. As a professional, you are responsible for ensuring that the appropriate decisions are made and that the appropriate actions occur. You have to solve the problem of written communication.

❑ 1.4 ❑
The Legal and Social Contexts of Written Communication

So far we have been discussing the function of written communication in its organizational context. We have assumed that your organization interacts with other organizations, such as with vendors who supply input and customers who purchase output, and that these organizations have complementary goals. Quality control in your organization meets the specifications of the customers in other organizations. However, your reports, memoranda, and letters exist in social contexts as well, especially a legal context. In these wider contexts, organizations do not necessarily have mutually complementary goals.

When you write reports to serve their organizational functions, you shouldn't overlook the potential legal implications of those reports. These involve, for example, product liability concerns. That test report you write to design engineering can become an important document in a lawsuit against your corporation, so you should write the test report with that additional context in mind. For example, you might test a truck jack and conclude that the user could have difficulty in getting the jack to cradle the axle. Your test report should be very specific on the problem and be clear and persuasive to design engineers as well as to a purchasing agent and the vendor of the jack. If that jack is adopted as standard equipment for the truck model, the instructions for use of the jack have to be very explicit on how to get the jack to cradle the axle. If you conclude that the jack design needs to be modified, then your report has to be explicit and persuasive so that it actually is modified.

In product liability cases, all "documents and writings" can be used as evidence in court. In most cases, these include all of the following:

correspondence	minutes	reports
notes	schedules	analyses
drawings	pictures	tables
graphs	charts	maps
surveys	invoices	receipts
purchase orders	pleadings	questionnaires
contracts	bills	checks
drafts	diaries	logs
proposals	recordings	telegrams
films	videotapes	computer disks

The term "documents and writings" can also be understood to include any preliminary drafts or notes for all of the preceding in whatever form, including:

printed	typed	longhand
shorthand	paper	paper tape
ribbon	blueprints	tabulating cards
magnetic	tape	microfilm film
videotape	phono record	movie film

Because of this legal context, you must be precise with all of your written communication, including working memos and drafts as well as final reports.[6] You should not treat any communication as incidental or as just passing some information along because it might be useful.

You must also think of the legal significance of your writing at all stages of a product cycle. That is, all of your documents must demonstrate that the product underwent well-considered design, testing, and manufacture and that the highest consideration was given to the safety of users and others. These documents therefore include the numerous working memoranda and reports related to design, testing, and manufacture as well as the instructions for product use, advertising and brochure materials, and warnings of dangers inherent or designed in a product. All of your written communication, not just warnings and instructions for the public, must be precise, accurate, and responsible in terms of the legal implications.

You and your organization must ensure careful record keeping as well. Your documents have an indefinite life span. They must be preserved as long as the products to which they refer are still in use. Even what you consider the most incidental and minor piece of paper must be carefully prepared with this in mind. In this respect, some good advice is contained in the following statement from an industrial journal: "the design engineer should discipline himself to organize his files so that nothing of his early thinking on a design project, or constructive comments of others relative to it is ever discarded. If it is worth sketching or discussing, it is worth filing."[7] In the same light, your company should develop a document retention policy and keep every item of communication necessary to document the entire life cycle of a product or service from initial concept to use or implementation. And all of these must be carefully and well written.

The legal and social context has other implications. Legal actions can be based on personnel issues, environmental issues, and intellectual property issues as well as product liability issues. Environmental concerns such as toxic waste disposal of any byproducts of a manufacturing process have an ethical dimension as well as a legal dimension. Even where legal concerns seem not to be an important consideration, the social context introduces your ethical concerns as a professional. Your reports are your imprint on the organization and therefore on society, indirectly and directly. When your written documents are preserved long after you have been promoted, changed jobs, or even retired, they are your legacy as a professional. The papers and reports you write in college self-destruct when you graduate, but the memos and reports you write as a professional will endure and must therefore serve their function whenever they are retrieved.

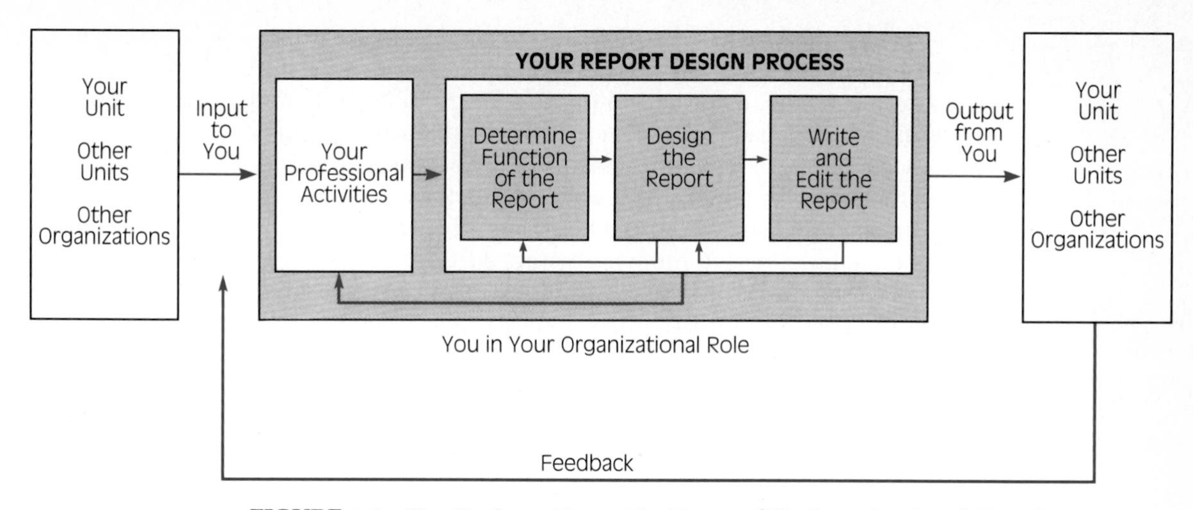

FIGURE 1.2 You Design a Report in Terms of Its Organizational Function.

❑ 1.5 ❑ The Report Design Process

The functional approach to written communication in organizations leads to the report design process that we present in this book, and it causes us to introduce many considerations which you might otherwise ignore when writing a memorandum, letter, or report.

You prepare your reports in the context of your role in the organization (Figure 1.2.). Input from the organization has stimulated your professional activities and results in output in the form of written communication that causes changes to occur in your organization. Your first consideration is to determine the function of your report. What do you want it to do? This consideration then leads to the actual process of design, writing, and editing a report. Accordingly, in this first section of the book we discuss the organizational context and the writing skills that are required for you to cope with it effectively. In this chapter we have introduced the function of written communication in organizations. Our purpose has been to make you aware of its importance and to provide a conceptual framework for report writing. In the next two chapters, we analyze this context in specific detail and explain how you analyze the needs of your readers and define your communication purpose. These are basic prewriting skills which you need to master before you outline and write a report. They enable you to design a report in terms of its function and the organizational context. That is, until you determine what you want your report to do, how do you know what to put in the report?

In the second section of the book, we explain how to design your report. That is, we explain how to establish the basic architecture of your report and how to arrange its various sections. Once you have determined the function of your report, you can outline its various parts. After you establish the basic structure, you can move on to outline the various sections of your report to determine what information to include and how to arrange it. You can determine whether each section is primarily persuasive or informative and outline it accordingly. During this design stage of report writing, you aren't putting many words on paper—you aren't doing the actual writing. That comes next.

In the third section of the book, we address the specific skills required to produce an effective report: writing, editing, use of visuals, and formatting. The paragraph is the building block of all written communication, so you must master the skills required for effective paragraphing. The sentences then provide the substance of your reports, and you must master the grammatical and stylistic skills required to write, rewrite, and edit sentences to be correct and effective. In addition, you need to know how to design and use visual aids because technical and professional documents require the use of visuals for effective communication more than do other types of written communication. You also need to know how to format your documents effectively because technical and professional documents depend on formatting more than other types of written communication usually do.

In the final section of the book, we introduce additional contextual issues. These involve considerations and skills that you will often need to call upon. Perhaps the most fundamental of these additional issues concerns group writing. Professionals write as members of groups as well as individually. Although in the first three sections of this book we discuss report writing as if you were doing all of the writing and editing yourself, the fact is that you will do these things with other people as well as alone. As a consequence, you need to supplement your individual writing skills with group-writing skills if you are going to produce coherent reports. Furthermore, at times you might have to write in a multinational context. That is, you will have to address people in organizations in other countries. The multinational context introduces considerations not present when you write within your own organization or to another organization or agency in the United States. For example, certain types of information you might typically include in a report should be omitted when you write to professionals in some other countries. You will be surprised at how a sales letter to a customer in France, Mexico, or Japan will differ from one to a customer in the United States. Also, you need to be aware of the specific written communication considerations required by the social and legal contexts. You need to know, for example, how to write adequate warnings and instructions. Typical report writing techniques often are inappropriate or insufficient in such communication situations.

We conclude with a brief discussion of electronic communication. This mode of communication (and tool for writing) has rapidly become a fundamental component of the organizational context. It is already affecting most aspects of organizational design and behavior. However, we believe that the electronic context will further develop to affect all aspects of report design and writing as well as to introduce additional forms of communication. In most of this book, we discuss report writing as if the only primary electronic consideration were word processing, a convenient extension of traditional written communication practices. In the future, however, we believe that the electronic environment will transform the very communication procedures and types of communication which professionals produce. In other words, we believe that writing is a rapidly changing technology, not a static thing. Accordingly, we believe that successful professionals must be able to adapt their written communication practices to changing and unknown situations. It is not enough simply to learn that there are x number of types of reports or *y* number of conventional communication procedures. Rather, you must be able to deal with the realities of change, of diversity, and of organizational and technological evolution. When you adopt a functional approach to writing for audiences in organizations, we are confident that you will have these abilities.

NOTES

1. Edgar H. Schein, *Organizational Psychology*, 3rd ed. (Englewood Cliffs, NJ: Prentice-Hall, 1980) 15.
2. Daniel Katz and Robert L. Kahn, *The Social Psychology of Organizations*, 2nd ed. (New York: Wiley, 1978) 45.
3. Katz and Kahn, *Social Psychology of Organizations* 52. They call these units "basic subsystems" of an organizational system.
4. J. C. Mathes and Dwight W. Stevenson, *Writing Shipyard Reports* (Ann Arbor: University of Michigan Transportation Research Institute, Marine Systems Division, Michigan, 1988).
5. Herbert A. Simon, *Administrative Behavior*, 3rd ed. (New York: Free Press, 1976) 14–15.
6. John E. Durst, Jr., "Use of Interrogatories in Products Liability Cases in New York," *Trial Lawyer's Quarterly* 15:2 (1983) 71.
7. R. Weber, "Product Liability: Some Ounces of Prevention," *Automotive Engineering* 92:2 (Sept. 1984) 54–55.

CHAPTER 2

Addressing the Needs of Your Readers

When you think about the implications of the organizational context for communication, which we discussed in the previous chapter, you can see that there are important implications for you as a writer.

First, most of your reports will follow unique paths—not predetermined ones—throughout the organization, and even beyond.

Second, because reports function as devices for coordinating among five types of organizational units and functions within organizations (production, support, maintenance, adaptation, and management), they inevitably communicate across departmental boundaries and address numerous readers. Your documents support activities in various departments and units of the organization. Therefore, almost any report that goes beyond the doors of your own department—and most do—will go to several other departments and units in your organization, and it will often address several individuals within those units to which it is distributed.

Third, individual readers within different departments have different roles rather than similar roles. This means that your readers are diverse as well as numerous.

Finally, your readers need information and advice from your report in order to do their jobs. The purpose of your report is to cause or enable others to make changes in the organization. To accomplish your purpose effectively, therefore, you must address the needs of individual people, not of faceless blocks on an organization chart.

Our input-output communication model can be modified to clarify this process. The organizational communication process consists of three stages whereby you interact with other persons in the organization (Figure 2.1). Some organizational problem or need causes persons in their organizational roles to provide input to you in your professional role. You undertake professional activities to analyze the situation to determine what organizational actions are required to solve the problem or meet the need. Then you write a report and send it to the various persons in the organization so that those actions will be taken. If you get your report to the right persons, and if those persons read and understand your report and then respond to the report with the required actions, the problem will be solved or the need met.

As you can see, the reports and memos which you write in an organization differ from other types of writing with which you are familiar in that they have different types of audiences. Many other types of writing assume a "homogeneous audience" rather than readers with different needs and uses for the information. For example, a scientific article in either a scientific journal or in a popular magazine will have readers with a common interest in the topic. In a scientific journal, such as *Science*, an article on the San Andreas Fault would be written for scientists. In the *National Geographic*, an article on the San Andreas Fault would be written so that an educated public could understand it. Although written at quite different levels, each article would have a homogeneous audience. The first article would assume a fairly high level of scientific knowledge on the part of all readers; the second would assume a moderately low level.

If we contrast the situation with that of a report on the San Andreas Fault written for the Los Angeles County government, however, the audience situation would be quite different. A report would have diverse readers in various agencies concerned with safety, public health, emergency planning, transportation, and other municipal functions. Different readers would have different needs from such a report as well as different levels of interest in the topic. They would also have significantly differing levels of knowledge about differing aspects of the report's content. In other words, such a report would address a "heterogeneous," or diverse, audience, not a "homogeneous" one. That, precisely, is a key factor of report audiences which you must understand and deal with.

But there is another aspect of difference to consider as well as diversity. There is also purpose to think about. A technical report differs in purpose from many other types of writing. The articles in *Science* and *National Geographic* are primarily informational. The writers don't expect the readers to do something after reading. Conversely, the report to the Los Angeles County government differs in kind because its purpose is to get decisions made and actions taken. The report contains information on the San Andreas Fault that persons need in order to evaluate current emergency evacuation plans, plan public health measures, prepare instructions for controlling traffic, modify

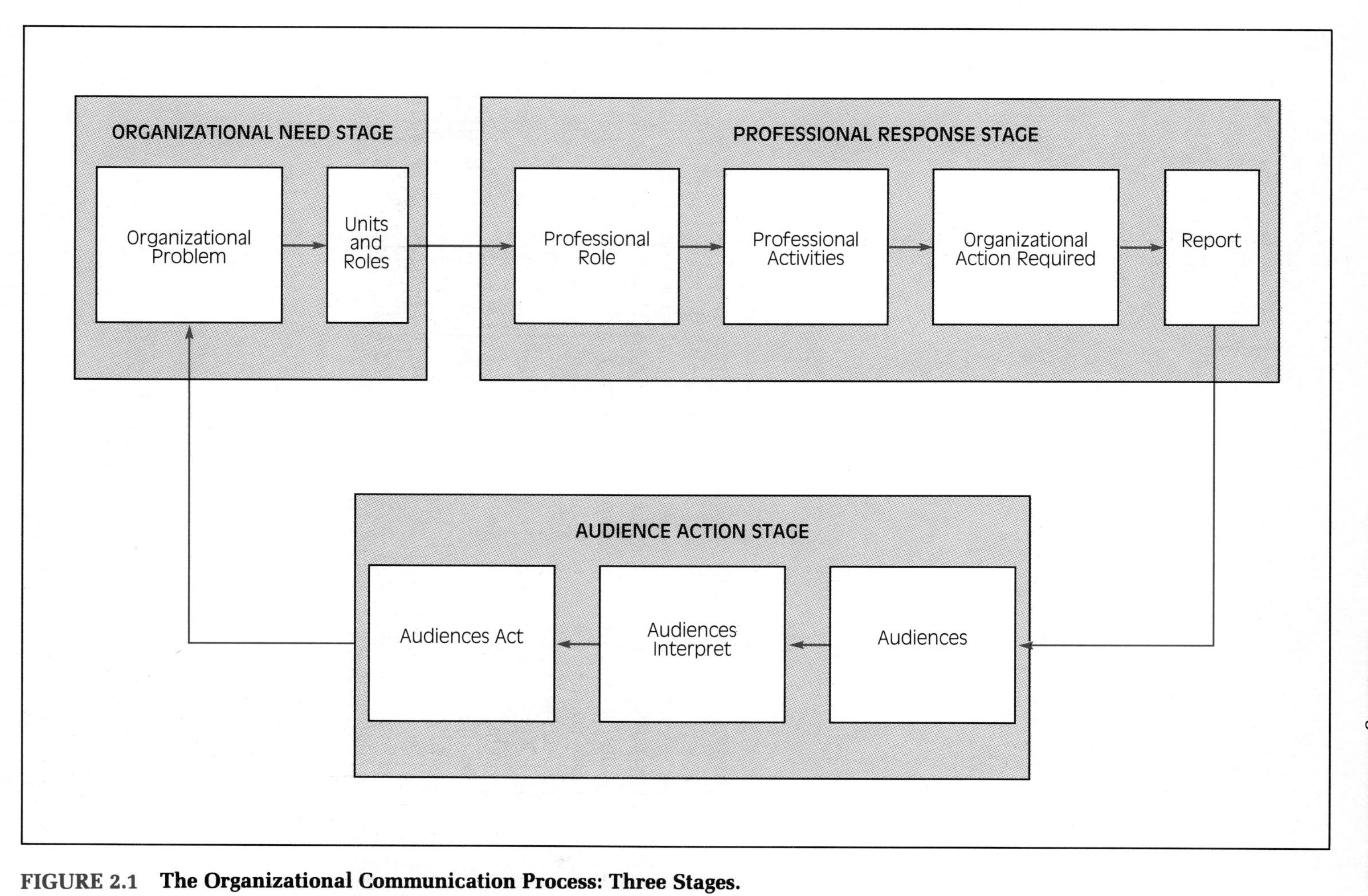

FIGURE 2.1 The Organizational Communication Process: Three Stages.

building codes, and make a host of other preparations for disaster. The writer of the report would expect the information to be used by the readers to whom he or she sends it, not to be read merely for its intrinsic interest. The report is a tool to produce something. Its value lies in its utility. A journal article, however, has intrinsic value quite apart from the actions and decisions which it might produce.

Designing a report to address readers' needs is not easy to do. To do so successfully, first you need to understand the problems that readers in organizations pose for a writer. Then, to cope with these problems, you need to appreciate how readers in organizations actually behave. Finally, you need to analyze your readers with a systematic audience analysis procedure. When you have performed this audience analysis, you are in a position to design the report to meet audience needs.

❑ 2.1 ❑ The Problems Readers Pose: A Specific Example

The problems readers pose were introduced by Herbert Simon. If you recall, he said you first must establish adequate lines of transmission for your report. This means identifying the persons in the organization who should receive your report and getting the report to them. You must then design the report so that when your readers receive the report they act in ways you intend if your report is to accomplish its purpose. These problems are illustrated by a communication process within a company called Babcock and Wilcox, a company connected with the 1979 accident at the Three Mile Island nuclear energy plant (Figure 2.2).[1]

Essentially, the accident at Three Mile Island was the result of a management communication failure within Babcock and Wilcox (B&W), the company that had designed the two nuclear reactors at Three Mile Island (TMI-1 and TMI-2).

The accident at TMI occurred on 28 March 1979, but the communication process began on 14 September 1977 at the Davis-Besse nuclear plant near Toledo, Ohio, also designed by B&W. At Davis-Besse an event occurred which was similar in many respects to the accident at Three Mile Island (TMI-2).[2] A valve in the core coolant water system stuck open, causing the pressure to drop and the water in the core to boil and escape through the open valve. Because of the continuous loss of coolant water, the emergency high pressure injection coolant water system automatically became activated. However, the nuclear plant operators at Toledo Davis-Besse misinterpreted a pressure gauge in the coolant system piping and prematurely shut off the emergency coolant water flow, even though the valve was still open and releasing coolant water. The operators again made a similar mistake on 23 October 1977.

Because the operators twice had mistakenly turned off the emergency core coolant water system, an engineer from B&W, Mr. J. J. Kelly, investigated the

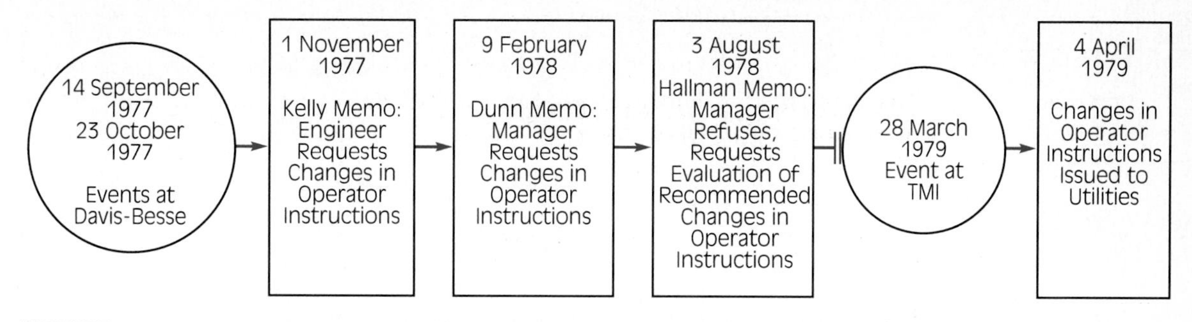

FIGURE 2.2 The Inefficient Management Communication Process in Regard to Changes in Operator Instructions.

situation. He decided that his company should issue changes in the instructions B&W had issued to the operators in the control rooms at the nuclear plants designed by B&W for public utilities. On 1 November 1977, as a result of this analysis, he wrote a memorandum in which he recommended guidelines for nuclear plant operators on when the high pressure injection system could be safely shut down. His memorandum received very little response and no action on the requested changes. Mr. Kelly therefore met with Mr. B. M. Dunn, the manager in charge of the Emergency Core Coolant System Analysis department at B&W.

On 9 February 1978, Mr. Dunn in turn wrote a memorandum (Figure 2.3) in which he recommended implementation of essentially the same guidelines for operator instructions that Mr. Kelly had recommended. The Dunn memorandum stated the problem rather explicitly: "I believe it fortunate that Toledo was at an extremely low power and extremely low burnup. Had this event occurred in a reactor at full power with other than insignificant burnup it is quite possible, perhaps probable, that core uncovery and possible fuel damage would have resulted." He recommended that "operating procedures be written to allow for termination of high pressure injection" under only two conditions. He concluded his memorandum with the statement, "I believe this is a very serious matter and deserves our prompt attention and correction."

About six months later, on 3 August 1978, Dr. B. F. Hallman, the manager of Plant Performance Services, responded to the Dunn memorandum. He wrote a memorandum (Figure 2.4) stating that he had not yet issued the guidelines recommended by Mr. Dunn because he had certain concerns about possible negative effects of the changes on some of the piping in the core coolant system. In his memorandum he asked Mr. B. A. Karrasch, the Manager of Plant Integration at B&W, "to resolve the issue" of how the high pressure injection system should be used.

He received no response. No decisions were made on the Dunn recommendations until the accident at Three Mile Island on 28 March 1979, when the valve in the core coolant water system stuck open and coolant water started to escape. Because of the continuous loss of coolant water, the emergency high pressure coolant water system automatically became activated.

THE BABCOCK & WILCOX COMPANY
POWER GENERATION GROUP

To
Jim Taylor, Manager, Licensing

From
Bert M. Dunn, Manager, ECCS Analysis(2138)

Cust.

File No. or Ref.

Subj. Operator Interruption of High Pressure Injection

Date February 9, 1978

This memo addresses a serious concern within ECCS Analysis about the potential for operator action to terminate high pressure injection following the initial stage of a LOCA. Successful ECCS operation during small breaks depends on the accumulated reactor coolant system inventory as well as the ECCS injection rate. As such, it is mandatory that full injection flow be maintained from the point of emergency safety features actuation system (ESFAS) actuation until the high pressure injection rate can fully compensate for the reactor heat load. As the injection rate depends on the reactor coolant system pressure, the time at which a compensating match-up occurs is variable and cannot be specified as a fixed number. It is quite possible, for example, that the high pressure injection may successfully match up with all heat sources at time t and that due to system pressurization be inadequate at some later time t_2.

The direct concern here rose out of the recent incident at Toledo. During the accident the operator terminated high pressure injection due to an apparent system recovery indicated by high level within the pressurizer. This action would have been acceptable only after the primary system had been in a subcooled state. Analysis of the data from the transient currently indicates that the system was in a two-phase state and as such did not contain sufficient capacity to allow high pressure injection termination. This became evident at some 20 to 30 minutes following termination of injection when the pressurizer level again collapsed and injection had to be reinitiated. During the 20 to 30 minutes of noninjection flow they were continuously losing important fluid inventory even though the pressurizer indicated high level. I believe it fortunate that Toledo was at an extremely low power and extremely low burnup. Had this event occurred in a reactor at full power with other than insignificant burnup it is quite possible, perhaps probable, that core uncovery and possible fuel damage would have resulted.

The incident points out that we have not supplied sufficient information to reactor operators in the area of recovery from LOCA. The following rule is based on an attempt to allow termination of high pressure injection only at a time when the

FIGURE 2.3 The First Dunn Memorandum (2 pages).

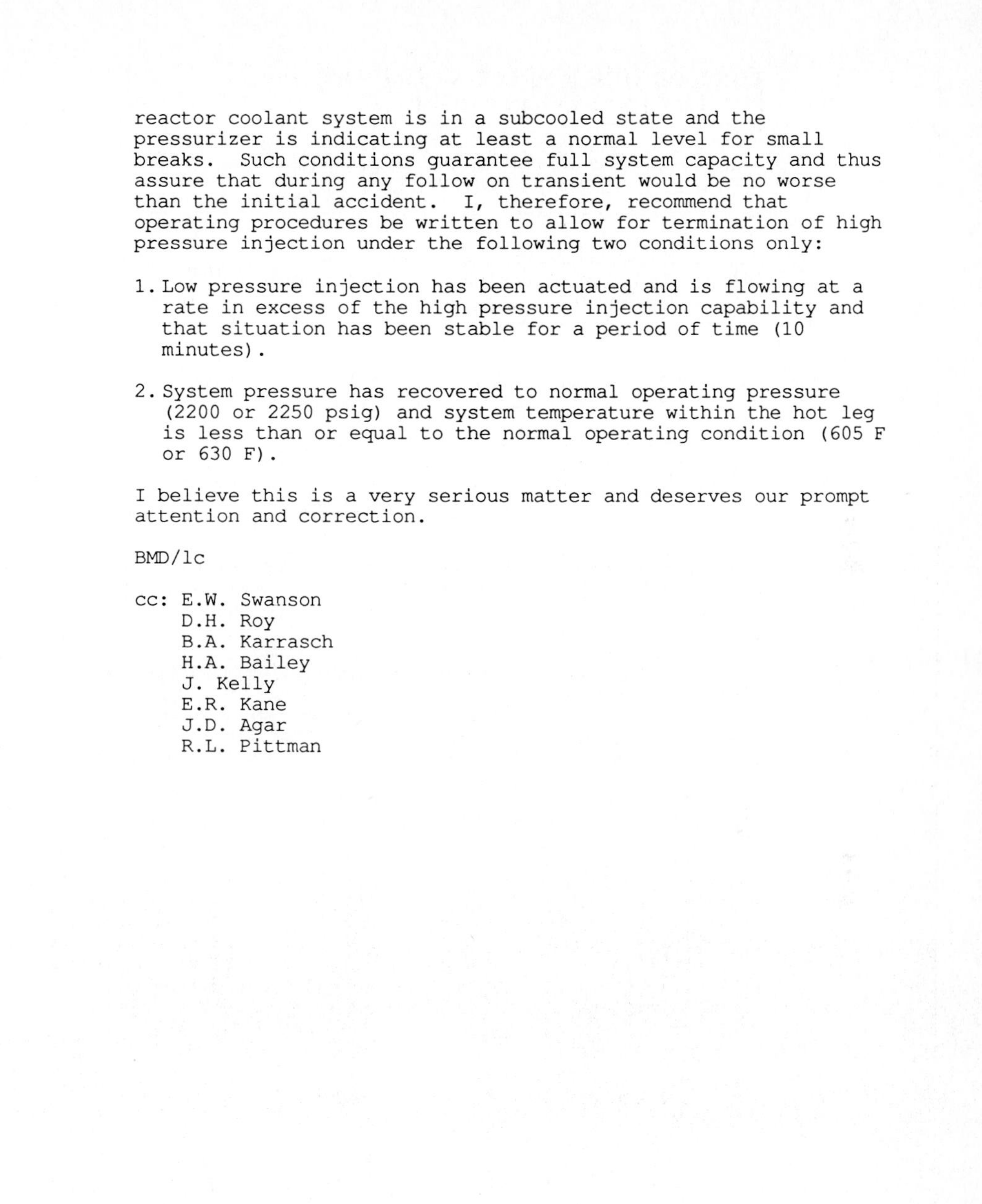

reactor coolant system is in a subcooled state and the pressurizer is indicating at least a normal level for small breaks. Such conditions guarantee full system capacity and thus assure that during any follow on transient would be no worse than the initial accident. I, therefore, recommend that operating procedures be written to allow for termination of high pressure injection under the following two conditions only:

1. Low pressure injection has been actuated and is flowing at a rate in excess of the high pressure injection capability and that situation has been stable for a period of time (10 minutes).

2. System pressure has recovered to normal operating pressure (2200 or 2250 psig) and system temperature within the hot leg is less than or equal to the normal operating condition (605 F or 630 F).

I believe this is a very serious matter and deserves our prompt attention and correction.

BMD/lc

cc: E.W. Swanson
D.H. Roy
B.A. Karrasch
H.A. Bailey
J. Kelly
E.R. Kane
J.D. Agar
R.L. Pittman

THE BABCOCK & WILCOX COMPANY
POWER GENERATION GROUP

To	B.A. Karrasch, Manager, Plant Integration		
From	D.F. Hallman, Manager, Plant Performance Services Section (1149)		
Cust.		File No. or Ref.	
Subj.	Operator Interruption of High Pressure Injection (HPI)	Date	August 3, 1978

References: (1) B.M. Dunn to J. Taylor, same subject, Feb.9, 1978
(2) B.M. Dunn to J. Taylor, same subject, Feb.10,1978

References 1 and 2 (attached) recommend a change in B&W's philosophy for HPI system use during low-pressure transients. Basically, they recommend leaving the HPI pumps on, once HPI has been initiated, until it can be determined that the hot leg temperature is more than 50 F below Tsat for the RCS pressure.

Nuclear Service believes this mode can cause the RCS (including the pressurizer) to go solid. The pressurizer reliefs will lift, with a water surge through the discharge piping into the quench tank.

We believe the following incidents should be evaluated:

1. If the pressurizer goes solid with one or more HPI pumps continuing to operate, would there be a pressure spike before the reliefs open which could cause damage to the RCS?

2. What damage would the water surge through the relief valve discharge piping and quench tanks cause?

To date, Nuclear Service has not notified our operating plants to change HPI policy consistent with References 1 and 2 because of our above-stated questions. Yet, the references suggest the possibility of uncovering the core if present HPI policy is continued.

We request that Integration resolve the issue of how the HPI system should be used. We are available to help as needed.

D F Hallman
D.F. Hallman

FIGURE 2.4 The Hallman Memorandum.

However, the nuclear plant operators at TMI, as had the operators at Davis-Besse, misinterpreted the pressure gauge and prematurely shut off the emergency coolant water injection system even though the valve was still open and releasing coolant water.

The difference between the Davis-Besse incident and the TMI accident was that TMI-2 was at full power. As Mr. Dunn had predicted in his memorandum of 9 February 1978, core uncovery and fuel damage occurred at TMI. Fortunately, no one was injured. However, the multi-billion-dollar plant was put out of service and a multi-billion-dollar cleanup was required. On 4 April 1979, the changes in operator instructions which Mr. Dunn had recommended were issued to the utilities with nuclear power plants designed by B&W. This completed the communication process that had been initiated on 14 September 1977. Seventeen months had elapsed.

An analysis of the inefficiency of this specific communication process within Babcock and Wilcox illustrates the problems that readers pose. Mr. Dunn failed to establish adequate lines of transmission for his memorandum and never realized that his recommended instructions had not been issued. Mr. Hallman never got a response to his memorandum, in part as a result of ineffective report design: he did not persuade his primary audience, Mr. Karrasch, to resolve the issue of how the high pressure injection system should be used.

Mr. Dunn did not establish adequate lines of transmission for his report. The purpose of his memorandum was to have a decision made to issue changes in operator instructions "to prevent premature operator termination of high pressure injection."[3] However, he did not send his memorandum to any person in a position to make the decision to issue changes in operating instructions (Figure 2.5). He sent his memorandum to numerous persons in B&W engineering departments and only incidentally to a person in a position to issue such instructions. He did not send it to Dr. Hallman, the Manager of Plant Performance Services, the person who could make the decision to act on his recommendations. He addressed his memo to Mr. Taylor, the Manager of Licensing in B&W, who testified that the memorandum had been misdirected "in the sense that operating instructions that are issued to the plant do not originate in the Licensing Section."[4] Thus, the lines of communication for the Dunn memorandum were completely inconsistent with the decision-making process. Not only was the decision never made; Mr. Dunn did not realize that it had not been made—before the TMI incident.

Dr. Hallman's memorandum was not persuasive to Mr. Karrasch, as Mr. Karrasch's testimony before the President's Commission stated:

> I recall glancing over it very quickly and keying on the two specific questions. I do not recall reading it very carefully at the time, but I do remember looking at the specific questions that Mr. Hallman was asking me. I remember thinking that they were rather routine questions from the Nuclear Service Department to the Engineering Department and that they could be answered in a routine fashion. I then am quite sure that I placed a note on top of the memorandum to

> one of two people who report to me in Plant Integration, with a message to him to please follow up on this and take any action that you deem appropriate, or something like, please answer the questions and get back with Mr. Hallman. I then, just that quickly, disposed of this piece of paper crossing my desk.[5]

Mr. Karrasch handled the Hallman memorandum in a "routine manner"; despite follow-up by Mr. Hallman, he essentially dismissed it. His reaction was to "simply forget it and to proceed with higher priority work." He "was busy doing something else." No one in Mr. Karrasch's department remembered his asking for them to follow up on the Hallman memorandum, and Mr. Karrasch "did not personally follow up."[6] Mr. Hallman contacted Mr. Karrasch informally by phone and in the hallway several times afterwards, but never received a written response from Mr. Karrasch.

The accident at Three Mile Island occurred about 8 months after the Hallman memorandum. Mr. Karrasch testified that the issue was "something that if it had had the proper attention could have been resolved within several months at most." Prior to TMI, however, Mr. Karrasch did not "take any fur-

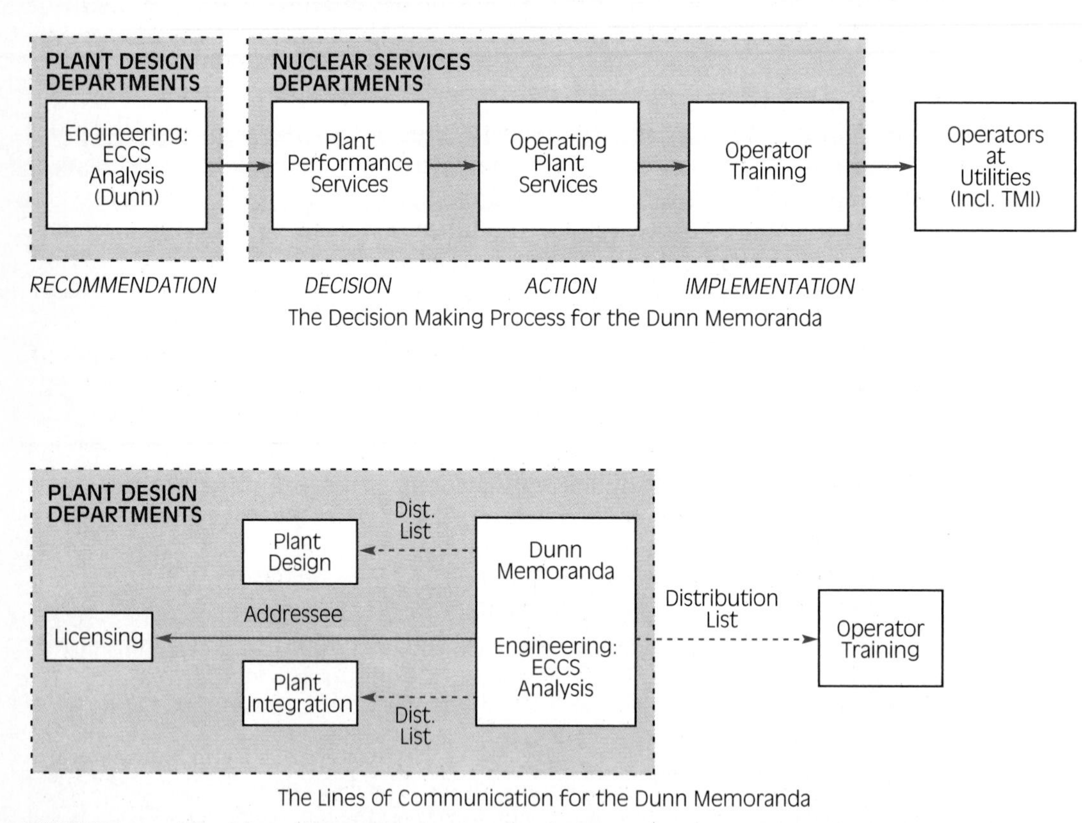

FIGURE 2.5 The Lines of Communication for the Dunn Memorandum Were Inappropriate for the Decision Making Process.

ther action at all on this matter."[7] Clearly, Mr. Hallman's memorandum was not persuasive. That Mr. Karrasch did not read it carefully and failed to appreciate the seriousness of the situation probably was the result of ineffective audience analysis and ineffective memorandum design which failed to take into account the behavior of report readers in organizations.

❑ 2.2 ❑ How Readers in Organizations Behave

With the benefit of hindsight, students in the classroom and persons in industry alike are usually perplexed by or even suspicious of the motives of the participants when we outline the communication process that led up to the accident at Three Mile Island. How could such a thing happen? However, they fail to consider how readers in organizations actually behave. They make a number of false assumptions about the persons to whom they send reports and memoranda. Having done so, they are unlikely to design a report that communicates effectively. In fact, we see nine false assumptions which report writers frequently make—several of which, as you can see, were mistakes made in the TMI situation.

1. It is false to assume that the person addressed is the real audience.

The person addressed often is not the most important person who should receive the report. Sometimes this occurs because organizational protocol requires reports to be addressed to various managers rather than to the real users. In the automobile test facility which we discussed in chapter one, test reports had to be addressed to the manager of the test engineer's own department rather than to the engineers in the design department who needed the information. By convention within the company, the test reports had to follow protocol lines of transmission rather than be sent directly to the persons who needed the information (70 miles away).

Sometimes this occurs when the writer has not determined who should receive the report, as the Dunn memorandum illustrates. Mr. Dunn addressed his memorandum to the Manager of Licensing, who had no need for the report and could not act upon its recommendation. He did not address it to the Manager of Plant Performance Services, who would have been responsible for acting upon the recommendations and implementing them. Nor did Mr. Dunn include the manager or anyone in Plant Performance Services on his distribution list. Although no protocol prevented Mr. Dunn from sending his memorandum to the appropriate persons, he just did not determine who those persons were.

2. It is false to assume that the readers are a group of specialists in the field.

Report readers are indeed specialists. However, almost never are they specialists in the subject matter of a report which is sent to them. That is, the writer is a specialist in some subject matter, and the readers are specialists in other subject matters. Precisely because the writer knows something that the readers do not know, the report is sent to its readers and has value to them. (If they already knew what was in the document, what would be the point of sending it to them?) Sometimes, of course, the writer is a nonspecialist asking a specialist for information. Indeed, this was the situation when Mr. Hallman wrote to Mr. Karrasch. However, most reports are written by people who have specialized knowledge and responsibility to other people who need information, decisions, and recommendations to carry out different specialized work and different functions.

For example, Mr. Dunn was the specialist in nuclear plant coolant systems writing to persons with different administrative and technical specialities and responsibilities. Mr. Hallman had administrative responsibilities, providing technical support to nuclear plants designed by Babcock and Wilcox. He and his staff misinterpreted the technical explanation in Mr. Dunn's memorandum.

3. It is false to assume that the report has a finite period of use.

Reports have long-term uses. Unfortunately, many writers assume that their memoranda and reports have a limited, short-term utility. The function of putting information into writing is to have it available for extended periods of time as well as for immediate use; it documents decisions and actions for future reference. For example, a contract administrator in a shipyard told us about an instance in which an engineer's inadequate documentation of a design change cost the company a major amount of money. It seems that the engineer had not documented the design change in sufficient detail; but some months later, the contracts administrator needed that type of information to negotiate the charges for the design change with the client. Unable to find it in the report, he was forced to reconstruct what should have been in the report. He told us that it took him over a month to reconstruct the reasons for the design change sufficiently to make a decision to bill the work against the contract rather than to treat it as shipyard overhead. A month's time was lost because of inadequate documentation.

In our consulting work for various companies, we have seen numerous other instances of this faulty assumption that reports have only short-term use, many of them involving legal uses of the information. For example, we have seen a product liability suit in a vehicle roll-over accident case which resulted in the death of a tractor driver. In this case, test reports written more than 10 years before the accident were the focus of attention. In another instance, we have seen a personnel action against an employer, a legal action which involved evaluations and correspondence written 5 years before a dispute which ended in the court trial. We have also seen that companies routinely keep all reports for a minimum of 5 years and that many of them keep reports 20 years or more because the information in them remains valuable and because it continues to have legal significance years after the reports first appeared.

4. It is false to assume that the writer and the readers always will be available.

The writer and the readers probably will be not available for long to clarify a report. In offices, businesses, and industries in the United States, the norm is for professionals to change positions in an organization and to change organizations. For example, about 50% of engineering graduates are with another organization 5 years after they graduate and take their first job.[8] Although organizational charts may remain unchanged, often for years, the names of the persons filling the roles on those charts constantly change. As an example, in one company we found that in only 3 months nearly one-third of the people on the distribution list for a report were either no longer with the company or not at the same site within the company. Three months: 30% turnover. Perhaps that was an extreme instance, but we think it shows dramatically that you cannot count on organizational stability.

5. It is false to assume that the readers are familiar with the situation.

Readers usually are not familiar with the situation. If you are lucky, your immediate supervisor may remember what you are working on and writing about. In a few cases even your second-level manager might know something about it. But usually you write your report primarily for persons beyond the doors of your own department, primarily to people who will not have the slightest idea why they are receiving your report. Usually those who requested it will have forgotten that they had asked for it.

6. It is false to assume that the readers have been involved in daily discussions of the material.

Almost no readers of any importance for your report will have been involved in daily discussions of the material in the report. Therefore, they can't "read between the lines" and must depend instead upon what you tell them in writing. Although your written reports and memoranda are certainly embedded in oral communication contexts, many of your readers, and certainly most of the important ones, will not have the benefit of these discussions. For example, we remember one report which we studied in a research project at a hospital supply company. The report had eighteen readers. Three could be classified as "important" in the sense that they could act and make decisions on what the report recommended. Of the three, none had ever heard any of the background discussion behind the project or the report; it was entirely first-hand news to them when they read the report. Of the fifteen other readers, fourteen had never heard about the project before. That leaves only one reader who knew about the report—the writer's supervisor—and her function was merely to sign off on the report and to transmit it to the seventeen other readers. That's a typical situation.

7. It is false to assume that the readers await the report.

Almost no one waits for a report to arrive. Readers who have no prior knowledge about a situation obviously can't be expecting a report dealing with it to arrive. But other readers who might have a reason to know about the situation get involved in other activities after they become aware of what you were writing about. By the time your report arrives, they have moved on to other things and have completely forgotten your work. When your report arrives, it is a surprise for everyone, interrupting their attention and dragging them away from other activities—or at least trying to drag them away.

8. It is false to assume that the readers have time to read the report.

Readers have no time to read reports. Their "in" boxes are full of paper which they want to dispose of as quickly as possible, and the last thing they want to see is another addition to the pile. Henry Mintzberg, a specialist in organizational theory, observes that "study after study has shown that managers work at an unrelenting pace, that their activities are characterized by brevity, variety, and discontinuity and that they are strongly oriented to action and dislike reflective activities."[9] As Mr. Karrasch said, he "just that quickly, disposed of this piece of paper crossing my desk" so that he could get on to "higher priority work." And don't assume that Mr. Karrasch was slipshod in his work; he—like most managers—was just a very busy man with a completely full schedule and an overloaded in-box. He was like the manager to whom we talked only this morning: his electronic mail in-box had forty messages in it when he arrived at work today; on his calendar for the next 4 days were meetings with sixty people coming in groups of two and three; and there was a 5-inch-high stack of mail in his box. In that environment, to have 10 uninterrupted minutes for reading a report is rare indeed.

9. It is false to assume that the writer and the readers share the same organizational values.

It is easy to think that values are absolute, but in an organizational context what we value is related to what we need to do our jobs. That is, what Ann does on the job is not what Dave does, so Ann needs and values types of information that don't interest Dave. At the same time, Dave needs and values information that Ann would find useless. Thus you cannot assume that your recommendations in a report, which perhaps assume a cost-savings premise, are going to be viewed in the same light by both Ann and Dave. Cost savings may be the most important premise to Dave because he works in the office of financial operations at the hospital. To Ann, quality control may be much the most important issue because she is the Quality Control Supervisor for the Immunopathology Lab. Thus, if your report simultaneously addresses both Ann and Dave, you have to take that into account in your report design. As organizational researchers Cyert and March say, assume that "an organization is a coalition of members having different goals."[10]

Sources of False Assumptions

These false assumptions have different sources. The most important source, probably, is the context and function of the papers that students write in their classes. Students learn to write for teachers and spend some 16 years practicing the skills involved. In so doing, unfortunately, they learn bad habits that will get them into trouble on the job. In writing for a teacher, students learn to write for one person; on the job there will be many readers most of the time. Students learn to write for a reader who knows more (or presumably knows more) about the topic than the student knows; on the job readers read precisely because they don't know something that the writer knows. And students in school learn to write for a reader who has no instrumental need for the information in a document and who will do nothing with it except to evaluate the writer; on the job, readers read reports because they must do so in order to get things done.

Subsequent sources on the job are several. Writers naturally try to simplify their communication tasks by writing for readers such as their own supervisors who, because they are familiar with the content of a report, can read between the lines. They also focus on their own technical material rather than on the needs of their readers because it is interesting to them. Also, at times, they use writing to attract the attention of "important" readers, managers and supervisors, rather than addressing it to the people who really need the information. The real users may be at any place or any level within the organization; they aren't all "important" people in the conventional sense of the term, but to the writer, the users of writing should be the most important people of all.

Primarily, however, we have found that writers simply are not aware of the need for audience analysis. They don't think about audience; they just think about themselves, what they have done, and about their subject matter. For them, a report is "about something," not "to somebody." To them, a report is an end, not a means. They are unaware that their reports address diverse readers with different needs and no time. They never think about a report having value over a long time or that it is legally significant. They forget that a conversation at the water cooler was not attended (or recalled) by everyone on the distribution list for a report. They can't imagine that readers wouldn't be interested in what interests them. They think that just because they spent 6 weeks working on something, readers will be sitting around, eagerly waiting and happy at the chance, finally, to read a 20-page, detailed analysis of some specialized topic. In other words, they don't think.

To be successful, report writers must consciously shift their focus from the technical material to the organizational functions of their reports. They must introduce the audience into the organizational communication process.

This conscious shift of focus is not easy, but we have found that a three-step audience analysis procedure can be helpful. In it you do three tasks:

- Identify your readers.
- Analyze your readers.
- Classify your readers.

The act of analyzing your audience will help you design your report in terms of your communication purpose.

❑ 2.3 ❑ Identifying Your Readers

Perhaps the most important advice we can provide is: identify your individual readers. Forget the faceless blocks on the organization chart and the great pyramid of the company's theoretical organization. Think instead of the communication situation and of the roles you and your readers have in it.

When you write a report you are the center of the organization—the point from which the lines of communication spread throughout the relevant departments in the organization. As a writer, you are not at the bottom with the great pyramid of the official organization chart pressing on your head. Instead, given the organizational problem your report is addressing, you are the most important person because you have some professional advice that several persons throughout the organization should have. Your lines of communication almost invariably will cut across the formal structure of the organization,[11] so if it has any use at all, the organization chart can only help you to identify the departments and organizational subsystems that you might overlook when you identify your readers.

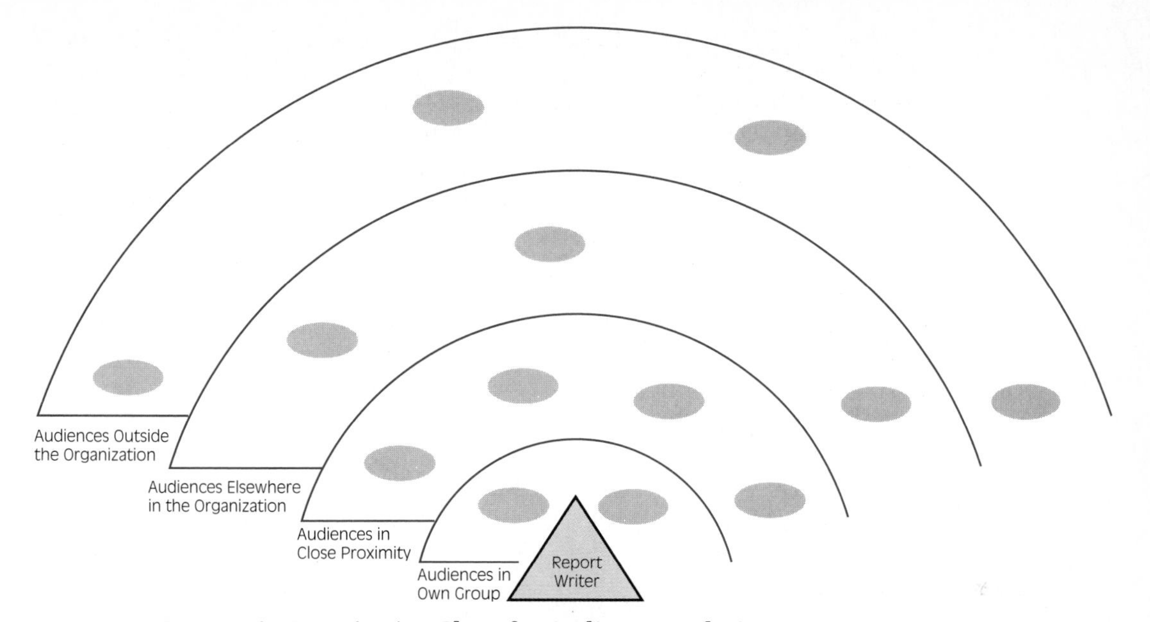

FIGURE 2.6 Egocentric Organization Chart for Audience Analysis.

For each report you write, try preparing what we call an "egocentric organization chart" (Figure 2.6). The egocentric organization chart puts you at the center of the communication network, not at the bottom of the pyramid. On this chart you need to identify all of the readers your report will have. The chart enables you to identify and establish lines of communication appropriate for your report.

In terms of organizational activities and familiarity with your technical activities, you will have readers at four degrees of distance from you, represented by the concentric rings on the egocentric chart:

- Readers in your own group or department.
- Readers in other departments in close proximity.
- Readers elsewhere in your organization.
- Readers in other organizations or other divisions of your organization.

Of course, you have readers in your own group or department, your ring on the chart. You interact with these persons constantly, usually orally. These persons, however, are not the persons for whom you write most of your important reports and memoranda. For written communication they usually are on your distribution list only for informational purposes. Sometimes they can even be considered as coauthors of your communication in the sense that they often review it before you send it out. Their names on a distribution list usually signal that they have been involved in the technical activities that form the background of your report.

Readers in the next ring, readers "in close proximity," are those persons with whom you and your group regularly interact according to your roles in the organization. These readers are in other departments of your organization with whom you work closely, not necessarily in close physical proximity. If you are in market research, you will be closely associated with persons in product development more than with persons in the other departments under the Vice President for Sales. If you are in technical service, often you will be closely associated with persons in production and, perhaps, design rather than with persons in sales.

Given the organization problem and units and roles involved, as identified in the "organizational need stage" of the communication process (Figure 2.1), for most reports of any importance you will have readers elsewhere in the organization. They are persons with whom you do not ordinarily interact as you perform your daily responsibilities. For example, a police officer assigned to oversee disposal of abandoned motor vehicles proposed a change in policy to reduce the holding time on abandoned vehicles and thereby reduce high storage costs. This particular report involved persons who did not often receive reports from the officer: the deputy chief of police, a city attorney, and the city finance director.

Finally, you often have readers outside the organization. The police officer's report, for example, was sent to owners of four towing services who did contract work for the city. Any report can address suppliers, vendors, clients, government agencies, and other businesses or industries.

At times, these "outside" readers can be your most important readers. A request by a design engineer for a change in a part specification has as readers a contract manager, the technical liaison person, a manager, and technical staff in the corporation with whom the engineer's corporation has the contract. That engineer routinely interacts with those persons in the other corporation as well as with persons in his own corporation. Thus, persons in other organizations also can be readers in "close proximity" in terms of your role.

Persons in other divisions of your own corporation usually can be considered as outside of your organization in a practical sense. We recall an engineer in one division of an automotive corporation who had to write a report to have a test facility in another division change its dynamometer testing procedures. The engineer had analyzed the discrepancies in test results between the two divisions and had found that the test facility in the other division was conducting the tests incorrectly. However, because the facility was in the other division, he had to write the report to get its test procedures changed without in any way implying that the other division's procedures were incorrect. Corporate protocol prevented one division from criticizing the operations of another division.

The contrast between a formal organization chart and the egocentric chart is illustrated by the audiences for a report written in a hospital consulting firm (Figures 2.7 and 2.8). The writer was a technician in the systems department. He was a systems analyst with a 2-year degree in data processing. His department was putting out work late and therefore having to rerun computer programs. The delay was caused when persons in the programming department apparently did not understand the specific information needs of the systems people. Consequently, his supervisor asked him to write a report to recommend a means of resolving the problem.

His report was sent to persons in the departments that are shaded on the company organization chart (Figure 2.7). His report went not only to members of his own unit, such as his supervisor, but to units throughout the operations wing of the organization. In addition, it went all the way to the CEO—it was read by both the president and the executive vice president. His specific readers are identified on his egocentric organization chart (Figure 2.8), and their relationships to him as the report writer are made clear. The official organization chart fails to identify these readers specifically and to establish their relationships to his report. His lines of communication are clarified by the egocentric organization chart.

Mintzberg discusses a corporate communication situation that illustrates the value of this egocentric perspective.[12] He quotes an assistant comptroller from a corporate headquarters:

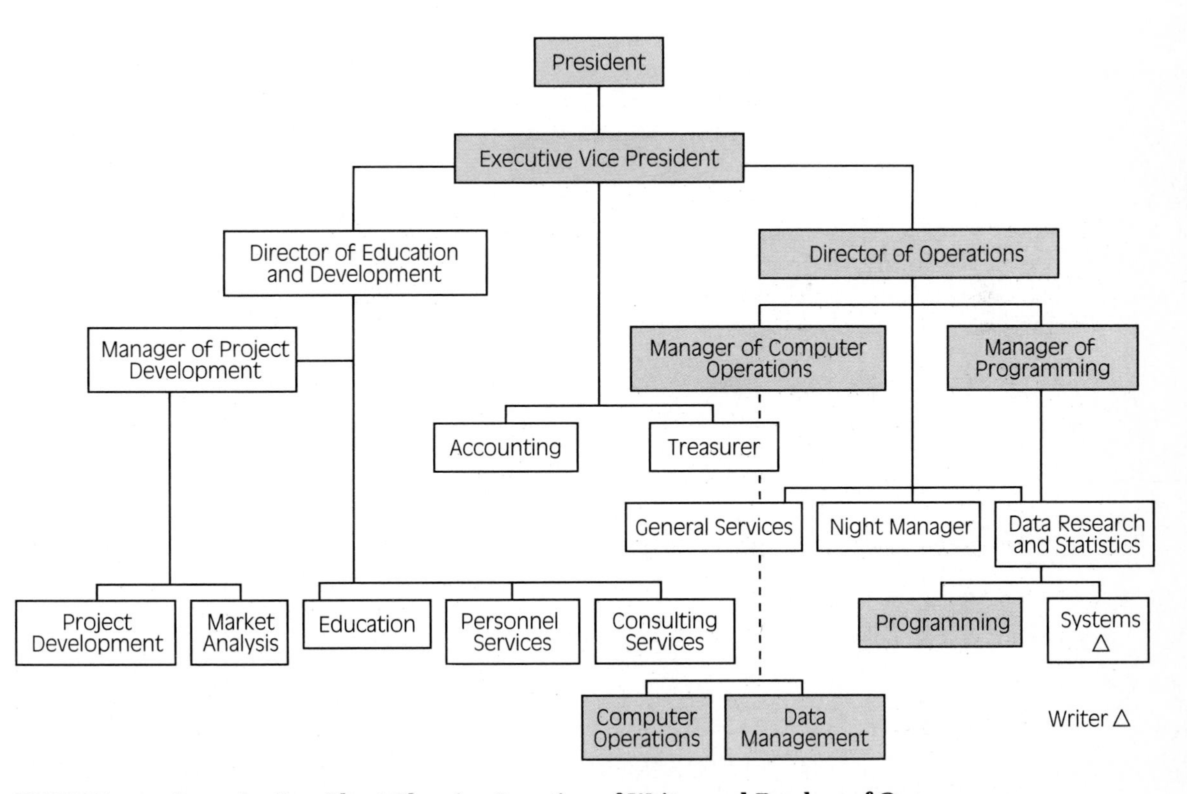

FIGURE 2.7 Organization Chart Showing Location of Writer and Readers of One Report in a Hospital Consulting Firm.

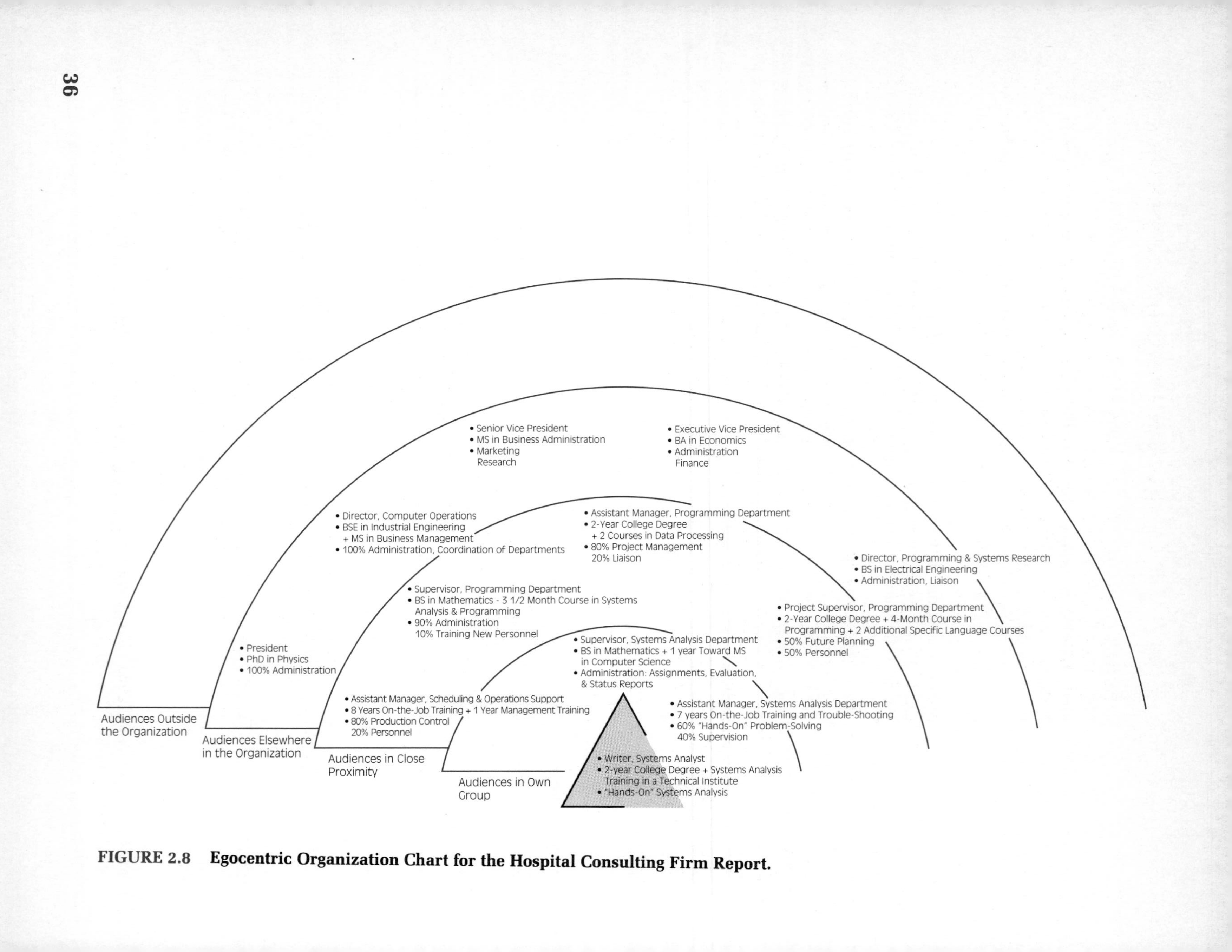

FIGURE 2.8 Egocentric Organization Chart for the Hospital Consulting Firm Report.

Our top management likes to make all the major decisions. They think they do, but I've just seen one case where a division beat them.

I received for editing a request from the division for a large chimney. I couldn't see what anyone could do with just a chimney, so I flew out for a visit. They've built and equipped a whole plant on plant expense orders. The chimney is the only indivisible item that exceeded the $50,000 limit we put on expense orders.

Apparently, they learned informally that a new plant wouldn't be favorably received, so they built the damn thing. I don't know exactly what I'm going to say.

The comptroller's first challenge is to analyze his audience by identifying his readers on an egocentric organization chart. He is faced with a situation where his audiences in the division clearly had thrown out the organization chart in their reports to him. They had viewed their audiences through an egocentric perspective. He now should do likewise; the official organization chart will not provide the perspective he needs to design his report.

Identification of your readers according to their distance from you in operational terms is an effective means of orienting yourself toward your readers rather than toward the subject matter of your report. The egocentric organization chart enables you to identify your readers because it provides a communication perspective rather than a chain-of-command perspective. It puts you in charge.

❑ 2.4 ❑ Analyzing Your Readers

Once you have identified the readers for a specific report, you need to characterize them in order to analyze their needs. Even if your audiences are familiar with your work, the diversity of their daily activities as well as their personal characteristics can inhibit effective communication if you do not take these into account when you write. You can characterize each of your readers in terms of

- Operational characteristics.
- Objective characteristics.
- Personal characteristics.

Characterizing your readers will go a long way in helping you to overcome any false assumptions you might have about your report audiences.

Operational Characteristics

Each of your readers has a unique role and responsibilities in your organization—unique operational characteristics. What that person does for the organization will influence what he or she wants or needs from your report and how he or she evaluates your conclusions and recommendations. Operationally, how does he or she spend time? Will daily concerns and activities enable that person to respond to your report easily, or will they make it difficult for him or her to grasp what you are talking about? What does that person know about your role and, in particular, the organizational situation and professional activities that have led to your report?

Furthermore, organizations are subdivided into units, each of which has its own distinctive norms, values, and jargon. As one article on management communication states: "These local languages make communication across boundaries difficult and prone to bias and distortion. The more differentiated the subunits, the greater the difficulty of communicating across these boundaries."[13] You need to be sensitive to the differences of norms, values, and interests among the departments in your organization if you are to communicate with them effectively. For each reader, you need to ask, what are his or her values as embodied in the organizational role?

Especially, you should concern yourself with the question, "How will my report affect that reader's role?" A former student of ours once told us of an experience he had had on one of his first jobs with a company. He had been asked to evaluate the efficiency of the plant's waste treatment process. Armed with fresh knowledge from his civil engineering courses, to his surprise he found that by making a simple change in the process the company could save more than $200,000 a year. He fired off his report with great anticipation of glowing accolades. None came. How had his report affected the roles of his important readers? The writer had not considered the fact that the plant manager and several supervisors were now faced with the problem of explaining their having spent $200,000 a year to do something that could be done less expensively. Small wonder that they were less than elated over his discovery, but the writer hadn't thought about it from their point of view before he wrote the report.

Objective Characteristics

Along with operational characteristics, each person has a unique set of educational background and prior work experiences that colors his or her perception of the information in a report. By objective characteristics we mean specific, relevant background data about each reader. Educational backgrounds often vary, so differences should be noted. When educational backgrounds are similar, do not assume that the person knows what you know. Remember that the half-life of much professional education today is about 5 years. Thus, anyone 5 to 10 years older than you, if you are recently out of college, probably will be only superficially familiar with the material and skills you have mastered. For example, few will be entirely comfortable with your regression analysis of statistical data. Ask yourself the specific question,

"Could he or she participate in a professional conference in my field of specialization?" You will eliminate almost all of your audience.

More important, perhaps, are the distinctions among professionals in terms of values and goals. Persons from different fields have professional goals and values that are inconsistent with each other and often inconsistent with the values and goals of the organization itself. For example, research chemists in chemical industries have considerably different values than chemical engineers within the same companies. From our own experience with these two groups in one company, we would almost say that the two can hardly talk to one another on occasion. Further, we could point to one survey in which 6% of the scientists thought it important to participate in decisions that affect the future business of the company, while 41% of the engineers thought it important.[14] Analogous differences might be found, perhaps, between the business office professionals and the professors in a university. Value systems such as these are inherent in professional training. Scientists are concerned for creating knowledge, engineers for developing applications of knowledge, social workers for helping people, environmentalists for preserving species diversity, teachers for educating students. These value systems will influence the way each responds to a report.

Personal Characteristics

Finally, each person has personal characteristics that influence the way he or she responds to a report. These can be personal idiosyncrasies. Many readers are fussy about minor details. We know of one supervisor, for example, who is a nut about tables. Your report can be superb technically and basically well designed, but if he spots an illogically formatted table, your report is in trouble. These also can be personal preferences for type and amount of detail. Some readers want results documented; other readers don't want to be bothered with documentation. Some readers are quick and decisive; other readers are plodding and methodical. To the extent possible, you should anticipate how they will respond to your report personally.

These personal characteristics involve values and goals as well. An individual professional's goals and the organization's goals are not necessarily compatible. Therefore, you have to take the attitudes and motivations of your readers into account when you write your reports. As Herbert Simon nicely states the point: "The function of the communication, after all, is not to get something off the mind of the person transmitting it, but to get something into the mind and actions of the person receiving it."[15] Other analysts introduce additional considerations, such as organizational "politics."[16] Thus, you should take personal characteristics into account whenever possible.

This diversity of roles, organizational responsibilities, and background is illustrated by the readers of the hospital consulting firm report (Figure 2.8). In his own group, the writer had two readers, a supervisor with a B.S. in mathematics who had an administrative role and an assistant manager with on-the-job training whose role was 60% "hands on" problem solving and 40% supervision. In close proximity he had four readers, a project supervisor, a supervisor, an assistant manager of programming, and an assistant manager in operations. Their training ranged from on-the-job to 2-year degrees with some data processing courses to a B.S. in mathematics. Their roles consisted almost entirely of administrative and managerial rather than technical responsibilities. Elsewhere in the company, the writer had two directors, a senior vice president, an executive vice president, and the president as readers—all upper management. Two had engineering backgrounds, while one had a background in physics, one in economics, and one in business administration. Again, all were involved 100% with administrative and management activities.

In addition to these operational and objective characteristics, the writer had to account for personal characteristics—the differences in attitudes between the programmers and the systems persons. As the writer said, tempers had been "flaring" in both departments because the programmers were not forwarding the information that the system analysts needed. Essentially, he was writing a report to open the lines of communication between programming and systems, and persons all the way up to the CEO became involved. Yet in the end the programmers and systems persons would have to cooperate to implement the recommendations in his report. Thus, personalities and organizational politics can be an important consideration when you characterize your report readers.

A convenient means of analyzing your report audience for a report is to use a form for characterizing your readers (Figure 2.9). Literally filling out the form for each reader, of course, is too time consuming. But you can use the form as a checklist to characterize each reader that you have identified on an egocentric organization chart for a report. Just the act of reviewing the form for each reader of your report will sensitize you to the needs, expectations, and probable reactions of those readers.

❑ 2.5 ❑ Classifying Your Readers

Once you have identified and characterized your audiences, you have to classify them in terms of their importance in regard to your purpose. Not only are your audiences diverse, they are of unequal importance. Particularly, be aware of these differences so that you design your report to address, primarily, the correct subgroup of readers among your total group of readers.

Report audiences are of three types (Figure 2.10):

- Primary audiences.
- Secondary audiences.
- Immediate or nominal audiences.

NAME: TITLE:

1. OPERATIONAL CHARACTERISTICS
    1. His or her role in the organization:
    2. What he or she needs from your report:
    3. His or her knowledge of your professional role and activities relating to this report:
    4. His or her daily concerns and activities and consequent value system:
    5. The persons who in turn will be affected by your report through him or her:
    6. How your report could affect his or her role:

2. OBJECTIVE CHARACTERISTICS
    1. His or her education——levels, fields, years:
    2. His or her past professional experiences and roles:
    3. His or her knowledge of your professional field:

3. PERSONAL CHARACTERISTICS
    1. Particular report writing interests and concerns:
    2. Personal goals, values, and motives:
    3. Organizational and professional commitments and values:

FIGURE 2.9 Form for Characterizing Individual Report Readers.

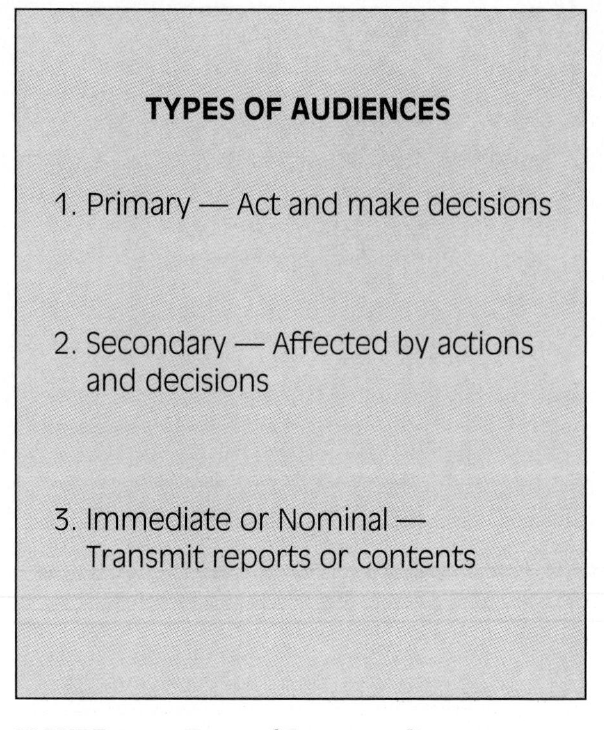

FIGURE 2.10 Types of Report Audiences.

The trick is to recognize that these three types of audience are not of equal importance, but successful report design still depends upon meeting all of their needs.

Primary Audiences

Primary audiences are those persons and groups who will act out and/or make the decisions necessary for you to accomplish your purpose. These persons will ensure that the appropriate actions are taken to solve the organizational problem or meet the organizational need that initially provoked the organizational communication process. Usually each report has a unique primary audience because the problems or needs themselves are unique, not routine.

But do not confuse the concept of "primary" with the concepts of "managerially important" or "single."

A primary audience for an individual report might be at any level within a company. The custodial staff might be the most important readers for a report which also addresses labor representatives and management representatives. It isn't the rank that determines the "primary" status; it is need and responsibility. Similarly, a primary audience might be one person or twenty. For example, a single decision maker might rely on staff input to respond to a report. That means the staff is really part of the primary audience along with their manager. In another situation, team leaders drawn from a dozen different projects might be the main readers for a document that would also be read by all of the the team members and the project management.

Also remember that primary audiences can exist in any ring on the egocentric organization chart and in fact can cut across rings. Your purpose defines who, among your many individual readers, can act and make decisions upon what you are writing about. You might be writing to one person in close proximity, or perhaps you are writing to many persons scattered throughout an organization. Each case requires individual analysis. For example, notice how very distinctive the primary audience was for the request by a shipyard for deviation from a contract equipment specification shown in Figure 2.11. The primary reader was a technical specialist at the Naval Sea Systems Command (NAVSEA) in Washington, DC. There were numerous other readers, but the report had to deal centrally and primarily with the very specific needs of the one person who could do what the writer wanted done.

In terms of the formal organization, the primary audience can be organizationally superior, equal, or subordinate to you. In terms of your communication purpose, those readers are of top priority; you need to identify your primary readers and then design your report to get their attention and the appropriate response.

Secondary Audiences

Secondary audiences are those persons who have to implement the recommended decisions or who will be affected by the recommended actions. They may not have the responsibility for making the decision or taking the action, but in their roles they are responsible for the changes that will occur. Thus, they need different types of information than your primary audiences need.

Because departments in an organization are interrelated, when you send a report to a primary audience, you automatically will involve a number of other persons as well. An analysis of the organization communication process model (Figure 2.1) indicates three types of secondary readers. The organizational problem or need stage invariably involves several organizational units and therefore several persons in their organizational roles. The report to NAVSEA, for example, had readers in Material Control in the shipyard, the unit that first identified the problem. The professional response stage introduces readers who are involved in your technical investigation to determine a solution to the problem. With the shipyard report, this specifically involved persons in Engineering who were the source of the proposed solution.

The audience action stage usually involves a number of secondary audiences in addition to the primary audience. The project engineer in the shipyard sent his report to Contracts and Purchasing, who will have to act after NAVSEA makes a decision on the request. The report also was sent to other persons at NAVSEA as well as a person with the Navy's Supervisor of Shipbuilding, its field department at the shipyard charged with monitoring the contract. When you determine the purpose of your report and analyze its communication paths, you will identify a number of persons to include on your egocentric organization chart.

Immediate or Nominal Audiences

Immediate audiences are those persons who transmit and route your report through an organization. They are typically the first-level and second-level managers who have a responsibility to review and sign or initial reports before they are released for distribution. These readers may have no other use for the reports themselves; they merely act as quality controls and as administrative gateways for information on the move.

Sometimes audiences are "nominal." That means their names are on a document because of protocol requirements. No one really expects them to do much with the report, but it is just standard practice to keep them informed. The report to NAVSEA, for example, had such audiences. By protocol the report was addressed to the Supervisor of Shipbuilding at the shipyard even though the primary audience was with NAVSEA in Washington. Also, the writer had to send a copy to his own program manager, who had reviewed the letter before it was sent. And he had to send a copy to his contract manager, who was responsible for handling this particular contract. These readers can read the report if they wish, but they really don't have to do so, and probably will not. (But the report will be in their files for future reference, an important point not to forget. The paper isn't necessarily wasted.)

Classification of Audiences in Various Organizations

This classification of report readers applies to all types of organizations—police departments, hospitals, insurance companies, public agencies, libraries, and schools as well as business and industry. Thus, the report to the Naval Seas Systems Command provides a prototype audience analysis for memoranda, reports, and business letters in most organizations.

The report which recommended adoption of a stress management program for the hospital (Figure 2.12) was written by a training supervisor in the Office of Hospitalwide Education. She had evaluated a proposal from an external consultant. Her report was transmitted by her manager, the Director of Training, to the Administrator of Human Resources and the Vice President of Human Resources for the hospital. The training director therefore served as an immediate audience; in this situation he was on the distribution list rather than the addressee. This particular report had a two-member primary audience: the Administrator of Human Resources and the Vice President of Human Resources. Both evaluated the recommendation in terms of the cost-effectiveness of the stress management program. That is, the Administrator of Human Resources first had to approve the recommendation; if he did, then it had to be approved by the Vice President.

The report had several secondary audiences. The clinical department and its director were at the head of the list because it had been designated for the pilot program. The nursing staff was next, as it might run the program on its own. The library also received the report, as it had to provide support for the program in the form of film purchase or rental. In addition, the report

Report Purposes:

(1) To request deviation from a contract equipment specification. The request is necessary because the vendor no longer is able to provide the specified part. (The decision probably has been reached unofficially as a result of oral communication.)

(2) To initiate documentation to modify the contract so that there is no future disagreement over contract provisions and cost implications.

Report Type:

One-page letter with attached documentation, including vendor drawings and calculations, to justify the replacement part recommended.

Report Writer:

Project Engineer in Program Management

Primary Audiences:

Technical specialist at NAVSEA responsible for analyzing and approving the request

Secondary Audiences:

NAVSEA:	Program (Contract) Manager Other technical specialists
SupShips:	Technical specialist, who will double-check the specifications
Yard:	Material Control, source of problem awareness and go-ahead to Purchasing Engineering (Machinery), source of alternative proposal Contracts, informational, but a change order might follow Purchasing, not copied but hold on purchase until authorized by Material Control

Nominal Audiences:

SupShips:	Supervisor of Shipbuilding, the addressee by protocol
Yard:	Program Manager, copied (the writer's Manager)
NAVSEA:	Program (Contract) Manager, responsible for routing the letter

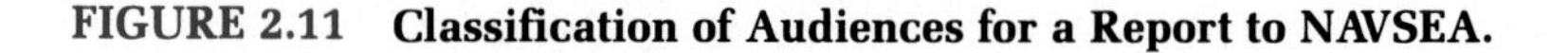

FIGURE 2.11 Classification of Audiences for a Report to NAVSEA.

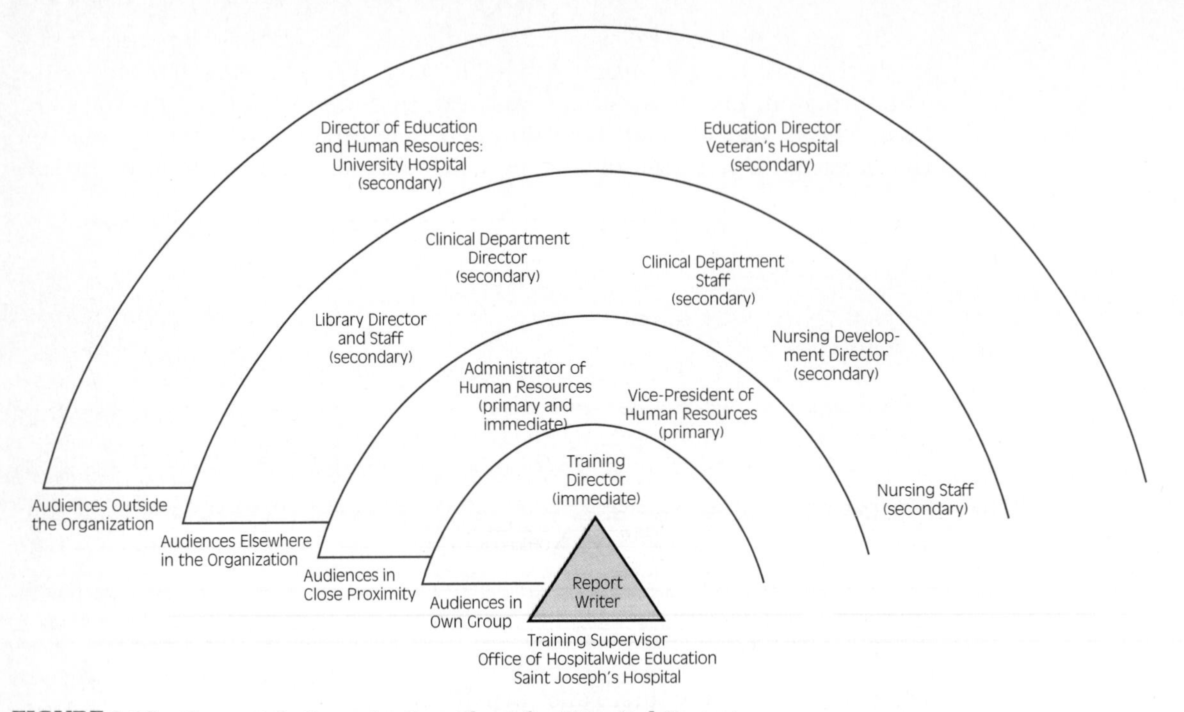

FIGURE 2.12 Egocentric Organization Chart for Hospital Report.

had secondary audiences in other organizations. Because the writer's hospital shared resources with other hospitals in the area, the directors of education of two other hospitals were on the distribution list for this report.

The Police Department Report (Figure 2.13) was written by a sergeant in a patrol division. He had investigated a traffic accident in which a person had been killed, and wrote the report to recommend prosecution of the driver at fault for manslaughter. For this report, therefore, the county prosecutor and chief trial lawyer for the county were the primary audience. Because this was a traffic accident, the city engineer was an important secondary audience. The report was transmitted by the sergeant's division commander; it had numerous secondary audiences within the police department itself. In addition to city and county personnel, this report had important secondary audiences outside the organization: the attorneys for the defendant and the victim, the insurance companies and their legal staffs, and the Records Office of the State Police.

The report to the manager of manufacturing in an industrial plant (Figure 2.14) was written by an industrial hygienist. She recommended partitioning off the employee lunch room and installation of a separate ventilation system. The report had for a primary audience the area manager of the laboratory which contained the lunch room, who would have to approve the request. An important secondary audience was the manager of manufacturing, who had to be notified of all issues that had Occupational Safety and Health Administration (OSHA) implications. Other important secondary audiences

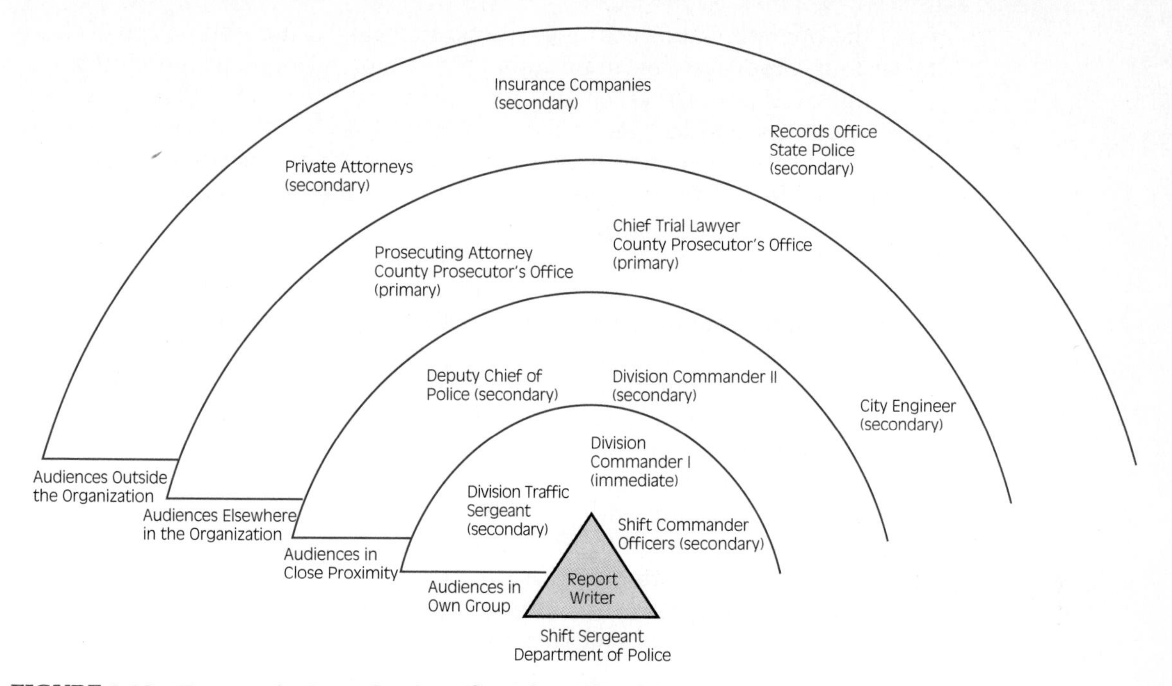

FIGURE 2.13 Egocentric Organization Chart for Police Department Report.

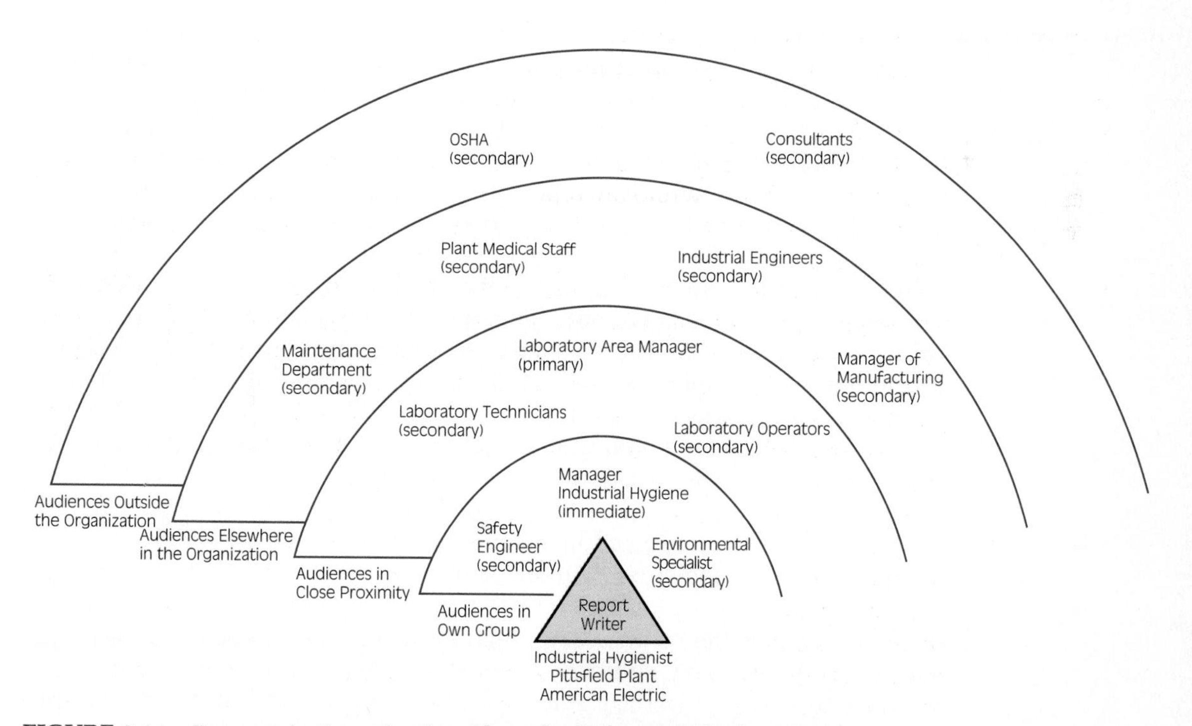

FIGURE 2.14 Egocentric Organization Chart for Industrial Hygiene Report.

were the operators and technicians who first raised the issue of potentially hazardous vapors in the lunch room. The report also went to maintenance personnel, industrial engineers, and the plant medical staff. OSHA personnel were another secondary audience; they would review the document during a site visit, although they weren't on the immediate distribution list. The Manager of Industrial Hygiene was the immediate audience who reviewed the report before distributing it.

Each of these types of audiences can pose a serious problem for a report writer. Primary audiences sometimes are difficult to identify, because persons actually responsible for a decision or action may be unknown to the writer. At Babcock and Wilcox, for example, Mr. Dunn did not know that the person responsible for acting on his request was a manager in Plant Performance Services, not a manager in Licensing. Furthermore, Mr. Dunn omitted important secondary audiences in Operating Plant Services as well as an additional person in Plant Performance Services who initially had been involved in discussing the situation with Mr. Kelly. At times, also, protocol may require the report to be addressed to an immediate audience rather than to a primary audience, as the letter to NAVSEA illustrates.

You cannot establish your audience priorities in general. Each report you write has a unique set of readers and therefore priorities among them.

❑ 2.6. ❑ Implications for Report Design

Serious problems can result when a report is not designed for its primary audience, even if it is sent directly to them. A primary audience has very specific—and usually very limited—information needs from your report. The Dunn memorandum illustrates this when one considers the purpose of the report. The report was recommending a change in operator instructions. Yet, rather than focus on the effects of the current and changed instructions, the memo presents detailed technical information. Most of this is background information, which was not only tangential but which actually confused the primary audience, Mr. Hallman and his staff. The writer of the Hallman memorandum, a supervisor, testified that he was "not aware that the two parameters, RC [reactor coolant] pressure and pressurizer level, would trend in opposite directions under a small LOCA [loss of coolant accident]."[17] Yet, that was the gist of the background information in the Dunn memorandum.

Secondary audiences usually are quite diverse in their needs. The efficiency of your communication—achieving your intended purpose—often depends on these secondary audiences receiving the information they need in order to take the appropriate actions triggered by the decisions and actions of the primary audience. For example, the report to NAVSEA makes a simple request to the primary audience, a request for deviation from the contract specification. The secondary audiences, however, needed much more information than the primary audience needed. Contracts would have to prepare a modification to the contract (a change order); Engineering would have to change all of the relevant drawings; technical staff would have to double-check the replacement vendor's specifications. For this particular report, the primary audience needed a 1-page letter; the secondary audiences needed the attached documentation, including drawings and calculations.

Immediate audiences often include the writer's own manager or his or her manager's manager. These managers need to know where the report is going; however, they usually know the technical information which the report contains. Often they have reviewed the report in draft form. When this is the case, unfortunately, the writer often writes to the manager rather than to the actual primary and secondary audiences (and omits information these other audiences need). This can be a "catch-22" situation, for your managers are indeed important readers for your reports. After all, your reports represent the department as well as you, and your managers are sensitive to that. However, your manager is often not the primary audience for your reports. You shouldn't let your managers' nearness, their power over you, or even their concern for the image of the department blind you to the fact that your real audience is usually elsewhere in the organization, beyond the doors of the department. Note: we said usually, not always. Sometimes immediate audiences can be both immediate and primary readers for some documents.

Your audience analysis ends with a classification of your audiences for a report by names, roles, and types. You conclude your audience analysis by determining the uses of your report by its readers and their information needs in order to use it (Figure 2.15). When you have identified your readers, determined how they will use your report, and specified the information they will need, you are ready to design your report.

You determine the uses and information needs of your report by anticipating the decisions or actions that you want to occur when the report is sent out. Your audience analysis will identify the persons who should receive it, both primary and secondary. You analyze the organizational roles of those persons in order to determine how they will use the report. Once you know their roles and therefore the uses to which your report will be put, you can determine the information that they will need to put your report to use.

As you can see, audience analysis is the first step in report design. You have considerable technical information at your disposal, but you don't know what to do with it until you assess audience needs. Effective audience analysis is not the solution to problems of report design, but it is a start. You have introduced considerations that you will need to take into account when you specify your communication purpose, select the information to put into the report, and arrange the information so that various readers can retrieve it and use it.

Your audience analysis will identify diverse readers for your report. Few of these will have any intrinsic interest in it. And the more important your readers are, the less time they will have to read it. Now you just have to write a report that doesn't have to be read but still communicates and gets the job done.

CHECKLIST OF REPORT READERS

PRIMARY

Name	Role	Uses of Report	Information Needs

SECONDARY

Name	Role	Uses of Report	Information Needs

IMMEDIATE

Name	Role	Uses of Report	Information Needs

FIGURE 2.15 Checklist of Report Readers.

NOTES

1. This explanation of the communication process is taken from J. C. Mathes, "Three Mile Island: The Management Communication Role," *Engineering Management International* 3 (1986): 261–68. An extensive account of the process is in "Three Mile Island: A Report to the Commissioners and to the Public," Nuclear Regulatory Commission Special Inquiry Group, Mitchell Rogovin, Director, II:1 (Washington, DC, January 1980) 157–61.
2. "Staff Report on the Generic Assessment of Feedwater Transients in Pressurized Water Reactors Designed by the Babcock & Wilcox Company," NUREG-0560, Office of Nuclear Reactor Regulation, U.S. Nuclear Regulatory Commission (Washington, DC, May 1979): 3–12. Other reports that analyze this aspect of the accident come to the same conclusion.
3. *Transcript of Proceedings*, President's Commission on the Accident at Three Mile Island, Public Hearing, July 18, 1979, and Public Hearing, July 19, 1979: 61. The two hearings are paged consecutively as one report. The memoranda are exhibits in the *Transcript* (retyped here for legibility).
4. *Transcript* 184.
5. *Transcript* 240–41.
6. *Transcript* 241, 242, 244, 249.
7. *Transcript* 265, 246.
8. Donald Peterson, Director, Engineering Placement, University of Michigan, reports that at a Placement Director's Conference at Westinghouse, June 1990, a consensus of corporate representatives and placement directors was that most corporations experience this rate of turnover with engineering hires.
9. Henry Mintzberg, "The Manager's Job: Folklore and Fact," *Harvard Business Review* July-August 1975: 50.
10. Richard M. Cyert and James G. March, *A Behavioral Theory of the Firm* (Englewood Cliffs, NJ: Prentice-Hall, 1963): 117. Some literature goes much further and defines organizations in terms of power and influence, consisting of systems of authority, ideology, expertise, and politics. See especially Henry Mintzberg, *Power In and Around Organizations* (Englewood Cliffs, NJ: Prentice-Hall, 1983): 111–242. This perspective introduces additional considerations that go beyond the scope of this book.
11. Jeffrey Pfeffer and Gerald R. Salancik, "Organization Design: The Case for a Coalitional Model of Organizations," *Introduction to Organizational Behavior*, ed. L. L. Cummings and Randall B. Dunham (Homewood, IL: Richard D. Irwin, 1980): 482. As Pfeffer and Salancik observe, "the pattern of interactions, responsibilities, and communication flows invariably differs from the formally specified pattern." Cyert and March distinguish between "through channels" communication, for which "the standard organization chart is viewed as a rule for communication," and "transmitting information across channels" when "it is necessary to coordinate the activities of subunits in the organization," for which "communication through channels is frequently quite inefficient." Our own research (see, for example, footnote 4, Chapter 1, and footnote 1 in this list) indicates that communication across channels is far more frequent in organizations than Cyert and March assume (Cyert and March, *Behavioral Theory* 109).
12. Henry Mintzberg, *Power In and Around Organizations* (Englewood Cliffs, NJ: Prentice-Hall, 1983): 198.
13. Michael L. Tushman, "Managing Communication Networks in R&D Laboratories," *Readings in the Management of Innovation*, ed. Michael L. Tushman and William L. Moore (Cambridge, MA: Ballinger, 1982): 357.
14. Richard Ritti, "Work Goals of Scientists and Engineers," *Readings in the Management of Innovation* 363, 369.
15. Herbert A. Simon, *Administrative Behavior*, 3rd ed. (New York: Free Press, 1976) 161, 164.
16. Mintzberg, *Power In and Around Organizations*, especially 420–66.
17. *Transcript* 160.

The beginning points for successful report planning and design are two fundamental issues: (1) Who are the readers and and what do they need? (2) What is the purpose of the document? In the previous chapter, we presented a procedure for thinking about your readers. In this chapter, we present an approach to thinking about your purpose. More particularly, we explain why you need to define your communication purpose. Then we outline a procedure for stating your communication purpose for problem solving reports, probably the most frequent and most difficult of the types of reports you will write as a professional. Finally, we discuss adapting the purpose statement to various other types of reports based upon the varied organizational functions of those different types of reports.

3.1 The Need to Formulate Your Communication Purpose

In part, defining your communication purpose is a cognitive act—an act of discovery—which enables you to focus on your readers and their needs and to switch your attention away from your own activities and the technical content of your report. If you are going to synthesize your data and to transform your analysis into useful form, you must start by coming to a clear understanding of why you are writing. But having done so, you must then communicate that purpose to your readers. You must attract their attention and succinctly orient them to the uses of your report. Otherwise, as we saw in the previous chapter, there is little hope that your readers will take the time to read your report, to understand it, or to act in appropriate ways.

First, for your own benefit, you need to formulate the communication purpose of every document which you produce.

Frequently, we find that technical professionals do not think very carefully about the purpose of their communications. Rather, they think about the purpose of the technical work lying behind the communications. For example, the purpose of your investigation might have been to test a vendor's part, to determine the cause of a loss of pressure, to analyze a portfolio, to design a sales inventory program, to survey a segment of the market. The purpose of your report, however, is to recommend rejection, acceptance, or modification of the vendor's part; to request installation of a new piping system; to authorize an investment; to present detailed system specifications for review; to propose a marketing strategy.

You should realize that the purpose of your technical investigation is not the same as the purpose of your report (as illustrated in Figure 3.1). Your purpose is not to tell your readers what your report is about; instead, it is to tell your readers what to do.

This distinction is nicely illustrated by two reports from automobile test facilities. One test report opened with this statement:

> Objective: To approve use of common Front End Sheet Metal mounting for all Conventional and Sport-Utility vehicles.

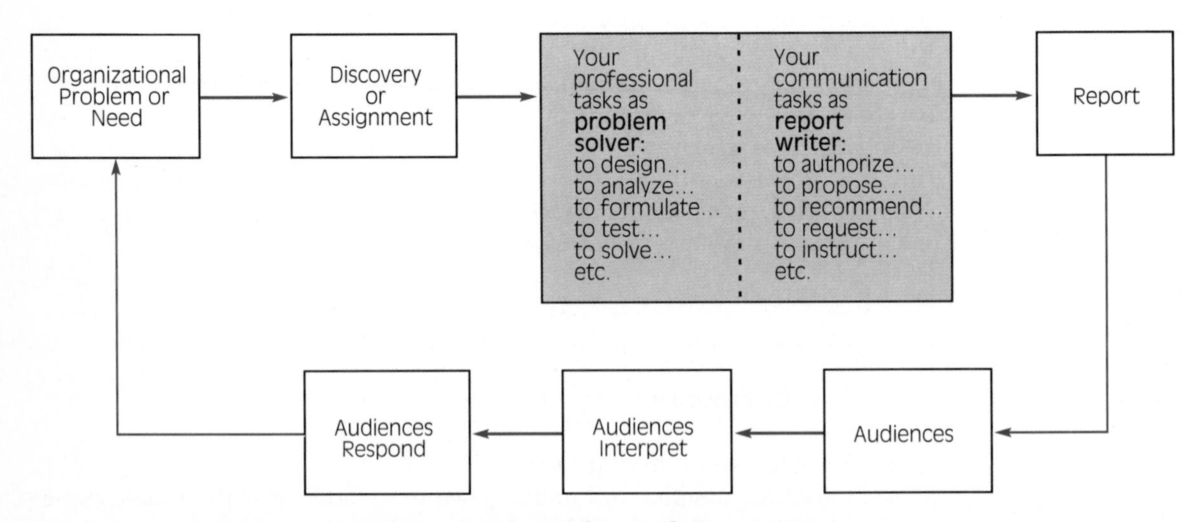

FIGURE 3.1 A Professional Is Both a Problem Solver and Report Writer.

Notice how this statement defines the purpose of the report, not of the test. It states that the report should result in action to revise manufacturing procedures.

A second test report opened with this statement:

> Objective: Test a new 1500 lb. screw jack submitted by General Equipment for use with the SB6-LB6 pickup line.

Notice how this statement defines the purpose of the test, not of the report. It states what the report is about, but in no way indicates the organizational function of the report. Unfortunately, that causes difficulty for both the writer and the readers in this report. The writer defined his purpose as "to test" something, but actually the testing had suggested that the thing tested—a truck jack—should be rejected because it failed the test. But the writer never made the recommendation—even though the truck fell off the jack three times during the test, the jack handle was twisted out of shape after the three lifts, and the handle was not long enough to protrude from under the body of the truck in the first place. In other words, there was no question about it: the jack failed the test, and failed miserably. Yet the writer was still thinking about the objective of the test, not the objective of the document, when he began to write. Accordingly, he had difficulty in understanding why he was writing: he was focused upon the testing, not upon the implications of the testing. Yet from the viewpoint of organizational action, the implications were the important thing.

Formulating your purpose is also very important for your readers. It is even important for those among your readers who are generally familiar with your work. As we discussed in Chapter 2, your readers are persons busy doing their own work, and they don't want to take time to read your report. Even though they asked for your report, they may well have forgotten that they asked for it, and even why they asked for it. Although your report may be very important to you, it is not important for them until you make it important. In addition, they have different concerns and values; they might supply a different purpose unless you tell them specifically what your purpose is. Finally, many of your readers will be unfamiliar with you, the organizational need, and the subject of your report. Unless you state your purpose, they won't know why they are getting the report. In all likelihood they will ignore it.

As you think about your purpose, both for yourself and for your readers, you need to distinguish between data and information. Data are facts; information is facts that have meaning and significance for an organization.[1] In a report, you present information that has meaning and significance. Through an investigation and analysis and then formulation of a purpose statement for a report, you interpret various types of data, giving them meaning and significance for the organization.

The need to formulate a communication purpose, to distinguish between information and data, is illustrated by an automotive engineer's report. This engineer went to Arizona to determine when would be the best period for road tests of a new truck model. He came back and wrote a 52-page report, consisting mostly of weather data and statistical analysis. His conclusion was that the second half of June would be the best time for road tests. Fortunately, his report was not distributed. With the assistance of his supervisor, he rewrote the report and ended up with a 3-page document. This engineer then was sent by his manager to take one of our in-house courses on report writing, primarily in order to learn how to distinguish between information and data. He had to learn how to formulate a communication purpose and then to present the relevant information persuasively.

The first Babcock and Wilcox memorandum, written by an engineer on 1 November 1977 (Figure 3.2), is another example of presenting information without clearly formulating a communication purpose for doing so. This writer designed the memorandum in terms of its subject matter, the operational problems at Davis-Besse, rather than in terms of B&W's organizational problem. He did not clarify the purpose of his communication. He speculates, "I wonder what guidance, if any, we should be giving to our customers on when they can safely shut the system down following an accident?" Then he recommends that "the following guidelines be sent." But he concludes with an invitation to a dialogue: "I would appreciate your thoughts on this subject." No wonder that one of the persons to whom he addressed the memorandum asked "whether this was an issue that should be addressed immediately or a technical evaluation of the issue raised"?[2] This primary audience raised the question of purpose: what action was required of him?

Contrast the B&W report writer's uncertainty with this naval architect's certainty:

> In my report (enclosed), I detail the intermittent welding requirements as imposed by the Navy. These requirements are for aluminum and steel. The purpose of this memo is to request that you prepare documentation that will permit the incorporation of U.S. Navy standards for intermittent welding into the production of the SWATH A-TSD.

Both reports have the same purpose: to get changes in instructions or documentation. The writer of the memo on welding, however, formulated and stated his purpose directly. No follow-up memo was required. The readers knew what was expected of them.

Formulation of a purpose statement, then, is an important cognitive and communication act. It requires you to determine the organizational implication of your professional activities—why you did what you did. It requires you to interpret the information you put into a report and to explain its organizational significance. You formulate your communication purpose in order to complete your professional activities. You have performed your professional activities in order to determine what organizational action is required

THE BABCOCK & WILCOX COMPANY
POWER GENERATION GROUP

To
Distribution

From
J.J. Kelly, Plant Integration

Cust.
Generic

File No. or Ref.

Subj. Customer Guidance On High Pressure Injection Operation

Date November 1, 1977

DISTRIBUTION

B.A. Karrasch	D.W. LaBelle
E.W. Swanson	N.S. Elliott
R.J. Finnin	D.F. Hallman
B.M. Dunn	

Two recent events at the Toledo site have pointed out that perhaps we are not giving our customers enough guidance on the operation of the high pressure injection system. On September 24, 1977, after depressurizing due to a stuck open electromatic relief valve, high pressure injection was automatically initiated. The operator stopped HPI when pressurizer level began to recover, without regard to primary pressure. As a result, the transient continued on with boiling in the RCS, etc. In a similar occurrence on October 23, 1977, the operator bypassed high pressure injection to prevent initiation, even though reactor coolant system pressure went below the actuation point.

Since there are accidents which require the continuous operation of the high pressure injection system, I wonder what guidance, if any, we should be giving to our customers on when they can safely shut the system down following an accident? I recommend the following guidelines be sent:

a) Do not bypass or otherwise prevent the actuation of high/low pressure injection under any conditions except a normal, controlled plant shutdown.

b) Once high/low pressure injection is initiated, do not stop it unless: Tave is stable or decreasing and pressurizer level is increasing and primary pressure is at least 1600 PSIG and increasing

I would appreciate your thoughts on this subject.

JJK:JL

FIGURE 3.2 The Kelly Memorandum.

to solve organizational problems and meet organizational needs. Until you have written a report that gets readers to respond, however, you haven't done your job.

❑ 3.2 ❑
A Procedure for Stating Your Communication Purpose

Many writers of technical and professional documents have a great deal of trouble getting started writing a report. Immersed in their professional activities, they find it difficult to step back and to decide what needs to be said. This difficulty causes many writers to work slowly, and it frequently produces reports which do not function effectively, confusing readers from the first paragraph. Moreover, in some extreme cases, the writer simply can't get started at all—a condition common enough to have a name: "writer's block" it is called.

To get started, you might simply state the purpose of the report in the first sentence. That is, "The purpose of this report is to ____." The test report (Figure 3.3) opens in this way with a direct statement of the communication purpose. Notice how such a direct statement of purpose enables the writer to focus on the important information; it immediately tells the reader why the report was written. The writer has put additional information in the "Background," which introduces the "Discussion," but right up front the report announces its purpose in one sentence.

Notice that in this particular instance, the report is a routine test report from an automotive proving ground presented in an established organizational context. For many of your problem solving reports—which are not routine and which do not deal with established organizational contexts—a different approach might be useful. Using this approach, you motivate your intended readers to read your report as well as tell them what the specific purpose is.

For these problem solving reports, we have developed a simple heuristic which you can use to begin documents of a variety of lengths and types. The heuristic consists of three basic elements:

- Statement of the Organizational Problem.
- Statement of the Investigation.
- Statement of Communication Purpose.

As Figure 3.4 shows, these three elements are derived from the stages of what we have called the organizational communication process. That is, first comes the problem. Out of the problem comes an investigation. And out of an investigation comes a communication. In using our heuristic, you develop a statement to reflect each of these three stages of the process. We have found that if these three elements can be succinctly communicated to readers right up front, their chances of understanding what and why they are reading will be enhanced. Also enhanced will be your own understanding of why you are writing.

The Organizational Problem

For every "what" there is a "why." That is, the organizational context or the research context for a document contains a problem and audiences who need to respond to that problem. As far as the organization is concerned, the problem may not be fully understood, much less solved. It may be a problem for which tentative solutions are just being considered. It may have been met already and partially resolved. The problem may be an ill-defined, future problem. It may be a partially defined, immediate issue. Or it may be a well-defined issue of past concern.

Regardless, organizations have problems and needs which require professionals to perform tasks and then to write about them. If there were no problem, there would be no need for a problem solving document. Thus, the introduction to a professional document, which establishes its purpose both for the writer and for the reader, cannot be effectively stated unless the organizational issue or problem behind it is explained. This is the first element in the problem solving process.

To state a problem clearly, you must understand what a problem is.

A problem is a conflict of perceptions.[3] A problem does not exist "out there"; it is an internal conflict which results when data or information which a person or organization receives does not correspond with a mental image of what the person or organization expects to receive or wants to receive. A problem is a gap, a conflict, a dissonance between the "ideal" and the "real," the "should" and the "is," the "expected" and the "actual." In an organization, a problem usually is some sort of conflict between an organizational goal or value and some actual state of affairs. It usually is a cost conflict, a technical conflict, or a performance conflict.

For example, here are the first few sentences from three different problem solving reports. Each of the examples points fairly well to the existence of a problem which required investigation, communication, and action:

> On August 8, 19xx, American President Lines authorized specification changes on five Seamasters through the issuance of Change Order No. 16 (ref. e). In making this economic decision, American President Lines [the owner] relied upon a preliminary cost estimate given by Doe Shipbuilding, the Contractor. The Contractor now submits a final estimate exceeding the preliminary estimate by 600%.

> Underwriters Laboratory has recently published a standard (UL-817) concerning the hook-up of AC power to chassis-mount connectors. However, this standard differs from our present wiring standard and creates a potential safety problem for users of instruments we fabricate.

Inter Company Correspondence

		File Code	Date	
			8/1/89	
To—Name & Department		Division	Plant/Office	CIMS Number
M. A. Bowen, Manager	Truck Engineering	E&RO	Chrysler Center.	415-04-31
From—Name & Department		Division	Plant/Office	CIMS Number
B. L. Ross, Test Engineer	Truck Development	E&RO	Proving Grounds	422-01-09

Subject:

TB-3056: APPROVE RELOCATION OF LIGHT DUTY CONVENTIONAL FRONT END SHEET METAL MOUNTING

OBJECTIVE

To approve use of common Front End Sheet Metal (F.E.S.M.) mounting, with Dayton A/C condenser, for all Conventional and Sport-Utility vehicles.

CONCLUSIONS

1. Standardized F.E.S.M. mounting should be satisfactory on Conventional and Sport-Utility vehicles with structural changes made as a result of endurance tests.

2. Dayton A/C condenser P/N 4039595 mounting is satisfactory for all Conventional and Sport-Utility vehicles.

RECOMMENDATION

Approve use of standardized F.E.S.M. mounting along with Dayton A/C condenser P/N 4039595 for all Conventional and Sport-Utility vehicles.

BACKGROUND

Sport-Utility and Conventional vehicles are now being built in the same plant. Standardization in location of F.E.S.M. mounts has been proposed to simplify plant handling. This standardization should also result in a cost reduction on some models. For these reasons, F.E.S.M. mount standardization was tested as outlined in the following "Discussion."

DISCUSSION

Two Sport-Utility vehicles, E501 (N8AW1) and E512 (NBAW1), plus Conventional vehicle E622 (N8W2) have been endurance tested at the Proving Grounds with standardized F.E.S.M. mounts installed. Both Sport-Utility vehicles have now completed testing, E501 with 23,338 miles of Schedule F-4 (Accelerated Off Road Endurance) and E512 with 33,739 miles of Schedule F-1 + G (Off Road Endurance plus Brake Steps). Conventional vehicle E622 has completed only 6,982 miles of Schedule F-4 and is awaiting design information on chassis brackets before continuing.

FIGURE 3.3 Test Report with Direct Statement of Purpose.

Bowen
Re: "...F.E.S.M. Mounting Relocation"

Standardized F.E.S.M. mountings should be satisfactory on Sport-Utility vehicles. Problems noted during tests on these vehicles have been corrected (as stated in replies to Reference 4, "Deficiencies." The most significant of these problems (Deficiency No. 1680002-3) resulted in the redesign of the radiator grille support extensions and of the lower outer reinforcements for the grille support.

Standardized F.E.S.M. mountings should be satisfactory on Conventional vehicles. This conclusion is based on previous feasibility tests on Proving Ground vehicles E124 (L803), E229 (M3D2), E230 (M8W2), and E981 (K6D1).

The standardized mounting location will be approved prior to final testing on Conventional endurance vehicle E622 (N8W2). Revised components (Reference 5, PCN 60823-400A) have now been installed on E622. Testing has not started, however. Any problems uncovered during subsequent testing will be reported by a Product Evaluation Report.

Use of the Dayton A/C condenser P/N 4039595 is acceptable for both types of vehicles. Change B of TB-3056 requested evaluation of this condenser mounting. This mounting was tested for 29,994 miles of Schedule F-1 + G on endurance vehicle E512. The Stress Lab also investigated the Dayton mounting and determined the mounting to be satisfactory for all light duty Conventional trucks with all configurations of F.E.S.M. mounting (see Reference 3).

With issuance of this letter, TB-3056 is considered closed.

References

1. Program Item 1680002.

2. Letter, B.L. Ross to M. A. Bowen, "Light Duty Conventional F.E.S.M. Mounting Relocation Test Program," dated 11/15/88.

3. Report, D. McKenzie, "L. D. Conventional Air Conditioning Condenser Fix," dated 11/17/88.

4. Deficiencies 1680002-1, -2, and -3.

5. PCN 60823-400A, "Revised Rod Grille Supt. Lower Reinf. and Mtg. Extensions."

cc: R. Chapman, Manager, Truck Development
W. M. Doerr, Supervisor, Truck Development
D. D. Freese, Test Engineer
R. W. Geisler, Engineering
R. E. Lutoesky, Engineering
D. McKenzie, Manufacturing

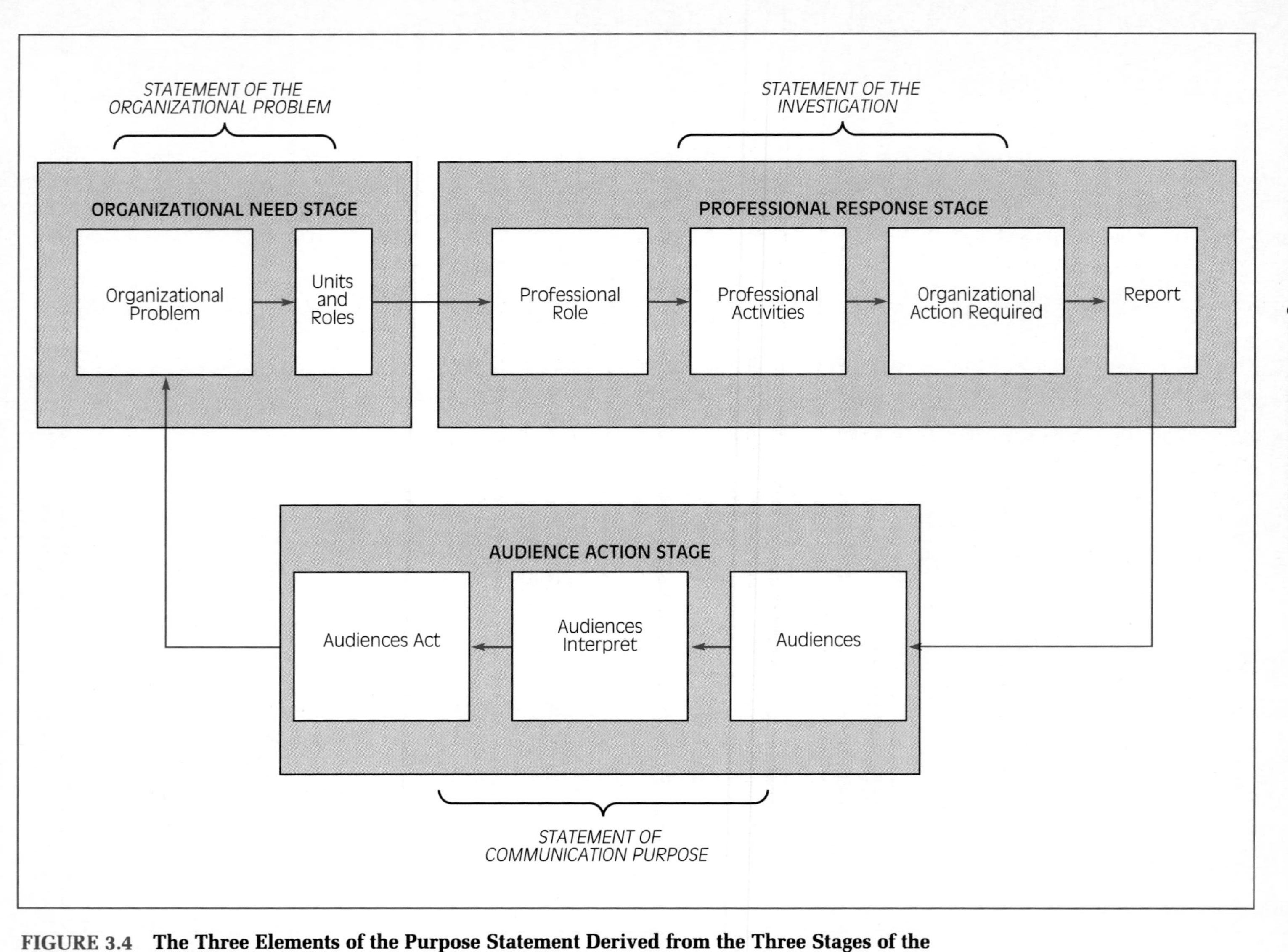

FIGURE 3.4 The Three Elements of the Purpose Statement Derived from the Three Stages of the Organizational Communication Process.

> Due to its high freeze point (158°F), biphenyl must be heated to near its flash point (223°F) before handling is possible. Even then its vapor pressure is low. Priming a pump to top-unload a tank car under these conditions is very difficult.

These statements present the "why" behind the investigation and the communication. They attract a reader's attention because they immediately place the report in an organizational context. The context clearly is not status quo. It presents a problem that needs to be resolved.

If you look closely at the examples, you can probably see a basic similarity among them. That is, all three statements—in admittedly different ways—identify the conflicts between two actual states of affairs. That can be made more evident if we slightly recast the statements into a simple and uniform syntax:

> (Fact A) We made a decision based upon an estimate given to us by a contractor. However, (Fact B) the contractor has now given us a revised estimate which is 600% over the original, and we are still expected to pay the bill.
>
> (Fact A) We must follow the new UL specification. However, (Fact B) if we follow the new specification, we will create a potential safety problem for users of our instruments.
>
> (Fact A) We are required to top-unload tank cars carrying biphenyl. However, (Fact B) we can top-unload biphenyl tank cars only if we heat the biphenal to dangerously near its flash point, and even then the pressure is too low for effective unloading.

As these examples indicate, in order to state a problem simply and directly, you can use a simple syntax: *Fact A; however, Fact B.* Many typical organizational problems can be stated in this way. For example, here is a statement with precisely the same syntax:

> The Shipping Department expected Warehouse C203 to be renovated by the date on which Martin Tool and Die shipped the Q-body dash. However, Spencer Construction will not have completed the job before the templates arrive.

We are not saying that there is a rigid formula to follow here. Naturally, the contrastive statements may take many forms and require quite different degrees of amplification. They might be stated as a compound sentence, a complex sentence, two simple sentences, or even a single simple sentence. Similarly, they might take only a few lines (as in the preceding examples) or in long reports they might require a paragraph or more for each of the contrasting elements. The point is that if you look for the conflict which lies behind your technical work, and if you attempt to state that conflict succinctly and immediately, you will help both your own understanding and the understanding of your readers.

Just ask yourself, "what is the conflict lying behind the technical work?" Then try to tease out that conflict by using a statement with this basic syntax: "A is a fact; however, B (which conflicts) is also a fact." What is the A side of your problem, and what is the B side?

The Investigation

The problem solving process is initiated when the organization discovers a problem. Perhaps you discover it for yourself; perhaps it is assigned to you. In any event, your role as a professional or manager is to be a problem solver. That is, in the second stage of the problem solving process you must perform some professional or administrative activity or investigation necessary to solve the organizational problem or meet an organizational need.

In simple terms, your function is to perform one, several, or all of these tasks:

to investigate the nature and consequences of the problem
to analyze problematic data
to identify causes of the problem
to determine feasibility of alternative solutions
to design an optimal solution
to perform tests documenting the problem and possible solutions
to establish effective ways of implementing a workable solution
to develop specifications for design, programming, or manufacturing
to prepare guidelines, procedures, instructions, or documentation for use

The nature of a professional's technical investigation is tied to the nature, degree, and status of the problem or need. Usually, it is what he or she has been trained to do as a professional—as an accountant, a forester, a social worker, or an engineer. The work that you spend much of your time on, answering technical questions and performing technical analyses, constitutes the second stage of the organizational problem solving process. Again, you want to convey a clear sense of what that work was, both for your own understanding and for the understanding of your readers.

Here are some examples of investigations required in this professional response stage:

We evaluated alternative methods of storing the templates.

Operations was asked to determine whether this was a part failure or an operator failure.

Consequently, Bob Liggett asked me to submit a plan on how the laboratory's existing instruments can be brought into conformity with the UL-817 standards.

Chief Thomas Brand asked me to determine if the Downtown Development Authority's request could be implemented within the Police Department's current budget structure.

We have completed the Comprehensive Test Package in accordance with your specifications.

I was asked to prepare a preliminary plan for the 1988–89 academic year with a proposed budget and implementation schedule.

These statements orient readers to the writers' professional activities and the subject matter of their reports. They are statements of the objective of the investigation that enable the readers to relate your activities to their needs.

Stating the investigation clearly is not difficult if you avoid getting bogged down in technical details. The statement often is best phrased in terms of the overall purpose or objective of the investigation rather than the methodology or specific technical tasks of the investigation. At times, however, even at the beginning of a report, you might conclude that a simple statement of the objective of an investigation might be insufficient to orient the reader to the report's purposes. In such a situation, you can introduce the more specific questions or tasks of your investigation. Often, you can do this most effectively simply by listing the questions which your investigation sought to answer:

Specifically we investigated three questions:

- What would be the potential advantages of the van-pool?
- How would the costs be shared between the employees and the company?
- Would company-sponsored van-pools be accepted by employees?

A short and uncomplicated list of specific questions (or tasks) often catches a reader's attention and provides a better sense of the scope of the investigation than a simple statement of overall objective. (Also, the list may set up a structure which you can use in the subsequent discussion.) But be cautious not to get bogged down in the excessive details, acronyms, and part numbers which, unfortunately, we see in so many introductions to technical and professional reports. Give your readers an overview of the nature and purposes of the investigation. Later on you can get into the more specific details.

One final point about statements of the investigation. Often, you will find it necessary to connect the investigation or subject matter of the report to the organizational problem, as some of our examples have illustrated. Readers understand the objective of an analysis when they understand its source:

Consequently, Operations hired RMT Inc. to conduct leaching tests on the foundry wastes.
Mr. Lanier asked me to investigate the feasibility of such a van-pool.

If your department or you in your professional role initiated the investigation in response to a problem, say so:

> We [or Safety Effects Laboratory] concluded that a percentage of the mutagenicity could have been an experimental artifact; therefore, we analyzed the Omaha test procedure.

The technique of stating the investigation or the assignment is to use the active voice, not the passive voice. ("Smith asked me to do something" not "I was asked.") The passive voice obscures the source of the assignment or the initiative and therefore omits organizational information useful for some of your readers.

Usually you will find that a general statement of the nature, objectives, or tasks of your investigation will be sufficient for your readers to understand your conclusions and recommendations.

❑ 3.3 ❑ The Communication Purpose

When you have finished your professional activities, you may have solved the problem conceptually and technically. However, you have not yet solved the problem or met the need organizationally. That is, you still must ensure that decisions are made and actions are taken to address the problem in the ways that your solution suggests.

The organizational need explained in the organizational problem indicates that you are not the only person with an interest in the problem or the only one for whom it has significance and consequence. Therefore, your role requires that you must interact with the organization and assure that the purposes that need to be accomplished are in fact accomplished. You must put on your writer's hat and tell people in the organization what must be done. That is, you must accomplish the third stage of the problem solving process, almost certainly using both written and oral communication paths to do it. Here you must establish a purposeful communication link between yourself and those readers in the organization who must act (primary audiences), those who will be affected (secondary audiences), and those who have supervisory need to know (immediate or nominal audiences).

The written documentation which you produce will function as a tool, a means by which your readers can respond as your purpose would have them respond. If you are successful, the understanding of the problem that your readers develop should roughly equal your own understanding. Their perception of your resulting technical work should be accurate and comprehensive. And on the basis of this understanding and your recommendations, readers should be able to act effectively. That is, your readers should be able to create the organizational responses required to implement a solution to the initial problem. The document which you write is only the means to that end. Its purpose is to effect that response.

Here are some examples of statements that signal the expected responses to the reports:

> This report recommends appropriate repair procedures.

Chassis Fuel Systems rejects the solder repair solution.

This report presents the actions necessary for Allen Foundry to obtain approval from DNR to use the Leaton landfill for waste disposal purposes.

After reviewing bid items #88-108 through #88-113, I recommend purchase of bid item #88-112.

This is a request for the United States Coast Guard to designate its representatives at the deadweight survey of Hull 324.

Accordingly, the contractor requests that these Quality Deficiency Reports be retired.

This report is designed to aid the Software Development Division in understanding and using the Comprehensive Test Package.

Your communication purpose, then, is understood and stated in the context of the problem of concern to the organization and the resulting investigation. It signals the organizational responses or needs which must be met if the issue is to be resolved. To state your communication purpose, use a verb that indicates the response you expect. Verbs such as "authorize," "propose," "recommend," "request," and "instruct" signal action on the part of the reader.

A Heuristic for Stating the Communication Purpose

In sum, a complete introduction or purpose statement for a problem solving document has three basic elements:

- It identifies the problem, conflict, or need as perceived by the organization.
- It specifies the objective of the investigation taken in order to solve the problem or meet the organizational need.
- It explicitly states the communication purpose, that is, the writer's reason for writing and how the writer intends others to respond to the organizational problem and the consequent investigation.

Each of these elements should be stated succinctly. (Often a single sentence will suffice for each element.) The following schematic suggests a possible order and syntax for the statements:[4]

- Organizational Problem
  (A fact or goal) ____;
  however, (a conflicting fact or goal) ________________________
  ____.
- Investigation
  Consequently, I did X (or Smith asked me to do X) ____________
  ____.

- Communication Purpose
  The purpose of this (document type) ______________________
  is to (writer's purpose) ______________________
  so that (reader's purpose) ______________________.

Here are some examples of three-step purpose statements that embody the entire organizational communication process:

> Recent customer demands indicate welding galvanized material on the #885 five-headed spot welder is essential. However, the welding parameters necessary to effectively weld galvanized material are different from the parameters we use on plain material. As a consequence, the shear strength values are below quality control standards. Jim Johnson, Manager of Manufacturing Engineering, assigned me to determine which welding parameters to change and how to change them. This report presents my recommendations for modifying the #885 Resistance Spot Welder to weld galvanized material.

> We are interested in developing more effective and efficient analytical methods on our adhesive base. Our objective is to obtain measures of physical properties on the base that will indicate with a high degree of certainty that the base will perform acceptably in production and yield an acceptable final product. We would like to employ Investigation Lab's services in assisting us with this problem.

> On November 24, 1988, at 1:15 a.m., Patrolman G. Adams was involved in a property damage accident at the intersection of Packard Ave. and Fourth St. Unit 241, driven by Adams, was damaged. I went to the scene to supervise the investigation. This report presents my recommendations on whether or not disciplinary action should be taken.

> We are in need of a new pumper for the Fire Department (see Request 32-88-0021) and have let bids for purchase of the same. After reviewing bid items #88-108 through #88-113, I recommend purchase of bid item #88-112.

These statements implement all three elements of the heuristic. If you master this heuristic, you will make writing as well as reading your reports much more efficient than perhaps they might now be.

Trouble Spots to Expect

Typically, writers' minds are full of the details of the investigation in which they have been immersed. Thus, they will tend to focus on this investigation, leaving out or glossing over both the problem and the purpose. On the basis of analyzing literally thousands of introductions to technical documents, we can generalize on the probable trouble spots to expect.

About the Problem You are likely to make any of these mistakes: you leave the problem out entirely; you present it, but dwell on the background, not on the immediate issue of concern in the investigation and document; or you present it in an overly technical way that is unclear to at least some of the readers who will use the document.

The following introduction, for example, seems broad and diffuse rather than sharply focussed:

> There is a need within the XXXXX Corporation to integrate all of the graphic materials into a common visual format. In today's competitive document management marketplace, where XXXXX was once the leader, it is essential that XXXXX visually project its image as the strong technological corporation it is. XXXXX is a decentralized corporation within which are six major organizations. These six organizations are then also divided within by smaller individual program groups. With few exceptions, each group has its own mark, symbol, or logotype. All of these images lend a confusing and overwhelming impact on the XXXXX logotype itself. The combination of a wide variety of symbols and the placement of these symbols in positions where emphasis is not on the XXXXX logotype creates a diffused image.

This statement contains background information that seems rather general for the situation. Certainly, readers within XXXXX (a major electronics corporation familiar to you) would be familiar with the decentralized corporate structure. However, some readers might find it difficult to identify the specific issue of concern.

The writer of this statement agreed with us that this introduction presented general background rather than clearly introduced the purpose of the report. She therefore rewrote it, using our three-step heuristic. Her rewrite seems much clearer:

> The XXXXX Corporation wishes to project a uniform corporate image through the design and placement of corporate logos on stationery and print materials. However, among the six major divisions of the corporation—and even within the divisions—differing logo designs and placements have developed. The objective of this investigation was to develop corporate-wide guidelines for standardizing logo design and placement. This report presents these guidelines for your review and approval.

The rewrite defines the problem succinctly and directly. Moreover, it makes clear the communication purpose of the report. The rewrite as a whole illustrates the usefulness of the purpose statement outline we have presented.

The organizational problem or need is the reason for your investigation. At times, you will be asked to do an analysis without being told what the problem is. When you come to report the results of your investigation, however, you need to first state the problem that initiated the investigation. To do so could require a few questions and phone calls. You should make the effort both for yourself and for your readers.

About the Investigation You are likely to make any of these mistakes: you define the investigation in such a way as to obscure the organizational responsibility for the work done; you state it in overly general terms; you state it in overly particular and technically unfamiliar terms.

The following introduction has all three of those problems:

> YYYYY Motors has recently initiated a significant corporate tire testing program. Improvement in the wet traction performance of the original equipment tires is one of the goals of this program. YYYYY Motors participation in wet traction studies has necessitated the development and construction of test surfaces which have controlled friction properties. This report concerns the design and building of one of these test surfaces.

As does the prior statement, this statement provides no hint of the organizational context. The reader has no idea who was responsible for the investigation presumably explained in the report. In addition, this statement is quite vague about the purpose of the investigation itself: "the development and construction of test surfaces" is either a background statement or an open-ended statement of the purpose of the investigation. The phrase "controlled friction properties" is overly technical for a management audience. After reading this paragraph, a management reader certainly would have questions about both the organizational problem and the purpose of the subsequent investigation. After all, at its Proving Grounds, YYYYY Motors has 162 miles of test surfaces. Why do they need more?

Perhaps the most common problem with stating the investigation is that writers go into far too much detail. They start to introduce their technical investigation rather than to introduce the report. For the purpose statement itself, you should limit yourself to only that information necessary for the reader to understand the purpose of the report. There are other places in the report to introduce the technical investigation itself.

About the Communication Purpose You are likely to make any of these mistakes: you do not state the communication purpose at all; you state it, but in a way that confuses the purpose of the report with the subject of the investigation; you state it in an overly general, vague way that does not answer the questions of readers.

The introduction of the report on the diffused image of the XXXXX Corporation has no statement of communication purpose at all. The introduction of the report on test surfaces confuses the purpose of the report with the subject matter of the report. The statement "This report concerns the design and building of one of these test surfaces" in a general way states what the report is about. However, it does not state the purpose of the report.

The failure to state a communication purpose is quite common. It indicates a failure of the writer to turn from the investigation and face the reader. For the following four statements, ask yourself, "why?":

> The purpose of this communication is to discuss the impact of this proposal on all units of the hospital. [Why? Why are you discussing this? Why should I read it?]
>
> This report presents the findings of the preliminary investigation into the use of company-sponsored van-pools as an alternative to private vehicle transportation. [Why?]
>
> The purpose of this report is to provide information on the progress of the installation and overall cost figures to date. [Why?]
>
> This report presents a structural analysis of the frame that was developed. [Why?]

As we have discussed, the purpose of written communication is to affect some organizational process. You should define your purpose in terms of your intended impact.

With this discussion as background, try analyzing a number of specific examples of your own writing to see if the introductions to your reports have any of these trouble spots. If so, read back over the preceding discussion and then give your statements another try. As was the case with the person from the XXXXX Corporation, after you know what to look for, it is often a simple matter to spot the difficulty in an ineffective introduction and to know how to fix it.

❑ 3.4 ❑ Presenting and Adapting Your Purpose Statement

An effective definition of the communication purpose for a report is a good starting place for both the writer and the reader. With practice, you should be able to use the heuristic that we have discussed to produce effective introductions quickly and efficiently. However, you should also be able to adapt the heuristic to meet the needs of your own style, different document types, and different writing situations. After all, not all reports are problem solving reports, and audience situations and purposes are always distinctive. Accordingly, you need to be both alert and flexible.

One important thing to recognize is that the functions of reports in organizations do indeed differ. That is, your reports vary in purpose according to the particular kinds of organizational interactions which they involve. We classify report purposes along a spectrum, with problem solving functions at one end and administrative functions at the other. In between are various coordinating functions.[5]

Problem solving functions, as we explained, support the decision-making process, either to enable a decision to be made or to implement decisions once they have been made. These types of reports are important for management purposes in all organizations. Typically, they resolve nonroutine problems, they involve decision making to modify an organizational process, and they can deal with any stage of the problem solving process.

Because these problem solving reports seek to modify organizational processes, they are not only the most important reports you will write, they are the most difficult. This difficulty is indicated by Herbert Simon, who observes that "the possibility of permitting a particular individual to make a particular decision will often hinge on whether there can be transmitted to him the information he will need to make a wise decision, and whether he, in turn, will be able to transmit his decision to other members of the organization whose behavior it is supposed to influence."[6] It is precisely because of the importance and the difficulty of these types of reports that we have concentrated until this point upon a heuristic for writing their introductions.

But there are also administrative and coordinating reports to be written as well, and their introductions sometimes can be handled somewhat differently.

Administrative functions support organizational routines and "become matters of organization practice, perhaps embodied in manuals of practice and procedure."[7] For example, monthly progress reports of the sort required in many organizations serve a routine administrative function. They are primarily informational, providing information for various administrative functions such as documentation for the record, budget planning, or analysis of production schedules or personnel needs. Admittedly, these reports also serve other functions at times, but frequently they are required parts of operational processes which stipulate certain reports at certain times. Other types of routine administrative reports include trip reports, maintenance records, minutes of meetings, space or equipment inventories, and annual evaluations of personnel.

In between problem solving and administrative functions are various sustaining or coordinating functions for written communication. The principal function of these various reports is to coordinate organizational activities, either by maintaining ongoing processes or by facilitating modifications of these processes. Many types of reports perform various coordinating functions. For example, progress reports or status reports might be used to coordinate activities among several individuals or units involved in different aspects of a project. Research reports, test reports, long-range planning documents, specifications, feasibility studies, and laboratory reports can also function as coordinating devices, although they sometimes also function as administrative devices as well. Again, the labels or names of these reports are not important; their functions are.

Unlike problem solving reports, many administrative and coordinating reports can open with very simple statements of purpose, omitting either a problem statement or a statement of investigation, or both. For some routine reports, in fact, even the communication purpose is implicit in the subject line of the report and can be omitted. As an illustration, the presentence investigation report (Figure 3.5) has no purpose statement. Notice that the standard form itself, labeled "Presentence Investigation Report," signals the purpose of the report. All audiences immediately know the purpose of the report when they see the form. Many other reports in standard forms and formats, such as Status Reports, Minutes of Meetings, and Engineering Change

MICHIGAN DEPARTMENT OF CORRECTIONS CF0145

PRESENTENCE INVESTIGATION REPORT

Honorable_______________County_______________Sentence Date ___________
Docket_______________Attorney_______________Appt.___________Retained_______
Defendant _______________Age_____________D.O.B. _____________

CURRENT CONVICTION(S)

Final Charge(s)	Max.	Jail Credit	Bond	Proposal B
1.___________	______	_____days	___	______
2.___________	______	_____days	___	______
3.___________	______	_____days	___	______

Convicted by: Plea____Jury____Judge____Plea Under Advisement____ Nolo Cont.____HYTA: Yes____ No ____
Conviction Date_______________Plea Agreement _______________
Pending Charges:_______________Where _______________

PRIOR RECORD

Convictions: Felonies_____Misdemeanors____ Juvenile Record: Yes_____ No _____
Probation: Active_______ Former_________ Pending Violation _______________
Parole: Active_______ Former_________ Pending Violation _______________
Current Michigan Prisoner: Yes_____ No_____ Number_______________
Currently Under Sentence: Offense_______________ Sentence _______________

PERSONAL HISTORY

Education_______________ Employed ___________ Where _______________
Psychiatric History: Yes____ No____ Physical Handicaps: Yes____ No____ Marital Status_______________
Substance Abuse History: Yes_____ No_____ What_______________ How Long _______________

RECOMMENDATION

Agent_______________ Caseload No._______________Date _________
Signature_______________Supervisor's Approval: _______________

FIGURE 3.5 Format of Presentence Investigation Report Signals Communication Purpose.

Notices, have no explicitly stated purpose statements. In these cases, the purposes of the reports are embedded in the terms "status report," "minutes of meeting," or "engineering change notice."

We want to stress that it is wise to be cautious in adapting the heuristic for the purpose statements previously discussed to fit different types of reports and different situations. Obviously, you don't want to confuse a routine administrative or coordinating report with an important problem solving report. But it is possible to adapt the purpose statement in a variety of ways, depending upon the function of the report, its contents, its audience, and the situation. With that thought in mind, here are some possible variations that do make sense at times.

You can put the purpose first

The preceding three-step heuristic has an order and a logic based upon the order of the organizational communication process. That is, the problem comes first. The problem leads to the investigation. And the investigation in turn leads to the communication purpose.

This order certainly would be appropriate in problem solving reports when your audience analysis indicates that you need to prepare your important readers before stating a recommendation that they might not expect. However, at times, you can attract immediate attention by putting the statement of communication purpose first and then providing the statements of the organizational problem and the investigation which explain or support it:

> This report presents my recommendations for modifying the #885 Resistance Spot Welder to weld galvanized material.
>
> Recent customer demands indicate welding galvanized material on the #885 five-headed spot welder is essential. However, the welding parameters necessary to effectively weld galvanized material are different from the parameters we use on plain material. As a consequence, the shear strength values are below quality control standards. Jim Johnson, Manager of Manufacturing Engineering, assigned me to determine which welding parameters to change and how to change them.

Placing the statement of communication purpose first usually is appropriate in design and manufacturing situations where modifications of ongoing processes are the rule, not the exception, and where the topics of discussion are generally familiar to all potential readers. It is also appropriate in routine administrative and coordinating reports. (For example, "This is a status report for the Publications and Marketing Department. It covers the period 1 July 1990 to 30 September 1990.") Putting the communication purpose first also makes sense when the the complexity of the problem or the investigation would require more than a few sentences of introductory explanation. You don't want to delay the reader from getting to the purpose statement quickly.

Naturally, you need to do your audience analysis in order to decide whether or not to put the statement of communication purpose first. Will your audiences know what you are talking about? Does the purpose statement make any sense standing by itself? How much space will it really take to define the problem and the investigation? Do you need to lead into your purpose diplomatically, or can you be up-front blunt? These are, of course, issues to consider with every document; but at least be aware that in some problem solving reports and in many more routine administrative and coordinating reports, you can start with your purpose statement.

You can expand the problem and investigation elements

Occasionally you have to write a longer report, sometimes a bound "formal report." These reports differ from the working memoranda, letters, and short reports which you will spend most of your time writing, and which we have been discussing and illustrating. Longer reports are the result of major studies, design projects, research projects, and other special technical investigations. Research and development units often produce reports of this sort, as do consultants who report to external clients.

These formal reports often have what we call "extended purpose statements" to provide an introductory framework for the reader. This is appropriate when you expect your readers to read the report closely. But note that such reports are usually broadly informational rather than for immediate decision and action. They are the basis for analysis and deliberation. An example of an extended purpose statement is the introduction to a consulting firm report on the Charlevoix, Michigan, Water Supply System (Figure 3.6). In this introduction, the problem is stated in the first paragraph, and the specific objectives or questions addressed in the investigation are stated in the second paragraph. The third paragraph explains the methodology. Finally, the fourth paragraph presents the communication purpose.

As you can see, essentially the same elements are present as in the shorter examples which we illustrated; however, each of the elements (particularly the problem and the investigation) is amplified into a paragraph-level treatment in the more formal report. For memoranda and letters, this degree of amplification would be excessive; but in formal reports, readers generally expect some sort of amplified background, not the very succinct memorandum form of introduction.

You can add a forecast

The introductory section for the Charlevoix, Michigan, Water Supply System report illustrates a possible addition to our three-part heuristic: a forecast. For many reports, this additional element is included to help prepare readers to deal with structurally complex or long discussions.

The forecast outlines the organization and the content of the report. It should be specific, precisely identifying the sections of the report. As you can see from the table of contents of the Charlevoix Water Supply System Report, the forecast items are identical to the headings of the three sections.

As we discuss in the next section, however, the forecast most often is included in the beginning of the discussion. The summary directly follows the purpose statement in short reports and memoranda. Thus, a forecast usually is unnecessary and, in fact, would interrupt the logical connection of the purpose statement to the summary. It comes after the summary, when it is included, in order to introduce the discussion to those readers who read the discussion.

You can modify the pattern for research reports and technical articles

To this point we have been discussing the purpose statement for reports whose purpose is to modify organizational processes or facilitate organizational decision making, which is our focus in this book. Occasionally, however, many professionals in organizations have to write documents which are primarily informational rather than primarily instrumental. Two common types are scientific and technical articles and research reports (Figure 3.7). Our discussion of the purpose statement provides some guidelines for writing introductions to these types of documents, but variations are appropriate.

Introductions to research reports and articles are essentially extended purpose statements that establish the objective and specific purposes of a scientific or technical investigation.[8] They begin by first establishing a research problem or need, often by reviewing the historical background of a particular research topic. This is analogous to a statement of the organizational problem, but research reports and articles often begin with extended explanations of some specific scientific or technical unknown or need. Unlike the more focused problems of problem solving reports, these needs or unknowns are open ended. They are conflicts between our desire to know something and the fact that we do not know it. Often we have no idea whatsoever about the potential values of knowing something. There may be no clear and immediate application of the knowledge. We merely wish to know it.

Research reports and technical articles also begin by spelling out the specific research objectives or technical questions to be addressed or answered in the article. Whereas problem solving reports tend to emphasize the problem and purpose, often technical articles and research reports pay relatively more attention to the methodology. This is analogous to the statement of the investigation.

Statements of the communication purpose of technical articles and research reports are often omitted, as the documents primarily are informational.

Introductions to technical articles and research reports often include an additional element: an hypothesis. The hypothesis usually is presented after the statement of the research problem. The formulation of the hypothesis

TABLE OF CONTENTS

FIGURE 3.6 The Table of Contents and Extended Purpose Statement from a Consulting Firm Report. (3 pages)

CHARLEVOIX, MICHIGAN WATER SUPPLY SYSTEM IMPROVEMENTS

INTRODUCTION

The purpose of this investigation was to review the adequacy of the present Charlevoix water supply system to meet present and anticipated future water demands of the community. The present system was installed in 1922, when the permanent population of Charlevoix was about 2200. The population since then has doubled, and recent projections indicate that the population will double again by the year 2000.

The following questions therefore arise:

1. Does the present system have a maximum safe capacity to meet projected water demands?

2. Are the pumping, distribution, and elevated storage facilities capable of meeting projected demands?

3. Are these facilities capable of meeting recommended fire flow for the projected population?

The present system also lacks the disinfection facilities now required by the State for all municipal water supplies.

To evaluate the adequacy of the present water supply system in the face of future demands, the study uses population projections to establish future requirements and analyzes the capacity, equipment, and facilities of the present system. A comparison of future requirements with the abilities of the present system to meet those requirements yields the recommendations for improvement of the Charlevoix water supply system.

We use existing records, field studies, and earlier engineering reports to determine the needs of the water system. We review all general areas of the present water system. This report recommends several improvements in the present system plus some additional construction to bring the total system up to levels sufficient to meet anticipated requirements. The report presents the estimated total cost of the projects and a breakdown of costs for each part of the project.

The report is divided into the following main sections:

I. WATER SUPPLY REQUIREMENTS. This section analyzes the trends of population growth, pumping patterns, and per capita consumption in order to project future design flow requirements.

II. THE PRESENT WATER SYSTEM AND RECOMMENDED IMPROVEMENTS. This section compares the future requirements with the abilities of the present system and recommends system improvements.

III. COST ESTIMATES. This section itemizes the estimated costs for the improvements recommended.

TYPE OF REPORT	PURPOSE
Organizational Reports	Decision and Action
Scientific and Technical Reports (Research and Development)	Information, as Basis for Further Research and Development Action
Scientific and Technical Articles (and Papers)	Information, Contribute to Scientific and Technical Knowledge

FIGURE 3.7 Scientific and Technical Articles and Reports Differ in Purpose from Organizational Reports.

dictates the specific research objectives and methodology and thus provides their rationale. Often, because the document is presenting the results of the research, the implication is that the hypothesis has been affirmed. Thus, the statement of the hypothesis provides an implicit summary or forecast of the conclusions that are presented in the document.

Introductions to many research reports and articles also forecast the content of the rest of the document. This usually is necessary because the purpose of the document is informational. Readers need a forecast to provide a framework for understanding the particulars, which are presented in scientific and technical terms.

The introduction to the scientific article, "The Attitude Gap" (Figure 3.8), provides an extended problem statement that introduces the specific questions addressed in the study, which is the subject of the article. Particularly illustrative is the detailed problem statement in the first two paragraphs. The first paragraph states the first situation (the A), that a number of whites support racial equality. The second paragraph states the second situation (but, B), that whites don't support policies to ensure equality.

With the problem clearly framed, the issues or hypotheses for this conflict or difference are raised in the third paragraph. This then leads to the objectives (questions) of the research being discussed. These questions and an overview of some of the results of the study are presented in the accompanying figure and legend.

The introduction to the management journal article, "Conflict in Organizational Decision Making" (Figure 3.9), also has an extended purpose state-

15

THE ATTITUDE GAP

White Americans endorse racial equality, yet show little support for policies to achieve it.

The right of Black children to attend the same schools as whites or of Blacks to live where they choose is widely endorsed by white Americans. Public opinion surveys dating back to World War Two show a steady increase in the number of whites who say they believe in principles of racial equality.

But many of these same surveys reveal much less support for governmental policies designed to insure equal treatment. In recent years, social scientists have been debating the meaning of this gap between white support for principles and support for the policies meant to implement them.

Some analysts suggest the gap shows that white support for the principles of racial equality is mostly talk meant to appease Blacks while maintaining the status quo. Others think that many more Americans truly believe in racial equality today than forty years ago, despite lagging support for affirmative action or busing.

In 1985, a group of social scientists at the U-M's Institute for Social Research (ISR) published *Racial Attitudes in America* (Harvard University Press), a comprehensive examination of the attitude trends suggested by survey data collected during the last four decades. Authors Howard Schuman (director of ISR's Survey Research Center), Charlotte Steeh, and Lawrence Bobo look at responses to questions about principles of racial equality, enforcement of desegregation, and other indicators of how whites (and, to some extent, Blacks) view matters of race.

Is Progress Real?

One goal of the book, explains Schuman, was to address the controversy over the significance of the principle–implementation gap. In four articles appearing in *Scientific American* between

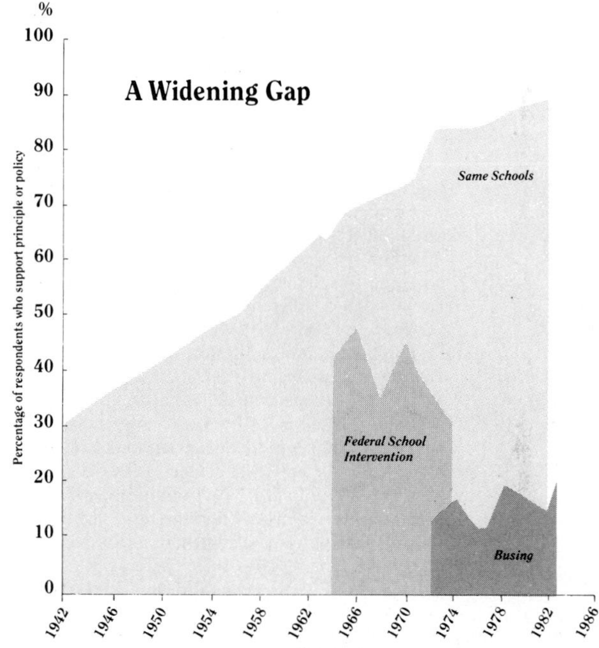

Surveys of white Americans show a growing gap between support for racial equality and for policies promoting equality. The University of Chicago's National Opinion Research Center (NORC) and U-M's Institute for Social Research (ISR) asked the questions shown here.

Same Schools (NORC): "Do you think white students and Black students should go to the same schools or to separate schools?" (Percent responding "same" shown on graph.)

Federal School Intervention (ISR): "Do you think the government in Washington should see to it that white and Black children go to the same schools, or stay out of this matter as it is not its business?" (Percent responding "government should see to it" shown on graph.)

Busing (NORC): "In general, do you favor or oppose the busing of Black and white school children from one school district to another?" (Percent responding "favor" shown on graph.)

FIGURE 3.8 Introduction to a Science Article.

MANAGEMENT SCIENCE
Vol. 36, No. 4, April 1990
Printed in U.S.A.

CONFLICT IN ORGANIZATIONAL DECISION MAKING: AN EXPLORATORY STUDY OF ITS EFFECTS IN FOR-PROFIT AND NOT-FOR-PROFIT ORGANIZATIONS*

CHARLES R. SCHWENK
Indiana University, Graduate School of Business, Department of Management, Bloomington, Indiana 47405

Though past research has shown that conflict may improve organizational decision making, business executives may have very different perceptions of the effects of conflict than executives of not-for-profit organizations. This exploratory study deals with executives' descriptions of the effects of conflict on their own organizations' decisions. Results show that high conflict is associated with high quality for the executives of not-for-profit organizations but with low quality for executives of for-profit organizations. Analysis of executives' written descriptions of these decisions suggests some reasons for this difference.
(CONFLICT; DECISION-MAKING; NOT-FOR-PROFIT ORGANIZATIONS)

Introduction

Research from a variety of perspectives shows that conflict and disagreement between decision makers can improve organizational decision making (Schweiger, Sandberg, and Ragan 1986; Schwenk 1982, 1984b; Tjosvold 1985). However, despite this evidence, research has shown that decision makers' dislike of conflict may lead to a reduction in conflict under stress (Janis and Mann 1977). Conflict may have very complex effects in decision making. Though it improves the quality of decisions, its aversiveness may cause decision makers to *perceive* it as harmful and to avoid it (Schweiger et al. 1986).

This study addresses several questions. First, is it true that executives perceive conflict as being aversive and harmful in organizational decision making? Second, are the perceived effects of conflict different for different types of organizations? Last, what may explain the differences in decision maker's perceptions of the effects of conflict? However, before these questions can be addressed, it is necessary that a more complete definition of the concept of conflict be provided.

Pondy (1967) has identified four definitions for the term "conflict" as it has been used in research. The term has been used to describe: (1) antecedent conditions (e.g. scarce resources), (2) affective states (e.g. tension or anxiety), (3) cognitive states (e.g. cognitive conflict or the perception of conflict), and (4) conflictual behavior (overt disagreement or resistance) (1967, p. 298). This study deals with conflictual behavior which occurs during decision making. When the term "conflict" is used in this paper, it refers to conflictual behavior in the context of a particular decision.

Previous Research on the Effects of Conflict

Several streams of research have dealt with the effects of conflict on the decision process. In this section of the paper I will discuss research which deals with specific aspects of decision-making and conclude with a summary of the probable effects of conflict on the decision process as a whole.

Maier (1970, p. 227) has shown that free discussion of divergent views on a problem may produce a larger number of correct answers to the problem if the discussion is

* Accepted by Arnold Barnett; received February 1988. This paper has been with the author 2 months for 2 revisions.

0025-1909/90/3604/0436$01.25

FIGURE 3.9 **Introduction to a Management Journal.**

ment with the first two elements very clearly stated. The first paragraph states the problem: A but B. Research indicates that conflict can improve decision making; however, decision makers avoid conflict because they think it is harmful. The second paragraph states the investigation in the form of three specific questions that the research addressed. The third paragraph is an additional element: a definition of the concept of conflict in order to specify the purpose of this investigation. In this article, then, the "Introduction" takes the form of what we call a "purpose statement."

Therefore, although research reports and scientific and technical articles seem to have detailed, technical introductions, they do more than simply present "background" or other types of introductory material. They are systematic, albeit detailed, purpose statements. They establish a problematic context and then the specific objectives of the research to be explained in the document. If you master the heuristic for formulating the purpose of an organizational report, you should be able to adapt it for writing the introduction in a research report or article.

Conclusion

In all likelihood, you will spend a lot of time on the job doing what might be called "bread and butter writing": letters, memoranda, short problem solving reports, and a host of routine administrative and coordinating documents. Perhaps you will also write research reports and technical articles. In either case, the basic heuristic which we have presented in this chapter is a good starting point for you to think about as you begin to write. By thinking carefully about your purpose, you will help yourself and certainly you will help your readers. Just remember, though, you need to appreciate the variety of functional objectives of communication that we have introduced in this chapter in order to formulate your communication purpose for any specific report. For each report, you need to determine your communication role. Our procedure for stating your communication purpose will enable you to do so.

NOTES

1. Peter G. W. Keen and Michael S. Scott Morton, *Decision Support Systems: An Organizational Perspective* (Reading, MA: Addison-Wesley): 56. Their distinction, made in order to clarify the function of data bases, seems applicable to report writing as well. Management Information Systems (MIS) and Decision Support Systems (DSS) are structured means of transferring information, with the professional's judgments built into the structure of the systems. Our approach to written communication can be seen as complementary, as a report presents the professional judgment that cannot be programmed.
2. *Transcript* 271.
3. Scholarship on the subject of problem statement is relevant here. See Richard E. Young, Alton L. Becker, and Kenneth L. Pike, *Rhetoric: Discovery and Change* (New York: Harcourt, 1970), especially chapter 5; also, Charles E. Lindbloom, *The Policy-Making Process* (Englewood Cliffs, NJ: Prentice-Hall, 1968): 13–14 is particularly relevant.

4. We are indebted to Dr. Carolyn Miller for suggesting this schematic for our purpose statement heuristic.
5. See Henry Mintzberg, *The Structuring of Organizations* (Englewood Cliffs, NJ: Prentice-Hall): 12, 13, and 63. Mintzberg characterizes an organization as consisting of a variety of systems: as a system of formal authority, as a system of regulated information flows, as a system of informal communication, as a system of work constellations, and as a system of ad hoc decision processes. The function of reports is to facilitate and specify their interactions.
6. Herbert A. Simon, *Administrative Behavior*, 3rd ed. (New York: Free Press, 1976) 88.
7. Simon, *Administrative Behavior* 154.
8. A particularly useful reference for writing research reports and articles is James P. Zappen, "A Rhetoric for Research in Sciences and Technologies," *New Essays in Technical and Scientific Communication: Research, Theory, and Practice*, ed. Paul V. Anderson, R. John Brockmann, and Carolyn R. Miller (Farmingdale, NY: Baywood, 1983): 123–38. His concept of "context-oriented rhetoric," illustrated by introductions to research articles and reports, is analogous to our purpose statement heuristic. An alternative to the Zappen model is presented in John Swales and Hazem Najjar, "The Writing of Research Article Introductions," *Written Communication* 4:2 (1987): 175–91. The Swales model is similar in fundamental rhetorical structure, as Swales acknowledges (note 1, 188). His order of subsections or "moves" also is analogous to that of our heuristic. However, Swales devotes considerable attention "to the question of whether announcements of principal findings (APFs) are also inserted at an introduction's close" (179), an issue of basic structure which we discuss in the next chapter.

SECTION II

Designing the Communication

CHAPTER 4

Planning the Basic Structure

A functional approach to report writing requires you to establish an effective basic structure before you get into the details. Your first concern is with the function of the report, providing information that readers need in order to respond or act so that the report accomplishes its purpose. The information that most readers need, however, usually is minimal—far less than the information you have available and indeed considerably less than you put into your report. Thus, you plan your basic structure according to reader needs rather than in terms of your investigation.

In this chapter, we introduce the principles of basic report design: your preliminary design considerations. We discuss the conceptual architecture of a report—how many levels or sections a report has and their relationships to each other. In the following four chapters, 5 through 8, we get down to detailed design. There, we explain how to design the particular aspects of the various levels or sections of a report.

In this discussion of basic structure, and indeed throughout this book, we are presenting principles and guidelines rather than talking about specific types of reports. We focus on the design and writing process that you should be able to apply to all types of reports you will have to write. Thus, we discuss basic structure in a generic sense. In actual practice, such as in a formal report, an informal report, or a letter, this generic structure can assume many distinctive forms. Our intent in this chapter is to introduce you to the principles supporting all of them.

The design of your basic structure is based on the behavior of persons in organizations. Our experience in industry has been that almost without exception professionals have no time to read a report and have very little interest in the content of a report. One management consultant said that, at most, managers have time for a "1-minute look" at any document. Another plant superintendent (and former design engineer and project manager) said, "If it's not on the first page, it isn't going to get read."[1] In part this is the result of the "unrelenting pace" of management activities; it is also that managers "are oriented to action and dislike reflective activities."[2] For almost all persons, however, the organizational context indicates that your readers have little interest in the content of your report. As we have discussed, you are writing to persons in terms of their roles in the organization. Their interest, therefore, is only in the information necessary for them to perform their tasks effectively and efficiently.

From the management perspective, researchers have long recognized that all managers read the summary or abstract of a report, but that few read the body or appendices.[3] Managers are interested only in the implications or impact of the report on their activities. Of course, not all readers are managers. Some people (for example, members of the manager's staff) must read into the body of a report and do need considerable degrees of detail. Some people read the body of a report, and some even dig into the details of the attached documentation. For example, a report on a part failure of a reactor baseplate (with a loss of over $50,000 and a potential loss of $2,700,000 if the reactor had been running at design capacity) concluded that the failure had been caused by welding slag plugging a valve. (The welding slag was produced during replacement of a cooling system pump.) Most readers of the report needed to know only that the cause of the failure had been determined and that revised repair and operating procedures were now in effect. A few readers, however, such as the supervisor in charge of the process reactor, had to read the details of the investigation of the accident. Furthermore, as you can see, the supervisor of the welding crew, and even the welders themselves, needed to be alerted to the fact that procedures used earlier had caused problems, and they needed details of new procedures so that similar problems could be avoided in the future.

Thus, reports in organizations differ considerably from the types of writing required of students in most of their professional courses in college. As we have indicated, in these courses students typically write for one person, a professor. The professor usually knows more about the subject matter of the report than the student does. The professor also knows the "answer" ahead of time; in fact, many well written student reports hold no surprises for their readers. In addition, the professor (as well as the student) focuses on the methodology, the analysis, and the calculations rather than on any conclusions that could be formed. The professor is concerned that the student has mastered the methodology and the skills he or she is training the student to perform. The student is intent on demonstrating mastery of those skills. Student laboratory reports therefore are arranged to focus on the particulars of the investigation, not on the organizational implications of the results.[4] Usually, an organizational context for the investigation or experiment is lacking in most student lab work.

Organizations then hire the professors' students. They do so because the students have mastered the skills taught by the professors. When students become professionals, however, they don't have to demonstrate this mastery in their reports; technical expertise is simply assumed as a basic requirement of their positions. They write their reports in terms of the organizational implications of their work, not in terms of their investigation. This organizational approach is illustrated by the report to "Approve Relocation of Light Duty Conventional Front End Sheet Metal Mounting" (Figure 3.3). Notice how the organizational information dominates all sections of the report, even the "Discussion." Details of the tests themselves and the data on the results of the tests—information usually expected in students' test reports—are omitted.

You plan your basic structure to address the diversity of your readers' needs. Essentially this means that you design your report to be read in parts by different people, not as a whole by the same person. You assume that no reader should have to read the entire report, and you plan it accordingly. For some of your administratively important readers—supervisors, managers, and upper management—you strive for the ideal in technical writing: you write a report that doesn't have to be read but still achieves its purpose.

To accomplish these objectives, two principles can be seen to underlie basic report structure: (1) reports have three parts; (2) reports move from general to particular. These basic principles are inherent in many types of reports which appear to be quite different from each other. In other words, behind the surface differences, these reports share a common basic architecture. You can even adapt these principles to writing technical articles and formal technical reports.

❑ 4.1 ❑ Reports Have Three Parts

Effective reports have three basic parts: an overview, a discussion, and documentation (Figure 4.1). Each of these parts should be considered as a self-contained unit. That is, each part should be written to be read independently and even selectively, probably by different readers. Only with a simple 1- or 2-page document might you expect your readers to read the entire document, although even with these you should establish an effective basic structure. You design your report in three parts so that different readers can read different sections of your report without having to read the entire document.

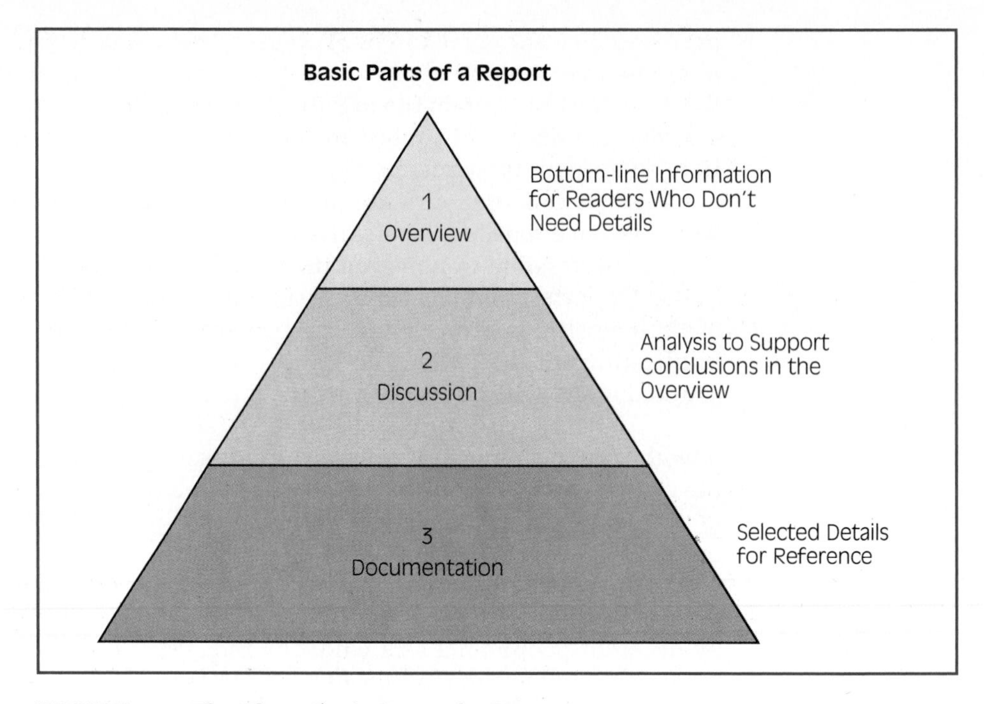

FIGURE 4.1 The Three Basic Parts of a Report.

The Overview

The overview of your document primarily addresses management readers, although it also meets the needs of some secondary audiences and immediate audiences. In the overview you present only the bottom-line information of your report, that is, the most important, action-oriented information. This information focuses on the organizational context rather than the technical content of the report.

Few readers, especially managers, need particulars or details to understand the decisions and actions that a report recommends. You present your conclusions and recommendations in organizational terms rather than in terms of your investigation, and thus do not need the technical particulars. Managers usually will read no further in a report. Many other readers also need to know only the organizational implications of your report. If you meet their needs in one page, you have communicated with them.

The overview provides a useful function for other readers as well. For readers who need to read other sections of your report, the overview indicates that there is information important for them to read and guides them to that information. For the few readers who will read most of your report, the overview provides a framework or road map so they don't get lost in the details. They know where they are going.

The overview also addresses nominal audiences. It indicates who should receive the report, helping to ensure that the lines of communication are adequate to its transmission. Because the overview has an organizational focus rather than technical focus, it helps get the report into the right hands for the appropriate decisions and actions to be taken.

The overview of a report goes by a variety of different names in different companies: "executive summary," "summary," "conclusions and recommendations," "opening component," "brief," "abstract" (actually a misnomer, as we will explain later), and even "introduction." Regardless of what it is called, this initial element of a report presents the most important organizational information.

The Discussion

The discussion of your report usually addresses technical and staff persons with an interest in some of the technical aspects of your report. Some will be technical persons who need the technical information for their own purposes, such as to revise a system component. Some will be other professionals who must act after a manager accepts a recommendation, such as persons who have to write specifications or deal with vendors. Some will be staff persons who must analyze the technical information in order to advise the manager who must make the decisions which your report recommends. The discussion itself, therefore, usually has diverse readers who read selectively. Often, it is segmented so that it can be read in this selective manner, especially in longer reports.

The purpose of the discussion is central to its role in the basic structure of your report. The function of your discussion is to present your technical analysis, to prove or support the conclusions and recommendations of your report. In a sense, what is in the overview must rest directly upon what is in the discussion. That is, your discussion is designed to present your thinking, not to present the technical documentation. It is quite selective in detail and has a conceptual focus. It doesn't present a mass of particulars or detail. Instead, it presents your reasoning—your analysis—and selects those particulars which support your reasoning and conclusions. The discussion is an intermediate level of your report in terms of detail.

Again, the discussion goes by several names in industry: "discussion," "body," "analysis." Among these, perhaps the term *analysis* is most meaningful in view of the function of this part of a report.

Documentation

The third part of your report, the documentation, presents selected data and other particular information to support the analysis in the discussion. Again, there is a direct relationship between what needs to be in the documentation and what is in the analysis or discussion section of the report. The analysis rests upon the details of the documentation.

The documentation serves two functions. First, in relation to the discussion, it serves as a repository for particulars that would clog the discussion and obscure the analysis. Second, it provides additional information for some readers without thrusting unnecessary information on other readers.

The documentation for a report has many forms: reference memos, computer printouts, calculations, parts lists, and so on. These are appendices or attachments to a report. Often, they are highly formatted with little prose. They usually address specialists who need little interpretation and explanation.

A final note on the documentation. It too should be selective. (We emphasize "should be." Often it is not. Unfortunately, many report writers sweep up all the paper available at the end of a project, fire a staple through it, and call it "appendix." This is an approach to avoid.) The complete documentation for a report remains in the file (cabinet or electronic) for future reference. Any documentation attached to a report should be selected for the convenience of some readers in terms of the purpose of the report and of the analysis in the discussion. Of course, many short reports contain no documentation; instead, they reference relevant documentation.

Meeting the Needs of Diverse Audiences

The purpose of this 3-part design is to solve the problems of writing for audiences in organizations. Your report has different types of audiences with different information needs. With a 3-part design, you can meet these different needs in such a way that your readers can skip irrelevant information. They can read selectively and, therefore, efficiently.

The need for a 3-part structure is illustrated by the report "Reflag of Interlakes Shipping Corporation Ships," which is a negative example because it lacks such a structure (Figure 4.2). All readers must read the entire report, and read it carefully, in order to understand the conclusions and to uncover the actions requested in the report. The report essentially lacks a summary: it is all detailed discussion, with documentation attached. It does not meet the needs of diverse readers.

In contrast, the letter on the "Clutch to disengage turbine starting pump" (Figure 4.3) illustrates how an effective basic structure meets the needs of diverse readers. It is addressed to a manager, but also calls the report to the attention of a supervisor. The overview of the report is completed at the end of the second paragraph. The manager has to read no further; he knows that the writer has a solution to his problem "at a total cost of $369.60." The supervisor, however, will read the details to verify the cost breakdown and evaluate the installation instructions. He might have to refer to some of the documentation attached (Bulletin 326-B). The documentation in addition addresses other secondary audiences—those who operate and maintain the equipment and those who need to know the specific details behind the selection of the recommended clutch accessories. The 3-part structure enables different readers to read selectively.

For many reports, the overview addresses a primary audience of managers and other decision makers (Figure 4.4). James Souther's article, "What to Report," indicates that all managers read the summary (i.e., overview) of a report. As that study indicates, these readers have little interest in the details of the analysis or discussion of the report. The analysis as well as the documentation is for the various other audiences who need to know details of the problem and analysis. Some need to be convinced of your conclusions, especially those readers involved in the problem you were analyzing. Some need the analysis in order to implement the solution. Others will be affected when the solution is implemented and need the analysis in order to determine exactly how their activities will be affected.

The test report on "Relocation of Front End Sheet Metal Mounting" (Figure 3.3) is designed for efficient decision making. The overview, after the heading information, states the objective of the report, the conclusions, and the recommendation. Readers of this report, both managers and most persons who need to know about the changes approved, need to read no further. The background and the discussion provide additional information for the few persons who need to know the reason for the tests, the type of tests conducted, information about the change, or the status of the change. The references provide additional but selective documentation on these items.

In other reports, the overview addresses secondary audiences, while the discussion addresses the primary audiences (Figure 4.5). Reports dealing with implementation rather than decision making often have this design. Managers and others in the organization just need a general idea of what changes the report will cause in the organization. Those who need to operate the equipment or to order a replacement part will read the discussion, which has the detail necessary for them to take action.

The report on "Procedure to Protect Fan Coil Units" (Figure 4.6) is designed for implementation. The cover memo provides an overview of the procedure—the need, purpose, and application. This is all that most secondary audiences need to read. Those who are to implement and supervise the procedure are the primary audiences for this report, and the discussion is designed for them. In it, specific responsibilities as well as actions are itemized. The report cannot accomplish its purpose without the actions in the discussion being taken. This report design contrasts with that of the "Relocation of Front End Sheet Metal" report, which can accomplish its purpose without any person reading the discussion.

Reports, therefore, are different from other types of writing with which you are familiar, such as essays. Essays traditionally also have three parts—a beginning, a middle, and an end. However, in an essay the beginning, middle, and end form one coherent unit. An essay's parts are not self-contained: an essay is a unified whole. The writer of an essay expects readers to

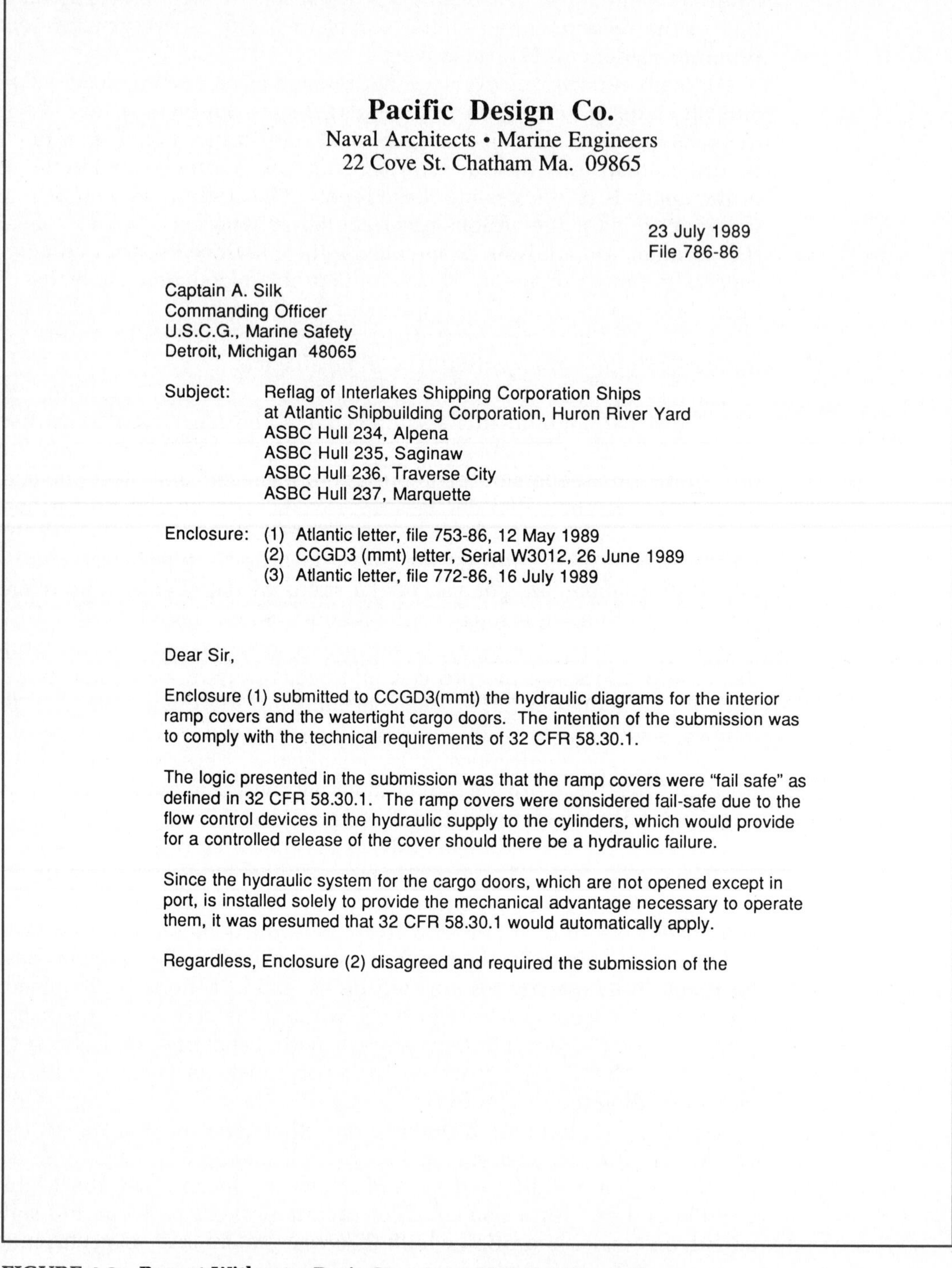

Pacific Design Co.
Naval Architects • Marine Engineers
22 Cove St. Chatham Ma. 09865

23 July 1989
File 786-86

Captain A. Silk
Commanding Officer
U.S.C.G., Marine Safety
Detroit, Michigan 48065

Subject: Reflag of Interlakes Shipping Corporation Ships
at Atlantic Shipbuilding Corporation, Huron River Yard
ASBC Hull 234, Alpena
ASBC Hull 235, Saginaw
ASBC Hull 236, Traverse City
ASBC Hull 237, Marquette

Enclosure: (1) Atlantic letter, file 753-86, 12 May 1989
(2) CCGD3 (mmt) letter, Serial W3012, 26 June 1989
(3) Atlantic letter, file 772-86, 16 July 1989

Dear Sir,

Enclosure (1) submitted to CCGD3(mmt) the hydraulic diagrams for the interior ramp covers and the watertight cargo doors. The intention of the submission was to comply with the technical requirements of 32 CFR 58.30.1.

The logic presented in the submission was that the ramp covers were "fail safe" as defined in 32 CFR 58.30.1. The ramp covers were considered fail-safe due to the flow control devices in the hydraulic supply to the cylinders, which would provide for a controlled release of the cover should there be a hydraulic failure.

Since the hydraulic system for the cargo doors, which are not opened except in port, is installed solely to provide the mechanical advantage necessary to operate them, it was presumed that 32 CFR 58.30.1 would automatically apply.

Regardless, Enclosure (2) disagreed and required the submission of the

FIGURE 4.2 Report Without a Basic Structure: Negative Example.

Pacific Design Co.

complete details of the hydraulic systems and imposed the requirements of 32 CFR 56.60.25 to the hydraulic hose.

It is understood that concurrent with Enclosure (2) an inspection requirement has also been issued either to show that the hydraulic hose in these systems is in compliance with 32 CFR 56.60.25 or to replace the hose.

We do not agree with these requirements and under normal circumstances would respond to Enclosure (2) with sufficient additional information to substantiate our position. We have, in fact, already responded to the letter regarding the stern ramps (Enclosure (3).

Due to the imminent delivery of the first two of the five vessels and the effect that these requirements could have on that delivery, it is requested that you have the installations on board inspected to verify the fail-safe design of the interior ramp covers, as described in this letter. Additionally, it is requested that you concur with our understanding that the cargo door system is not a regulated system.

If you are in agreement, it is requested that the inspection requirement for the hoses to be in conformance with 32 CFR 56.60.25 be cancelled, as only those hydraulic systems in 32 CFR 59.17.1 require compliance with 59.17.5 through 59.17.9 which, in turn, invoke compliance with 32 CFR Part 56.

Thank you for your prompt attention to this problem. If additional information is needed, feel free to contact me.

Sincerely yours,

Thomas Weber

Thomas Weber
Atlantic Design

2

W.A. KRAFT CORP.

ENGINE POWER SYSTEMS-GEN SETS-INDUSTRIAL TRANSMISSION EQUIPMENT

45 SIXTH ROAD
P.O. BOX 2189
WOBURN, MA 01888-0389
617/938-9100
TLX 94-0391
FAX 617/933-7812

January 6, 19XX

Mr. Ronald Adams, Manager
Industrial Mfg. Corp.
P.O. Box 263
Centerville, Ohio

Attention: Mr. Andrews, Supervisor, Maintenance Department

Subject: Clutch to disengage turbine starting pump

Gentlemen:

Thank you for the courtesy extended our representative, Ed Driscoll, on his recent visit to Centerville, at which time he discussed with you your requirement of the clutch to be used to disengage the turbine starting pump from the main generator engine.

We understand you wish to mount the clutch on a shaft which would be turning 720 RPM, and that the duty of the turbine pump is 117 HP. Based on this, the torque requirement is 853 pound feet, and a clutch having 16.3 HP per 100 RPM is indicated. Model #CL-310 therefore would seem to fill the bill quite nicely, at a total cost of $369.60.

We refer you to Bulletin #326-B, enclosed. On Page 10 you will find the description, capacities, etc., of this clutch. You will also note, it is available in 2-1/4" and 2-7/16" bores. Accessories are described on Page 11. Two possible spider drive arrangements are suggested, as described in Figs. 2 and 3 on the back cover of the bulletin. The spiders and their dimensions are indicated on Pages 18 and 19.

We are pleased to quote as follows:

1. XA5752 Model CL-310 Clutch in standard bore of 2-1/4" or 2-7/16" $185.60
2. Part #3507 Throwout Yoke 2.25

Diesel and Gas Engines • Generating Sets for Prime and Standby Power • Cogeneration Systems
Clutches • PTO's • Universal Joints • Fluid Couplings • Torque Converters •
Marine Transmissions • Pumps • Variable Speed Drives

ENGINEERED SYSTEMS — SERVICE — SALES — PARTS

PRINTED ON RECYCLED PAPER

FIGURE 4.3 Letter Report Illustrates Basic Report Structure. (3 pages)

W.A. KRAFT CORP.

45 SIXTH ROAD PO BOX 2189 WOBURN, MA 01888-0389 617/938-9100 FAX 617/933-7812

Mr. Ronald Adams, Manager
Industrial Mfg. Corp.
Mr. Andrews, Maintenance Dept.
January 6, 19XX

3. Part #3039 Hand Lever..$6.70
4. Part #1144-E Operating Shaft 7.95
5. X8152-3-B Spider (maximum bore 2.439/2.441") 167.10

Please note, the X8152-3-B Spider quoted is the through shaft spider described on Page 18, which , you will note, has a hub outside diameter of 4-1/4". On this hub, you would mount the driving sheave. In short, the driving sheave would have to be of such construction that it could be bored out 4-1/4". If this were not possible, then the alternate arrangement (Fig. 2 on back cover) would be applicable, in which case we would recommend Spider Flange A-5932 at $78.65 in lieu of the above $167.10.

Thank you for our opportunity to quote. We look forward to being favored with your order.

Very truly yours,

W.A. KRAFT CORP.

Leo Fitzpatrick

Leo A. Fitzpatrick
Sales Engineer

LAF/jmf
ENCLOSURE

PRINTED ON RECYCLED PAPER

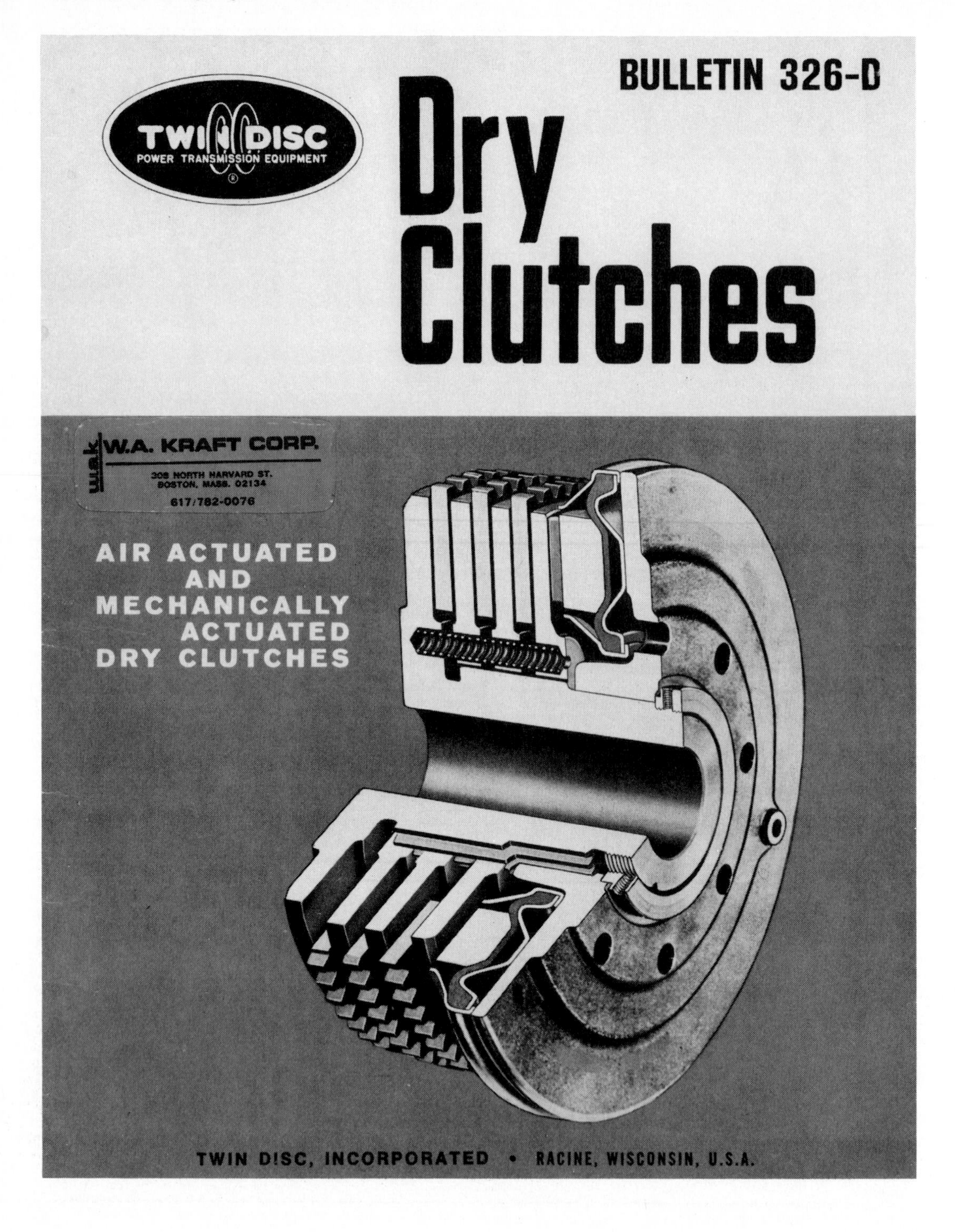
BULLETIN 326-D
TWIN DISC
POWER TRANSMISSION EQUIPMENT
Dry
Clutches
W.A. KRAFT CORP.
308 NORTH HARVARD ST.
BOSTON, MASS. 02134
617/782-0076
AIR ACTUATED
AND
MECHANICALLY
ACTUATED
DRY CLUTCHES
TWIN DISC, INCORPORATED • RACINE, WISCONSIN, U.S.A.

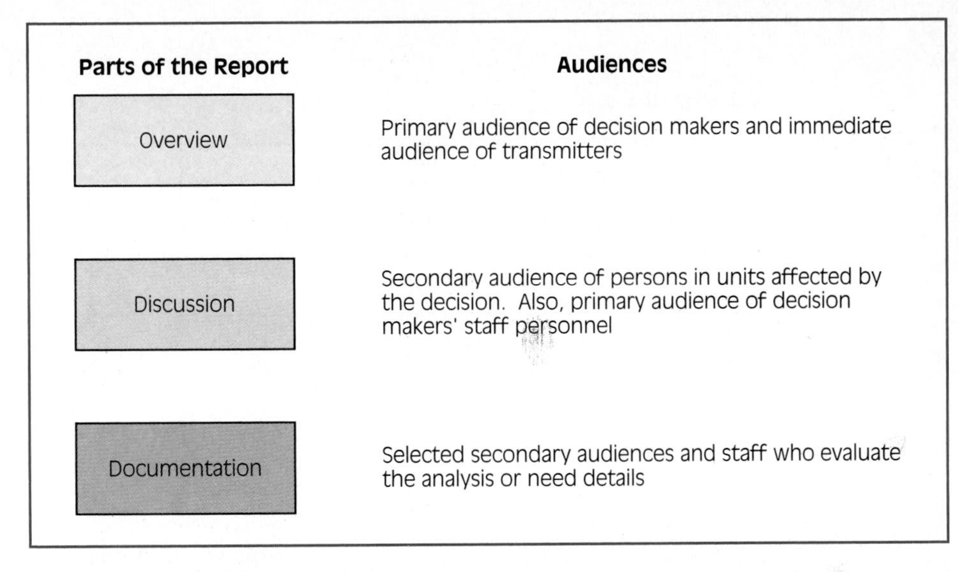

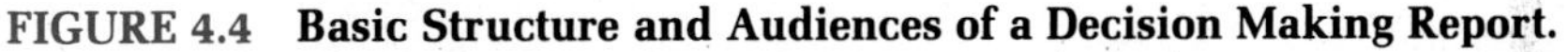
FIGURE 4.4 Basic Structure and Audiences of a Decision Making Report.

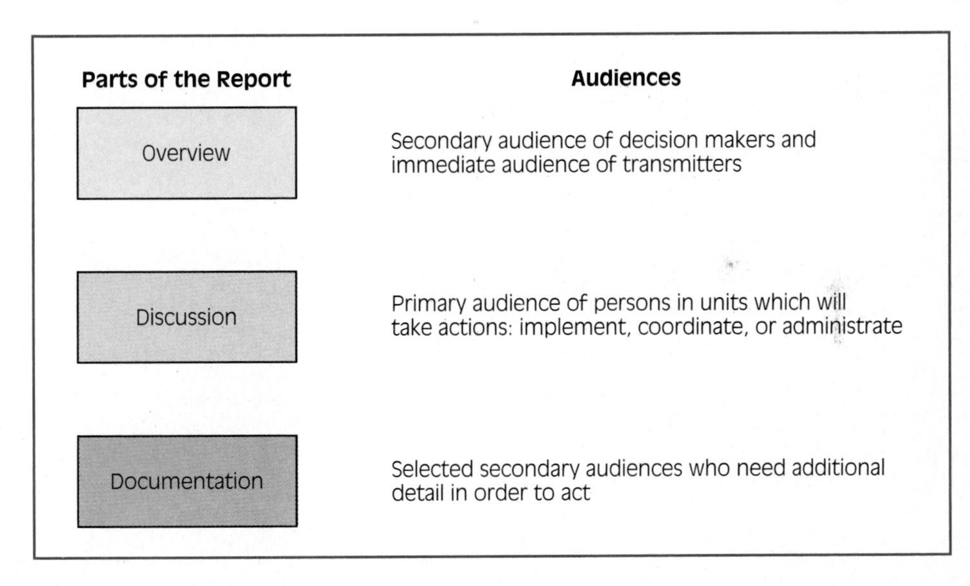

FIGURE 4.5 Basic Structure and Audiences of an Implementation Report.

read the entire essay, so the beginning is designed to lead into the middle, and the end wraps up the entire essay. A report is different because it is action oriented: it causes change in the operations of an organization or it ensures that operations can continue efficiently. If it is written to be read from beginning to end, as was the Dunn report recommending changes in operator instructions (Figure 2.3), it might not accomplish its purpose.

Atlantic Shipbuilding
MEMORANDUM

To: Construction Engin. Material Control	Trades and Shops Quality Assurance	**Date:** 8 June 1986
From: L. Hernandez *L.H*	**Dept.** Production and Construction	**File:** 10-28E **Page** 1 of 1
Subject: Procedure to Protect Fan Coil Units		**Contract:** Generic

Recently, corrosion in fan coil units in the production staging area has caused concern. Henceforth, all units stored in the open should be protected from rain and moisture. The attached Work Standard has been adopted for protection of fan coil units throughout the yard.

Work Standard #A5-23/10 is issued to alert all shops and trades to the procedures for protecting Fan Coil Units. It explains when and how to protect these units. Material Control is responsible for assuring that all units released to Construction are adequately protected.

Supervisory personnel should review this WS with the affected trades and shop personnel at the next Work Procedure meeting.

Attachment: Work Standard, A5-23/10

Distribution

S. Gutherie (Asst. General Manager)
Shipfitters
Joiners
Sheetmetal Workers
Riggers
Laborers
Painters
Welders
Electricians
Machinists
Pipefitters
Tool Room
Transportation
Crane Operators

FIGURE 4.6 Report with Discussion Aimed at Primary Audiences.

form ST-78

Atlantic Shipbuilding **Work Standard**	**Subject:** Fan Coil Unit Protection **No.** A5-23/10 **Date:** 3 June 1986 **Page:** 1 **of** 1

1.0. Purpose

This work standard presents a procedure for protecting fan coil units from corrosion.

2.0. General

2.1. The fan coil units are designed for indoor use. They should not be exposed to rain and moisture without proper protection.

2.2. Material Control is responsible for assuring that all units released to Construction are adequately protected against moisture damage.

3.0. Action

3.1 All fan coil units to be transported and stored in the open must be covered to protect against rain and moisture.

3.2. Units must be covered prior to removal from the warehouse. Plastic (4 mil min.) or similar wrapping material should be used.

3.3. Temporary storage sites at assigned production staging areas must be above standing water levels.

3.4. Units must be examined during regular preventive maintenance to assure continued protection. Any damaged covering must be replaced.

3.5. Units must remain covered against moisture until they no longer are exposed to weather conditions. UNITS MUST REMAIN COVERED WHEN REMOVED FROM THE PRODUCTION STAGING AREA AND TAKEN TO THE CONSTRUCTIN AREA.

3.6. Quality Assurance is responsible for inspection of fan coil unit protection in the production storage area.

Issued by: L. Hernandez, Production and Construction

Approved by: V. White, Quality Assurance

❑ 4.2 ❑
Reports Move from General to Particular

The 3-part structure essentially embodies a general-to-particular movement in a report. The overview presents the most general or "bottom-line" information, although it can contain some precise information such as cost. The discussion presents an analysis that uses particulars and results of calculations in order to formulate generalizations and conclusions. The documentation presents the particulars that support the analysis.

A report outline for a Presentence Investigation Report for the Michigan Department of Corrections nicely illustrates the general-to-particular structure.[5] A probation agent writes the Presentence Investigation Report to make a recommendation to the judge, who will use the document at sentencing. The writer therefore designs the report to present and support the recommendation. The report has this structure and content:

1. (Form CFO-145, a 1-page Overview): a summary containing the basic facts, background, and specific recommendation.

2. Evaluation and Plan: support for recommendation, with conclusions based on judgments and interpretations of facts in terms of relevant criteria.

3. Investigator's Description of the Offense: selective presentation of facts to support the judgments and conclusions in the Evaluation and Plan section.

4. Defendant's Description of the Offense: Facts

5. Prior Criminal Record: Facts

6. Marriage and Family: Selected Facts

7. Basic Information Report Form (CFO-101): Facts

Background Sections:
- A. Employment and Economic Status: Selected Facts
- B. Education: Selected Facts
- C. Substance Abuse: Selected Facts
- D. Mental and Physical Health: Selected Facts

The report thus is arranged according to the structure of the reasoning behind a recommendation:

Recommendation
  based on

Conclusions
  based on

Judgments, Interpretations, Inferences
  based on

Application of Criteria to Facts
  based on

Analysis of Facts
  based on

Selected Facts
  based on

Facts, Facts, Facts

This general-to-particular design can apply to organizational reports in all fields, regardless of length.

The Importance of General to Particular

Even with 1-page reports, the general-to-particular design is important. Although most writers assume that anyone will read a 1-page report or letter all the way through, that isn't necessarily so. For example, look again at the Hallman memorandum in the Three Mile Island case (Figure 2.4). That memorandum was written with the most general and important information coming at the end. The design of the memorandum makes the topic seem to focus on some minor technical issues that have no urgent priority. The addressee and primary audience for the memorandum reacted accordingly. He testified that he recalled "glancing over it very quickly and keying on the two specific questions" which "were rather routine." He did not notice he was being asked to make a decision. Because the important general information of the report was at the end, he failed to appreciate the seriousness of the situation and did not respond.[6]

General-to-particular design also is important from the standpoint of readability. Studies of how persons read information indicate that they pay more attention to general information or what is called "high-level" information.[7] Readers don't read a report in a linear fashion from beginning to end; they read a report as a "hierarchical" structure, sorting out details within a framework of generalities. General-to-particular design for the entire report establishes a basic framework for the various subordinate general-to-particular structures in the report, which we discuss in later chapters on sections and paragraphs.

An effective 1-page letter with a general-to-particular design is the letter proposing to lease some city property for storage of broken concrete (Figure

CENTRAL CONCRETE PRODUCTS

Producers of Sand, Gravel and Ready Mix Concrete

PHONE 772-3695 W. HIGH STREET BOX 389
773-5270
MT. PLEASANT, MICHIGAN 48858

October 22, 1990

Mr. Blaine R. Hinds, City Manager
The City of Alma
525 E. Superior
Alma, MI 48801

Dear Mr. Blaine,

Alma Concrete Products is interested in leasing 4 or 5 acres of the former airport property located across from our plant on Bridge Street. We propose to use this property for storage of broken concrete and asphalt. If this is of interest to the City, please contact us to discuss terms of the potential lease.

The property would be used to store broken concrete, broken asphalt, and clean fill sand and gravel. These discarded products would then be recycled into usable construction materials. This should benefit the City economically and environmentally.

The plan would be to build a secure fence around the leased premises. In addition to Alma Concrete, others could use the area to dump broken concrete and asphalt. They would first have to stop at Alma Concrete to have their load inspected and to gain access to the disposal site. Only concrete, asphalt, and sand and gravel would be accepted.

This type of operation would benefit the City by:

A. Generating a source of revenue to the general fund.

B. Providing an inexpensive and convenient disposal area for this type of salvage for the citizens of Alma.

C. Preventing the dumping of these materials along roads and on vacant lots.

D. Preventing waste by recycling an otherwise lost resource.

E. Conserving precious space in other disposal areas by the elimination of these bulky but recyclable items.

If this proposal is of any interest to you, please contact me at our Mt. Pleasant office, 764-1426.

Sincerely yours,

R. J. Fisher, Jr.

R. J. Fisher, Jr.

FIGURE 4.7 Letter Report with General to Particular Design.

4.7). Unlike the Hallman memorandum, this letter states the purpose in the first paragraph. The second paragraph then summarizes the proposal in general terms. The following paragraphs explain the proposal and its benefits. The technical specifics are clearly subordinated by the letter's design. The general-to-particular design embodies the writer's priorities and provides a framework that focuses the reader's attention on the high-level information.

The report of Investment Portfolio activity (Figure 4.8) also illustrates the general-to-particular structure very clearly. The "Summary of Activity" section at the beginning summarizes the details in the two sections of the discussion. The generalizations are:

> The $110 million leverage program approved at the May Corporate ALCO was completed.
>
> Sixty million dollars in Asset-Backed Securities were sold for reinvestment in Agency-issued CMO product.

These generalizations pull out the important implications of the particulars found in the discussion:

> The purchases were done in three months. . . . the purchases averaged 9.65% yield. . . . the average life of the purchases is about 3.6 years. . . . the maturity predicted for the purchases is expected to be not less than 2 nor more than 7 years.
>
> MBS, at $2.5 billion, represents nearly 63% of the portfolio securities, up dramatically from $1.2 billion or 36% of the portfolio of August, a year ago. . . .

At the beginning of the report, these details are not presented because they might obscure the important generalized information: a program which had been approved earlier has been completed; $100 million in securities had been sold for reinvestment; the on-balance portfolio has reached an all-time high. These general points then establish the hierarchical structure for the report, leading into the details of the discussion. The important information is in the generalizations, however, not the particulars.

Nevertheless, generalizations can be precise in the form of "bottom-line" numbers, and often should be. In the report on the "Clutch to disengage turbine starting pump" (Figure 4.3), the cost total in the summary is $369.60. This is the sum of the itemized particulars in the discussion, the important number which the writer wants the primary reader to see as early as possible in the report. The writer wants the reader to know that for only $369.60 his problem is solved.

Thinking from General to Particular

Writers often fail to think in terms of general-to-particular movement. They think in terms of the chronology of an investigation. In other words, they think particular-to-general, in terms of what they did. To think general-to-

NBIn

MEMORANDUM

NATIONAL BANK OF INDIANAPOLIS

October 11, 1990

TO: Frank E. Stewart, Vice-President, NBIn Wisconsin
John S. Turner, Vice-President, NBIn Indiana
Lee A. Carlson, Vice-President, NBIn Ohio

FROM: Frank B. Angelo, Municipal Trader, NBIn Corporation FBA

SUBJECT: August 1990 Investment Portfolio Activity

SUMMARY OF ACTIVITY

Current Activity:

- Completion of the 110 million leverage program approved at the May Corporate ALCO.
- Sale of 60 million in Asset-Backed Securities for reinvestment in Agency-issued CMO product.

Portfolio Review:

- The on-balance sheet portfolio surpassed $4.0 billion for the first time. MBS represents about 63% of portfolio securities. 76% of this is in CMO/REMIC product.

CURRENT ACTIVITY

August saw the completion of the 110 million leverage program initially approved at the May Corporate ALCO. The 110 million in purchases done in the three months following the May ALCO were primarily in Agency-issued CMO product, except for approximately $28 million in seasoned 15-year FNMA and FHLMC paper described in last month's activity memo. In total, the purchases averaged 9.65% yield, with an average life of about 3.6 years. The maturity sensitivity of the purchases is not expected to shorten to less than 2.0 years or extend much beyond 7.0 years.

The other major activity during the month centered around the sale of just more than $60 million in Asset-Backed Securities (ABS), with reinvestment in Agency-issued CMO product. The swap was done for three reasons. First, CMO spreads to AMB spreads were relatively attractive. Second, we were able to pick up yield, 5-10 basis points,

FIGURE 4.8 First Page of Report on Investment Portfolio Activity Illustrates General to Particular Structure.

particular is to adopt the reader's perspective and to think in terms of the function of a report. Yet, as a professional, you always work from particular-to-general, so shifting to the reader's perspective is often difficult. In your actual investigation, you work from setting up an apparatus to collecting data, to interpreting results, to formulating conclusions, and to deciding upon recommendations. You work from particulars to generalizations. Of course, this is the norm of professional activity, an inductive process analogous to that of scientific discovery. When you come to write a report, but fail to think of its function, you find it tempting to write the report in terms of what you did—in terms of the chronology of your technical investigation rather than in terms of your communication purpose. You write the way you think: particular-to-general. Unfortunately, this tendency is reinforced in college courses when student reports focus almost entirely on experimental procedures, design processes, algorithms, and calculations because those are the appropriate educational concerns of the professors.

In addition, as a professional, you are often interested in technical details, the particulars, because you are the expert. In his report on Operator Interruption of High Pressure Injection (Figure 2.3), Mr. Dunn devoted most of his attention to the theory behind the operation of the emergency core coolant system and to the technical details of the operator actions at the Toledo nuclear plant. Much as this technical discussion might have interested Mr. Dunn, however, it was unnecessary for most of his readers. Actually, it confused his most important readers, those in a position to act on his recommendation. In general, you can assume that most of your readers don't have the same technical enthusiasm for your subject that you have. Most of the time, they are interested in outputs and implications; as a technical expert, you naturally tend to be interested in inputs and details.

Our discussion of the needs of report readers, however, indicates the tendency of readers to prefer and even demand a general-to-particular basic structure. Their first question upon receiving your report is, "Do I read it, route it, or skip it?" If you want them to read it, put the "bottom line" first: think general-to-particular.

An additional point about basic structure and thinking general-to-particular: your reports will be somewhat selectively redundant. The general-to-particular design requires you to repeat certain key information—and you shouldn't be reluctant to do so, as long as the repetition is functional. That is, some information will be repeated in the summary and the discussion, sometimes word for word, but usually in differing degrees of amplification. For example, a conclusion stated in the summary needs to be repeated in the discussion so that the analysis supporting that conclusion is clear to the reader. A recommendation in the summary, which can meet the needs of a decision maker, also needs to be explained in greater detail in the discussion for those who have to implement decisions. The same holds for information in the purpose statement. Generalizations dealing with the organizational problem and the investigation in the overview often need to be repeated and further explained in the discussion.

The petrochemical department progress report (Figure 4.9) illustrates this selective redundancy. In this case, most phrases in the "Brief" are taken directly from the discussion, with very little amplification:

From the "Brief":

> Productivity during the first quarter exceeded prior records. Despite a slight decrease in the second quarter, we project a record year for production.

From the "Productivity" section of the report:

> During the first quarter, production exceeded previous production records. . . .
> During the second quarter the production rate did decrease slightly. . . .
> Despite . . . 19XX will be a record year.

The paragraph in the "Brief" on the Expanded Day Program has a similar selective redundancy.

This redundancy is a consequence of designing your report so that it can be read selectively. Because you don't expect most of your readers to read your report straight through, you have to repeat some information in different parts of your report. The repetition inherent in relating generalizations to particulars enables you to meet the needs of diverse audiences.

Distinguishing General Information from Particular Information

The concept of basic structure requires you to formulate generalizations in order to establish general-to-particular relationships among your material. To do so, you must be able to distinguish between generalizations and particulars. Also, you must be able to arrange information according to differing levels of generality.

Once when we were preparing to teach a technical writing course in industry, a manager told us that he was convinced it was not sufficient just to tell his staff to put the conclusions and recommendations up front in a report. He pointed out that his company's standard format had these headings but that often the statements which his staff put under these headings were not conclusions or even generalizations. He said that we had to be sure to teach his staff what a generalization is before telling them where to put it. And he was right. We've since seen the same phenomenon in many organizations.

To us, the ability to formulate generalizations is what separates the professional-level employee from the technical support employee. As professionals, your primary role is to formulate generalizations, judgments based on interpretations of data and other types of evidence. In one sense we can say that the purpose of a report is to present these generalizations. Data themselves can be transferred by various means, especially electronically, without being translated into reports as we define them in this book.

Generalizations are interpretations of particulars such as sensory obser-

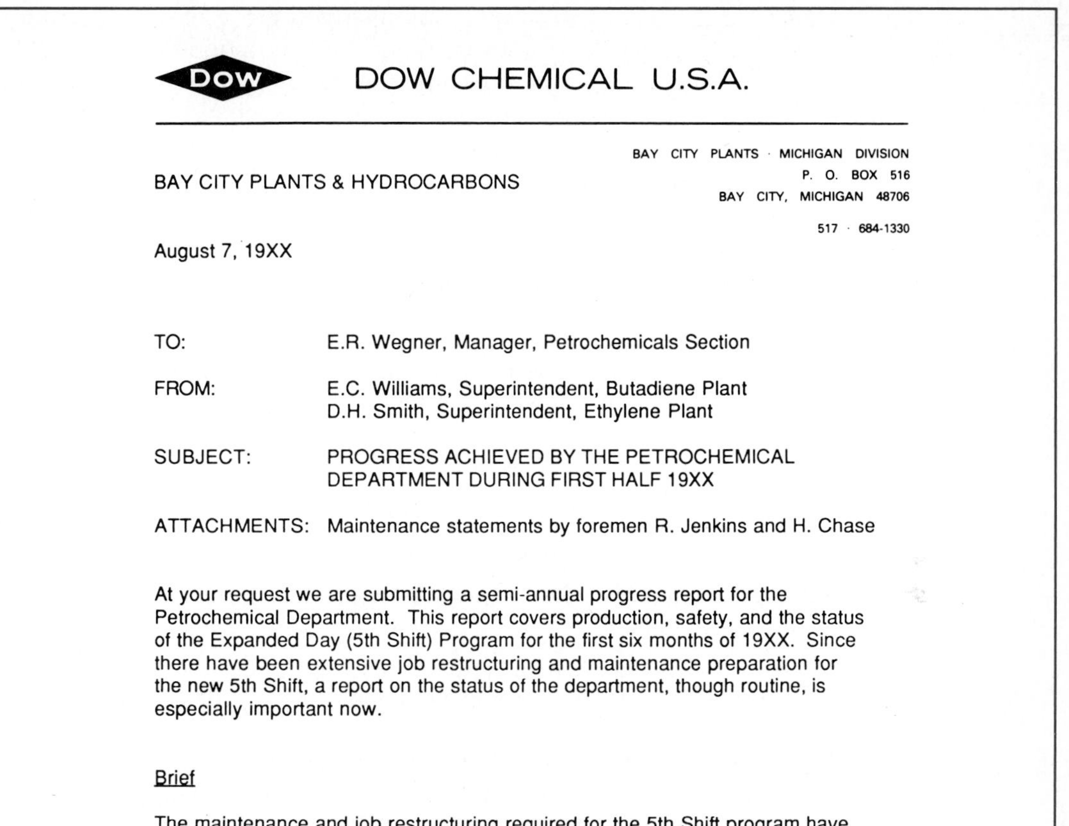

DOW CHEMICAL U.S.A.

BAY CITY PLANTS & HYDROCARBONS

BAY CITY PLANTS · MICHIGAN DIVISION
P. O. BOX 516
BAY CITY, MICHIGAN 48706
517 · 684-1330

August 7, 19XX

TO: E.R. Wegner, Manager, Petrochemicals Section

FROM: E.C. Williams, Superintendent, Butadiene Plant
D.H. Smith, Superintendent, Ethylene Plant

SUBJECT: PROGRESS ACHIEVED BY THE PETROCHEMICAL DEPARTMENT DURING FIRST HALF 19XX

ATTACHMENTS: Maintenance statements by foremen R. Jenkins and H. Chase

At your request we are submitting a semi-annual progress report for the Petrochemical Department. This report covers production, safety, and the status of the Expanded Day (5th Shift) Program for the first six months of 19XX. Since there have been extensive job restructuring and maintenance preparation for the new 5th Shift, a report on the status of the department, though routine, is especially important now.

Brief

The maintenance and job restructuring required for the 5th Shift program have not harmed either productivity or safety in the department. Productivity during the first quarter exceeded prior records. Despite a slight decrease in the second quarter, we project a record year for production. Safety also has been good. We had no disabling injuries in the department during the reporting period.

The Expanded Day Program development continues to progress as scheduled. Necessary maintenance has been done because the men have adapted well despite the obstacles. The first step of job restructuring has already been accomplished. To continue this progress we now need to begin hiring, to provide equipment, and to provide an adequate building. The Unifiner also needs a minor maintenance building of 400-500 square feet.

Productivity

The 19XX productivity of the department has been excellent. During the first quarter, production exceeded previous production records by 7%. During the second quarter the production rate did decrease slightly (3% below the first

FIGURE 4.9 First Page of Progress Report Illustrates General to Particular Structure.

vations and data. For example, suppose you are a planner for the County Parks Commission. The Commission needs population-density statistics, and you are asked to report the population density in Webster Township. You tabulate the exact number of persons living in each of the thirty square miles of the township. This tabulation represents the census takers' direct observations of the persons living in Webster Township: particular information. When you average these statistics, you arrive at general information. You report that the population density in Webster Township is 32.7 persons per square mile: a generalization. You present information that cannot be directly observed.

Particular information takes the form of empirical data and related observations: measurements, counts, direct impressions of the senses, details, and calculations. Here are some examples:

The planned park is 6 miles from Dexter.

All fractures occurred on test units run less than 5,000 miles.

The dry density was 102#/ft.^{3}

With the add-on heat shield in place, engine housing temperature was reduced from 475°F to 292°F.

The drive-in window line, between 10:35 and 11:00 A.M. on Saturday, May 2, extended onto Packard Street by up to four cars.

When you read a meter, of course, you are not directly observing temperature. However, you are observing a specific number on the meter.

Generalizations come from averaging, comparing, interpreting, inferring, evaluating, judging, and concluding. These activities are conceptual skills: you are processing particulars and producing generalizations—mental concepts. Here are examples of various types of generalizations:

The planned park will be accessible to Dexter residents.

The planned park is close to Dexter.

The planned park location meets legal requirements. It is more than 5 miles from a section with a population density of more than 100 persons per square mile.

All fractures occurred in low-mileage test units.

Fractures will occur before 5,000 miles or not at all.

All units with mileage under 5,000 miles should be tested.

The relative density was close to 100%.

The soil is very dense.

Virtually no settlement of the soil will occur under the projected loads.

The carburetor changes account for a 70°F increase in inside surface temperature.

> The G-22 engine housing design has deficiencies.
>
> The line extended several cars onto Packard for about a half hour.

Your role as a professional is to formulate generalizations such as these. Your role as a report writer is to present them.

When you present information by moving from general to particular, you are presenting a generalization and then explaining or supporting it. You separate the generalizations from the particulars and put the generalizations first. If your analysis shows that the planned park is 6 miles from Dexter, and if your conclusion is that at that distance the park would still be accessible to Dexter residents, you state your conclusion before presenting the particulars that support it:

> The Dexter Town Council asked the County Parks Commission if the planned park would serve the citizens of Dexter effectively. I checked our guidelines and the legal requirements. Here is the information for your May 3rd meeting with council members.
>
> The planned park is accessible to Dexter residents according to our standard guidelines for park planning. It will service the town of Dexter according to state tax laws as well. The relative accessibility of the park to the residents of Onsted in York Township might be the real issue.
>
> *Criteria for accessibility*. Commission guidelines state that . . .

To be able to write this way, you must be able to distinguish between a generalization, "The planned park is accessible to Dexter residents," and a particular fact, "The planned park is 6 miles from Dexter." You present generalizations that are your interpretations of specific data rather than simply state the specific data themselves.

In addition to interpreting particulars and formulating generalizations, you need to distinguish among levels of generality. The statement that "Charlevoix should anticipate a significant increase in population by the year 2000" is a generalization or conclusion based on interpretations (projections) of population trends. The statement that "the Charlevoix water system will be unable to meet the projected water demands for the year 2000" is another generalization, a conclusion based on interpretation of several types of information, on intermediate judgments, and on assumptions. Both statements are generalizations, but the two statements differ in their degrees of generality. The first is a single value judgment interpreting projected data or a curve on a graph (how much increase is significant?). The second is a conclusion based on a comparison of the projected population increase with the capacity of the present system, which involves important assumptions about water use by the population. The second statement is more general than the first in the sense that it is a deduction which rests upon a number of supporting intermediate premises or generalizations.

In order to develop your reports, you need to distinguish among at least four levels of generality: empirical data, results, conclusions, and recommendations. These require different degrees of interpretation and judgment by you. Some statements require you to interpret particulars objectively. Other statements require you to exercise varying types of critical judgment. To help distinguish among these levels, it is useful to distinguish among four different types of statements (Figure 4.10):

- Empirical data
- Results
- Conclusions
- Recommendations

Empirical data are directly observed or measured facts which require only recording, not mental processing: The time is 08:11:24. The temperature at the beginning of the test run was 475°F. Cost for insurance for the office during FY 1990 will be $24,150.

A *result* requires routine mental processing of particulars or data. The statement that the "soil had a dry density of 102#/ft.^{3}" is a presentation of empirical data. It is a measured fact. The statement that "the soil had a relative density of close to 100%" is an interpretation of that data. It is a low-level generalization requiring objective interpretation of data. One cannot measure or observe "a relative density of close to 100%," yet that statement requires no value judgment on the part of the writer. A low-level generalization is a statement of information that cannot be observed directly, but which at the same time requires little judgment or evaluation. The lowest-level generalizations are the results of calculations and objective interpretations of empirical data.

A *conclusion* requires interpretation of results. When you state that the "soil is in a highly densified condition," you are interpreting data and results by making a value judgment. You are formulating a conclusion, a subjective concept of "high density." When you state that "virtually no settlement will occur at this site under the proposed loading" you are making a deduction, which is an even higher level conclusion. It requires you to make intermediate judgments concerning the limits of acceptable settlement for the proposed loading, which are assumed but unstated premises in your argument. Mental processes such as *concluding, deducing, inducing, inferring,* and *judging* result in high-level generalizations because you interpret empirical data and results in terms of criteria and standards.

A *recommendation* requires evaluation of a conclusion in an organizational context. It is an even higher level of generalization than a conclusion because it is a second order generalization. For example, when you assert that "we consider this site an excellent location for the placement of vibratory machinery" (because virtually no settlement will occur) you are evaluating a conclusion in an organizational context that requires a decision and action. Implicitly, you are saying that you recommend the site for the placement of vibratory machinery.

Examples of levels of generality are presented in Figure 4.10, Levels of

Examples of Four Levels of Generality

1. Empirical Observations

- With the add-on heat shield in place, the engine housing temperature was reduced from 475°F to 292°F.
- The planned park is 6 miles from Dexter.

2. Results

- The carburator changes account for 70°F increase in inside surface temperature.
- The planned park location meets legal requirements. It is more than five miles from a section with a population density of more than 100 persons per square mile.

3. Conclusions

- The G-22 engine housing design has deficiencies.
- The planned park will serve the citizens of Dexter effectively.

4. Recommendations

- The add-on heat shield should be released for production.
- The bonding proposition to develop the Dexter County Park should be put on the spring ballot.

FIGURE 4.10 The Ladder of Generality.

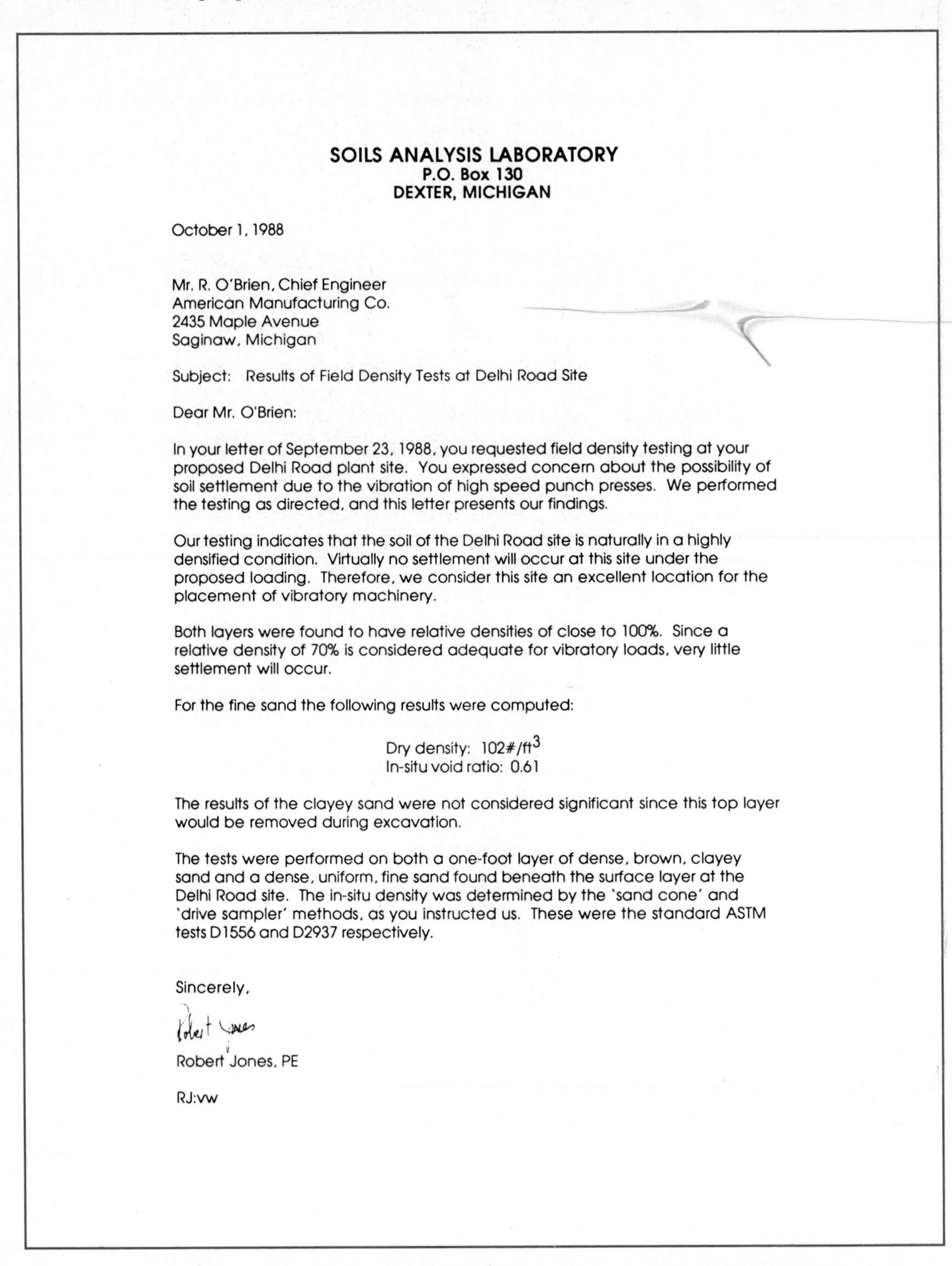

SOILS ANALYSIS LABORATORY
P.O. Box 130
DEXTER, MICHIGAN

October 1, 1988

Mr. R. O'Brien, Chief Engineer
American Manufacturing Co.
2435 Maple Avenue
Saginaw, Michigan

Subject: Results of Field Density Tests at Delhi Road Site

Dear Mr. O'Brien:

In your letter of September 23, 1988, you requested field density testing at your proposed Delhi Road plant site. You expressed concern about the possibility of soil settlement due to the vibration of high speed punch presses. We performed the testing as directed, and this letter presents our findings.

Our testing indicates that the soil of the Delhi Road site is naturally in a highly densified condition. Virtually no settlement will occur at this site under the proposed loading. Therefore, we consider this site an excellent location for the placement of vibratory machinery.

Both layers were found to have relative densities of close to 100%. Since a relative density of 70% is considered adequate for vibratory loads, very little settlement will occur.

For the fine sand the following results were computed:

Dry density: 102#/ft^3
In-situ void ratio: 0.61

The results of the clayey sand were not considered significant since this top layer would be removed during excavation.

The tests were performed on both a one-foot layer of dense, brown, clayey sand and a dense, uniform, fine sand found beneath the surface layer at the Delhi Road site. The in-situ density was determined by the 'sand cone' and 'drive sampler' methods, as you instructed us. These were the standard ASTM tests D1556 and D2937 respectively.

Sincerely,

Robert Jones

Robert Jones, PE

RJ:vw

FIGURE 4.11 **Soils Test Report with Logical Relationships among Levels of Generality.**

Generality. The statements move from particulars ("with the add-on heat shield in place, the engine housing temperature was reduced from 475°F to 292°F") to results of objective interpretations of data ("the carburetor changes account for a 70°F increase in inside surface temperature") to conclusions based on interpretations of results ("the G-22 engine housing design has deficiencies") to recommendations on actions to be taken ("the add-on heat shield should be released for production").

When you design your basic structure and arrange information within the various sections of your report, you need to establish clear and logical relationships among these levels of generalization. Report design essentially consists of establishing these relationships or, in other words, a hierarchical structure. The Soils Test Report (Figure 4.11) illustrates how you arrange information to move logically among levels of generalization. After the purpose paragraph, the writer summarizes his conclusions and recommendations: this paragraph moves from a value judgment to a higher-level conclusion to an implicit recommendation. These are followed by a paragraph of results, interpretations of the data that lead to the conclusion restated from the summary: "very little settlement will occur." This interpretative results paragraph in turn is followed by a paragraph presenting the results of calculations from data—very close to empirical observations. Although data themselves are not presented, the report ends with a paragraph of facts. In sum, the report effectively moves from general-to-particular. The relationships among the various types of statements are clear and logical.

Reports which are primarily informational can have an implicit general-to-particular basic structure, but hierarchical relationships are deemphasized. A status report, for example, often consists entirely of factual information and low-level results such as percentages (Figure 4.12). In the sample status report, four types of information are presented. "Work Action Status" comes first and presents data in aggregated form: if there is an overview for the report, it is this section. "Delivery Date Status" has both aggregated and specific data. "Critical Work Actions" and "Comments" have data and particular information. Usually, such reports are in standard forms which enable the reader to spot the important information immediately. The forms embody standard procedures that essentially require writers to select information rather than design and write a report.

Most of the time, however, you have to design a report rather than fill out a form. Your first task should be to establish a 3-part basic structure.

❑ 4.3 ❑ Different Types of Reports Have Similar Basic Structures

Reports have a similar basic structure despite a diversity of formats. One common distinction is between the formal report and what we usually call an informal report. A formal report is distinctive because it is bound with a

Atlantic Shipbuilding
Project Summary

Planning

L. Duke, Supervisor
Scheduling

Project: SWATH A-TSD

Date: 17 June 1986

Contract: TY201-78

Work Action Status
As of 17 June 1986

	actual	scheduled	percentage
Work Actions (started)	101	140	72.1
Work Actions (completed)	185	220	84.1

Delivery Date Status
As of 17 June 1986

	actual	scheduled
Current progress	60%	65%
Delivery date	October 9	September 28

Critical Work Actions
As of 17 June 1986

Work Action No.	Description	Sch. Comp.	% Comp	Req. Manpwr
380	Outfit block 1790	7-20-86	25%	32
291	Install heat exch.	6-19-86	78%	10
894	Rework piping	6-23-86	43%	15

Comments:
Reworking piping, WA 894, required some redesign. This was accomplished by the Design Engineering Outfitting Section.

Distribution:
S. Holmes (General Manager)
S. Gutherie (Asst. General Manager)
Contracts
Program Office Navy
Production Engineering and Planning
Design Engineering
Quality Assurance
Production and Construction

FIGURE 4.12 Status Report with General to Particular Structure of Tabular Data.

cover, has a title page and table of contents, and sections that start on separate pages as chapters in a book do. Superficially, it looks different from the letters and memoranda that we have used in our figures so far. Formal reports are not as frequently written as informal memoranda and short reports in most organizations, but they are common in consulting organizations and some administrative organizations. The table of contents for a formal report from a hospital management consulting firm is shown in Figure 4.13. As you can see, it is from a long report with five major sections, including the appendices, and many subsections. Although we do not focus on such reports in this book, we want to make sure that you realize that our guidelines apply to them as well.

Despite the superficial differences in format, the formal report is similar in its basic structure to the informal report. Both have the three major parts of a report: overview, discussion, and documentation. Although the formal report has some additional subsections, it has the basic structure we have been discussing. The hospital management consultant report has an overview with two sections: The Purpose of the Study and Synopsis of Recommendations. The discussion has four major sections, I–IV, which are separated rather than included under a main heading such as "Discussion." The documentation consists of Appendices A–G. Despite its 27 pages, however, this formal report has the same basic structure as the report to Approve Relocation of Light Duty Conventional Front End Sheet Metal Mounting (Figure 3.3).

We have shown this basic design in other types of reports also. You have seen it in memoranda, letters, minutes of meetings, and procedures. You will find it in effectively written manuals, proposals, and design reports. The function of most documents written by professionals requires that they have the basic structure we have outlined.

❑ 4.4 ❑ The Basic Structure of Scientific and Technical Reports and Articles

Our guidelines for designing the basic structure of an organizational report are also relevant when you must write a scientific or technical report or article. Scientific and technical reports and scientific and technical articles (sometimes called "papers") are distinctive forms of technical writing. As we discussed in Chapter 3, these are written in research contexts rather than in organizational contexts. Although primarily informational, technical reports have basic structures similar to those of organizational reports. Although scientific and technical articles usually do not, they sometimes use an alternative strategy to achieve some of the functions of basic report structure.

TABLE OF CONTENTS

FIGURE 4.13 Consultant's Formal Report Illustrates Basic Structure.

The Scientific and Technical Report

The American National Standards Institute (ANSI) standard for *Scientific and Technical Reports* stipulates the appropriate content, organization, and sequencing of information to be included in scientific and technical reports.[8] The basic concept of the ANSI standard is that the technical report should be designed to permit readers to read selectively. That is, the text should contain an introductory summary which briefly compresses the report into a manageable scope. That is followed by a selective discussion which explains how the information and conclusions in the summary were developed. And that in turn is followed by appendices with details and data.

The ANSI standard presents the scientific and technical report as a formal report, as we discussed it previously. Although the outline has many elements that are not included in informal reports, primarily in what is called the "Front Matter," the report itself has the same basic structure. The structure suggested by ANSI consists of the following recommended elements in this order:[9]

Front Matter

Cover
Title Page/ Report Documentation Page
Notices
Abstract
Contents
List(s) of Figures and Tables
Foreword
Preface and Acknowledgments
List(s) of Symbols, Abbreviations, and Acronyms

Text (Body)

Summary
Introduction
Methods, Assumptions, and Procedures
Results and Discussion
Conclusions
Recommendations
References

Back Matter

Appendices
Bibliography
Glossary
Index
Distribution List

The text or body of the report has the basic structure that we have explained. The summary briefly states the "principal results, conclusions, and recommendations." Although some introductory and methodology information can be summarized, the emphasis is "on the results and accomplishments of the research." As in the basic structure of the organizational report, the summary precedes the other elements of the text, which we collectively call the discussion, in the basic structure of a report. (The standard uses the term *discussion* for one of the elements.)

As in an organizational report, the elements of the scientific and technical report are selective. The introduction "provides readers with general information that they must have to understand detailed information in the rest of the report." The methods, assumptions, and procedures section succinctly describes these "so that the reader can evaluate the results without referring extensively to the references." The results and discussion section is a selective interpretation of the results, with "supporting details not essential to an understanding of the results" placed in the documentation. The conclusions present "findings that have been substantiated in the discussion of results and discusses their implications."

The standard emphasizes that the report should be written so that readers do not have to read the entire text. The summary "is independent of the text from the reader's point of view." Furthermore, the conclusions section "should be written so that it can be read independently of the text."

The appendices "contain information that supplements, clarifies, or supports the text." Typical contents of the appendices are:

Detailed explanations and descriptions of test techniques and apparatus
Texts of other documents (for example, standard test procedures, laws, and management instructions)
Extensive data in the form of figures or tables
Computer listings of programs, input, and output
Mathematical analyses

These contents are those you often attach to organizational reports as well.

The scientific and technical report described by the ANSI standard has an abstract in the front matter. (We discuss abstracts in the next chapter.) This is an element of basic structure in many formal reports. It is an additional redundancy designed for other readers and must "be understandable independent of the rest of the report."

The Scientific and Technical Article

The scientific and technical *article* is different from the scientific and technical *report* because the article tends to have a homogeneous audience and to be primarily informational, whereas the audience for the report is often diverse, and the purposes of the report are sometimes more applied than theoretical. The audience for an article is homogeneous in that it consists of some community of interest. The community of interest, identifiable from the journal in which the article appears, can range from the very specialized, such as members of the Association of Military Surgeons, to an audience interested in use and application, such as an audience of health administrators, to a broad audience interested in a specialized field, such as hospital architecture and design, to a public audience, such as the readers of a university research news magazine. When you write an article, you need to define the interests and needs of your audience and determine your purpose in writing to that audience.

Because of its audience and purpose, the text of a scientific and technical article often has a 1-part structure. That is, rather than being designed in discrete parts to be read selectively by different types of readers, as is the technical report, the technical article is designed as a unified whole to be read—or at least skimmed—from beginning to end. The technical article therefore lacks most of the redundant structural characteristics of the report and lacks additional features such as appendices and attachments (a bibliography of references instead can send readers to search for these if they wish additional information).

Many scientific and technical articles thus lack some of the elements of the ANSI standard for reports. Their structures are outlined in the ANSI standard for the *Preparation of Scientific Papers for Written or Oral Presentation*[10] and in the *Publication Manual of the American Psychological Association* (APA)[11]:

ANSI	**APA**
Title	Title
By-Line	Abstract
Abstract	Introduction
Body of the Paper	Method
Organization	Results
Introduction	Discussion
Methods	References
Materials	Appendix [seldom used]
Results	
Discussion	
Summary [sometimes]	
References, etc.	

These are typical outlines for a scientific or technical article.

As you can see, the article typically lacks the 3-part basic structure. The outline has no opening summary and usually has no documentation section, although the APA manual provides for an occasional appendix. Instead, the summary, if there is one at all, comes at the end. (If there is no summary, the conclusions are interwoven in the discussion.) The ANSI standard even mentions that in an article you do not expect the readers to read selectively. It says the summary "is for the convenience of those who have read the paper and is usually intelligible only after this has been done."[12]

The abstract of a scientific and technical article, however, does serve a basic structure function. Although the abstract is different from a summary, as we explain in the next chapter, it also serves some of the functions of a summary. It provides an overview of the article and "may be the only part of an article that is actually read."[13] The ANSI standard says "each abstract must be self-contained, fully intelligible without reference to the body of the paper." Although the abstract is not written to cope with audience diversity, it serves several important functions of an overview. It enables some readers to read the article selectively and efficiently; it enables other readers to get the important information without having to read the article itself.

The outlines in the ANSI standard and APA manual illustrate the general architecture of technical articles. They are not prescriptive. In fact, our guidelines for the 3-part basic structure of a report can be applied to the design of scientific and technical articles as well. Some articles do have a summary in the beginning to provide an overview. As the ANSI standard observes, "other formats, such as the one in which major conclusions are presented at the very beginning of the paper, are permissible and in many types of journals may be preferable." When you analyze your readers' needs, you might be well advised to adopt such a format. (Also, be sure to look at the journal's advice to contributors and at other articles published in the journal. This will alert you to the expectations of readers and help you to decide whether an up-front summary would be a good idea in a particular article. In some cases you will find that the journal does not encourage the practice; in other cases you will find that the journal requires it. Just be careful to check your design against the general design expectations signaled by the journal's own guidelines and by the example of other articles published there.)

Conclusion

To a professional first entering an organization, the experience of encountering the many apparently "new" types of reports to be found there can be a bit overwhelming. For example, a new employee of the New York City Police Department will suddenly find that there is such a thing as an "Unusuals" report—an odd name for a frequently written type of report describing an incident such as a shooting. But there are also Complaint Reports, Investigation Reports, Precinct Memos, Letters, Field Operation Critiques, Management or Planning Reports, and a host of other types of documents which the organization uses to do its daily business. A classification of the department documents lists them in three broad categories: Forms, of which there are 440; Precinct Records and Files, of which there are 43; and Directives, of which there are numerous varieties, which may be issued by ten designated categories of personnel.[14]

Other organizations have their own array of different report types. To the outsider, these arrays can suggest that the architecture of each is unique and that its design must be approached individually. To a certain extent, of course, that is true; indeed, a writer must learn the audiences, purposes, needs, and conventions of individual documents in any organization. Nonetheless, we believe you can approach an array of different technical document "types" with a sense of their underlying similarity. They do not need to be seen as architecturally different. That is, you can look for the opportunity to exploit the principles we have discussed here: (1) "Layer" the document so that it consists of three parts. This will make it possible for different readers to get what they need without reading the document as a whole. (2) Make the document move from general information to particular information—both between sections of the document and within those sections. This hierarchical design will increase reader comprehension and facilitate selective reading. Whether you are writing a letter, a proposal, a report, a memorandum, a technical article, or an "Unusuals" report, these generic design principles can help to assure that your professional communication is functional.

NOTES

1. Comments by Dr. Edward Gilbert, a management consultant, in a seminar presented to the University of Michigan Program in Technical Communication annual Conference on Teaching Technical and Professional Communication (1975–86) and Dr. Howard Klee, a Superintendent with Amoco Oil Company, in a seminar presented to the University of Michigan Program in Technical Communication annual Conference on English Technical Writing for Japanese Engineers (1981–present).
2. Henry Mintzberg, "The Manager's Job: Folklore and Fact," *Harvard Business Review* July–August 1975: 50.
3. James W. Souther, "What to Report," *IEEE Transactions on Professional Communication*, PC-28, 3 September 1985: 5–8. Reprinted from an article in the Westinghouse *Engineer*, based on his study at Westinghouse in 1959–60. Our own research supports Souther's findings, as does the research of numerous other management communication researchers. In a sense, the surprising thing is how little attention writers in organizations seem to pay to this information, despite the degree to which it is known and the length of time we have known it.
4. J. C. Mathes and Dwight W. Stevenson, "Completing the Bridge: Report Writing in 'Real Life' Engineering Courses," *Engineering Education* November 1976: 154–58.
5. Based on *Michigan Department of Corrections Procedure*, OP-BFS-71.01, Effective Date March 1, 1983.
6. *Transcript of Proceedings*, President's Commission on the Accident at Three Mile Island, Public Hearing, July 18, 1979, and Public Hearing, July 19, 1979: 240–41.
7. Thomas N. Huckin, "A Cognitive Approach to Readability," *New Essays in Technical and Scientific Communication: Research, Theory, Practice*, ed. P. Anderson, R. Brockmann, C. Miller (Farmingdale, NY: Baywood) 90–108. See especially 95–96.
8. *Scientific and Technical Reports: Organization, Preparation, and Production*, American National Standards Institute (ANSI), Z39.18-1987.
9. ANSI, *Scientific* 12. The quotes on the next page are from the standard.
10. *Preparation of Scientific Papers for Written or Oral Presentation*, American National Standards Institute (ANSI), Z39.16-1979.
11. *Publication Manual of the American Psycho-*

logical Association, 3rd ed. (Washington, DC: American Psychological Association, 1983 and 1984).

12. ANSI, *Preparation* 11. The following quotations are on page 9.
13. *Publication Manual* (APA) 23.
14. We are indebted to Arthur S. Pfeffer, Director of the Police Management Writing Project, New York City Police Department, for sharing with us the information summarized here. It is included in the Preliminary Report of the project, issued in December 1980.

CHAPTER 5

Planning the Overview

Designing and writing a report is a process, but it is certainly not a rule-governed process. In this and the following three chapters we present detailed outlines of different parts of a report, usually with suggestions and guidelines for implementing them. However, these outlines are not meant to be followed mechanically. As in our discussion of basic structure, for which we provided a simple 3-part outline, these outlines and guidelines usually are derived from the function of reports in organizations. In many cases they are based on either traditional or experimentally tested communication principles as well. But they are not prescriptive. They are merely points of departure which we have found to be helpful to writers in organizations. After all, you define the function of your report in terms of its unique audiences and your specific purpose. Until you have done that, you cannot even outline your report. Certainly we can't outline it for you. Yet we can give you some general planning strategies to consider once you have developed a clear idea of your audiences, your purpose, and the information you will present. These strategies, if you use them appropriately, can help you to start fleshing out a plan for your document structure and writing the preliminary drafts of the document.

In this chapter, then, we divide the overview into three parts—the heading, the purpose, and the summary—and explain the characteristics of each in informal reports. Then we discuss the overviews of technical reports and formal reports, which vary somewhat from those of informal organizational reports and memoranda. Finally, we discuss writing an abstract, which you should not confuse with the summary of a report.

❑ 5.1 ❑ The Three Parts of the Overview and When You Write Them

The overview has three parts, each of which serves an important function: heading, purpose statement, summary of important information (Figure 5.1). The heading establishes the lines of communication and the significance of the report: it attracts the attention of the appropriate audiences to the report. The purpose statement, sometimes called a "foreword" or "introduction," establishes the organizational context and function or purpose of the report. The summary, often called "conclusions and recommendations," presents the most important information so that the audiences can act or respond to the report.

Writing the overview spans the entire report writing process. You have to identify your audiences and formulate your purpose for writing. When you have done so, you can start to write parts of an overview. However, you cannot complete the overview until you have written the report. In the overview you

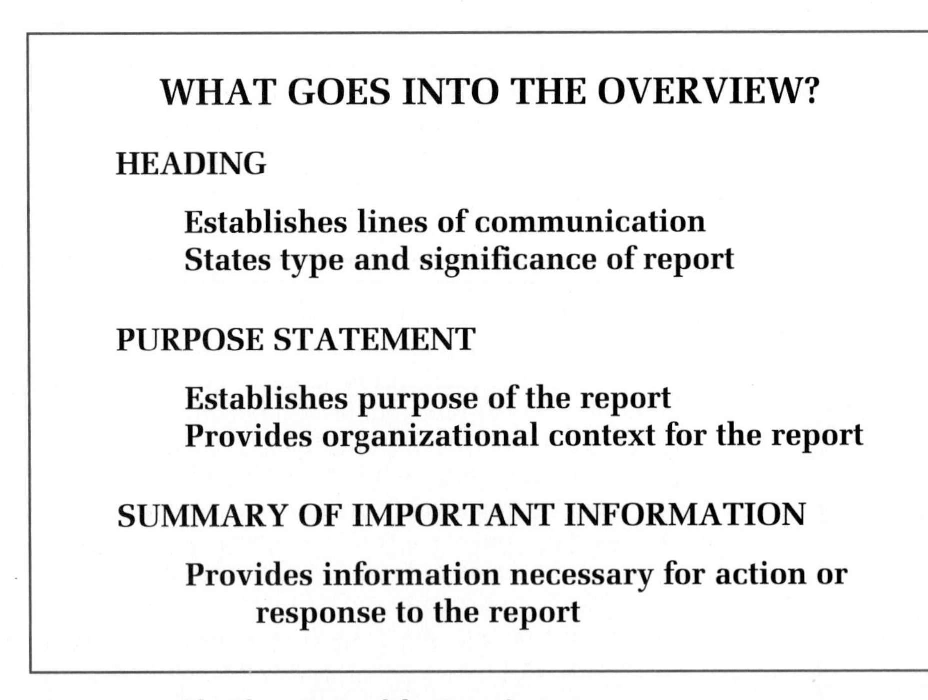

FIGURE 5.1 The Three Parts of the Overview.

summarize the most important information in the report, so you have to write the discussion before you can summarize it.

Once you decide essentially what you want to accomplish with a report, you can begin to plan its design. You can start and end that planning process with what we have labeled the "overview" of a report. You can start there because the overview gives you a compressed sense of your problem and purpose as well as of your essential conclusions and recommendations. You—like your eventual readers—can think about the overview without getting bogged down in all of the details. But you need to end there too. That is, by the time you have planned the report in detail and have actually written it, you will have discovered how "loose" your initial plan for the overview might have been. Your first version of an overview almost certainly will be overwritten, unselective, and perhaps imprecise. That is, when you are planning, you do the best you can to anticipate the final version of the overview; when you write the report, however, you learn that your plan needs modifications. Our advice, then, is to think about and perhaps draft the overview first, as a planning act. Later, when you are done writing the report, rewrite the overview and tighten it up. Never let your first version of an overview be your final version.

If we look at the process of doing the actual writing of a report (which—remember—follows an analysis and planning process), it might be divided into eight stages of actual writing, in which you do the following:

1. Prepare the heading.
2. Draft the purpose statement.
3. Prepare a rough draft of the summary.
4. Outline the discussion or analysis.
5. Write the discussion.
6. Prepare the attachments.
7. Rewrite and revise the heading, purpose, and summary.
8. Revise the discussion and edit the report.

This report writing process can apply to most types of reports that you write, but is particularly descriptive of the process of actually writing a short, informal report.

As we mentioned in the previous chapter, different types of reports have similar basic structures despite a diversity of formats (Figure 5.2). Formats vary considerably among organizations, and terminology often is a matter of convention within an organization rather than generally applicable across

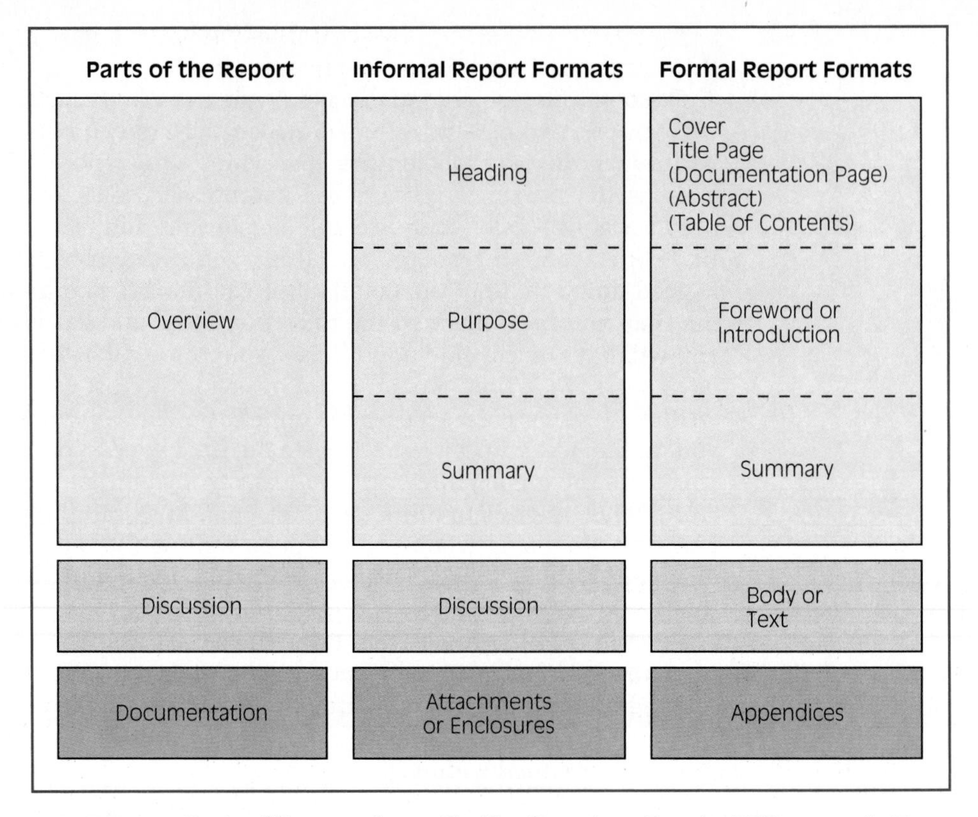

FIGURE 5.2 Parts of Reports Serve Similar Functions Despite Differences in Format and Terms.

organizations. Despite differences in format and labels, however, the various parts of a report serve similar functions. The summary of a formal report, for example, is on a separate page; the summary in an informal report might be only a paragraph without a heading, or even just one sentence of conclusion tacked onto the purpose statement at the end of the introductory paragraph. Both summaries, however, serve the same function.

This diversity of format is most apparent in the overviews of different types of reports. Formal reports are distinguished by such format items as covers, titles pages, and tables of contents. Informal reports lack covers and title pages, but come in a variety of heading formats (memorandum format, letter format, and the combined letter-memorandum format are typical variations discussed below). These heading formats present much of the same information that the cover and title page of a formal report do.

We devote most of our attention in this chapter to the overviews of various types of informal reports, which you will write much more often than you will write formal reports. We do so because, from an organizational point of view, the informal report, such as a memorandum or a letter, usually is more important than a formal report. In most organizations, decisions are made and actions are taken on the basis of informal reports. Formal reports often present material after the fact—for the record or for "PR" purposes. The term *informal report* should not suggest that the report is casual or unimportant. The term *formal report* should not suggest that it is more important than informal reports. The overview of an informal report is important because of the report's organizational function and usefulness.

❑ 5.2 ❑ The Heading in Informal Reports

The heading has four types of information (Figure 5.3):

- Subject Line
- Audiences
- Source
- Reference Information

The heading introduces your report into the organization. It establishes effective and appropriate communication paths and calls attention to your report. It is your first, and sometimes only, chance to grab a reader's attention. It answers the first question all persons in organizations ask upon receiving a piece of paper: "Should I pay attention to this?"

The heading is an important section of the report, not just a series of blanks you fill in mechanically. You have to compose a heading just as you do other sections of the report. You have to think out the specific heading for each report, then you have to write it. The only difference from the other sections of the report is that in the heading you write in a telegraphic style, not in complete sentences.

To write the heading requires you to to analyze the entire organizational communication process for your report. You need to identify the organizational problem and the persons involved in the problem and specify how your activities are related to them. You have to determine the specific subject of the report and your precise purpose in writing. And you have to identify the audiences for the report and the actions they will take.

We suggest that you at least draft your heading before you write your report. Doing so will force you to think about your communication rather than about your subject matter. Of course, after you write the report, you should revise your heading so that it represents the report which you actually ended up with.

Report headings are formatted so that the four types of information are clearly separated. In many organizations, the headings are on standard forms. Each organization and sometimes each department can have several heading formats—one for internal correspondence, for example, or one for a specific type of report the department routinely issues, such as a test report.

THE HEADING

1. **Subject Line. States the topic and suggests the purpose of the report.**
2. **Audiences. Identifies primary and secondary audiences by role and department.**
3. **Source. Identifies the writer and writer's role in the organizational unit.**
4. **Reference Information. Identifies relevant previous reports and communications. Gives project and file numbers or retrieval codes. States date of issuance.**

FIGURE 5.3 **The Heading in Informal Reports.**

Informal reports have three general types of headings (Figure 5.4). These vary considerably according to organizational style standards, as the various sample reports in figures throughout this book illustrate. The standard memorandum format, the "To:/From:" format, sometimes is typed on the memorandum by the writer and at other times is printed on a standard form. The letter report format also usually includes items, such as a subject line, which the non-business letter format does not include. Because of the importance of the heading information and the efficiency of the memorandum heading format, the letter format is often combined with the memorandum format. In the combined format, the heading itself is in memorandum format, but then the salutation and close of the letter format are also included.

Several of our examples illustrate an additional feature of heading formats: some of the "heading" information actually appears at the end of the report. Extensive distribution lists, for example, and some of the reference information often come at the end rather than the beginning simply so that the report does not begin with a whole page of preliminary information. Ideally, page 1 of the report contains at least the heading and the rest of the overview—the explanation of the problem and purpose as well as the summary of important conclusions and recommendations. In short reports, the first page may include a fair amount of the discussion as well. Sometimes, how-

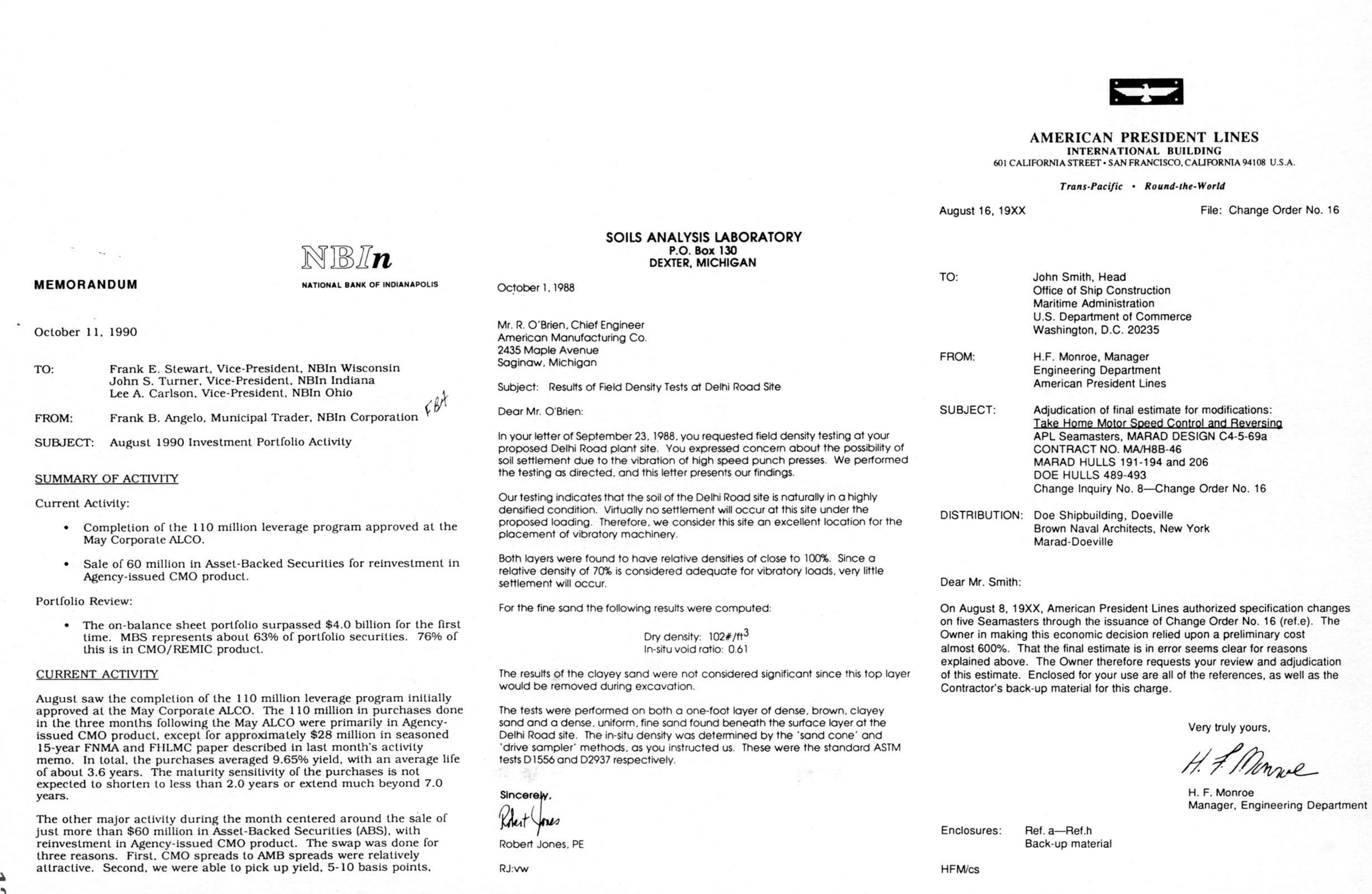

NBIn
NATIONAL BANK OF INDIANAPOLIS

MEMORANDUM

October 11, 1990

TO: Frank E. Stewart, Vice-President, NBIn Wisconsin
John S. Turner, Vice-President, NBIn Indiana
Lee A. Carlson, Vice-President, NBIn Ohio

FROM: Frank B. Angelo, Municipal Trader, NBIn Corporation FBA

SUBJECT: August 1990 Investment Portfolio Activity

SUMMARY OF ACTIVITY

Current Activity:

- Completion of the 110 million leverage program approved at the May Corporate ALCO.
- Sale of 60 million in Asset-Backed Securities for reinvestment in Agency-issued CMO product.

Portfolio Review:

- The on-balance sheet portfolio surpassed $4.0 billion for the first time. MBS represents about 63% of portfolio securities. 76% of this is in CMO/REMIC product.

CURRENT ACTIVITY

August saw the completion of the 110 million leverage program initially approved at the May Corporate ALCO. The 110 million in purchases done in the three months following the May ALCO were primarily in Agency-issued CMO product, except for approximately $28 million in seasoned 15-year FNMA and FHLMC paper described in last month's activity memo. In total, the purchases averaged 9.65% yield, with an average life of about 3.6 years. The maturity sensitivity of the purchases is not expected to shorten to less than 2.0 years or extend much beyond 7.0 years.

The other major activity during the month centered around the sale of just more than $60 million in Asset-Backed Securities (ABS), with reinvestment in Agency-issued CMO product. The swap was done for three reasons. First, CMO spreads to AMB spreads were relatively attractive. Second, we were able to pick up yield, 5-10 basis points,

SOILS ANALYSIS LABORATORY
P.O. Box 130
DEXTER, MICHIGAN

October 1, 1988

Mr. R. O'Brien, Chief Engineer
American Manufacturing Co.
2435 Maple Avenue
Saginaw, Michigan

Subject: Results of Field Density Tests at Delhi Road Site

Dear Mr. O'Brien:

In your letter of September 23, 1988, you requested field density testing at your proposed Delhi Road plant site. You expressed concern about the possibility of soil settlement due to the vibration of high speed punch presses. We performed the testing as directed, and this letter presents our findings.

Our testing indicates that the soil of the Delhi Road site is naturally in a highly densified condition. Virtually no settlement will occur at this site under the proposed loading. Therefore, we consider this site an excellent location for the placement of vibratory machinery.

Both layers were found to have relative densities of close to 100%. Since a relative density of 70% is considered adequate for vibratory loads, very little settlement will occur.

For the fine sand the following results were computed:

Dry density: 102#/ft^3
In-situ void ratio: 0.61

The results of the clayey sand were not considered significant since this top layer would be removed during excavation.

The tests were performed on both a one-foot layer of dense, brown, clayey sand and a dense, uniform, fine sand found beneath the surface layer at the Delhi Road site. The in-situ density was determined by the 'sand cone' and 'drive sampler' methods, as you instructed us. These were the standard ASTM tests D1556 and D2937 respectively.

Sincerely,
Robert Jones
Robert Jones, PE

RJ:vw

AMERICAN PRESIDENT LINES
INTERNATIONAL BUILDING
601 CALIFORNIA STREET • SAN FRANCISCO, CALIFORNIA 94108 U.S.A.

Trans-Pacific • Round-the-World

August 16, 19XX File: Change Order No. 16

TO: John Smith, Head
Office of Ship Construction
Maritime Administration
U.S. Department of Commerce
Washington, D.C. 20235

FROM: H.F. Monroe, Manager
Engineering Department
American President Lines

SUBJECT: Adjudication of final estimate for modifications:
Take Home Motor Speed Control and Reversing
APL Seamasters, MARAD DESIGN C4-5-69a
CONTRACT NO. MA/H8B-46
MARAD HULLS 191-194 and 206
DOE HULLS 489-493
Change Inquiry No. 8—Change Order No. 16

DISTRIBUTION: Doe Shipbuilding, Doeville
Brown Naval Architects, New York
Marad-Doeville

Dear Mr. Smith:

On August 8, 19XX, American President Lines authorized specification changes on five Seamasters through the issuance of Change Order No. 16 (ref.e). The Owner in making this economic decision relied upon a preliminary cost almost 600%. That the final estimate is in error seems clear for reasons explained above. The Owner therefore requests your review and adjudication of this estimate. Enclosed for your use are all of the references, as well as the Contractor's back-up material for this charge.

Very truly yours,

H. F. Monroe

H. F. Monroe
Manager, Engineering Department

Enclosures: Ref. a—Ref.h
Back-up material

HFM/cs

FIGURE 5.4 Three Typical Informal Report Heading Formats.

ever, that won't be possible unless lists of references or distribution are delayed until the end of the report. Similarly, memos put the "from" line at the beginning (where you should initial the report to release it), but the same report in letter form would put a signature line at the end, where you would sign the report to release it.

But don't pay too much attention to these superficial variations; look at the functions behind the information which headings present. That is the important point.

In this section we discuss the appropriate parts of a heading in terms of the informal report. How they appear in a formal report is discussed in a following section.

Subject Line

The subject line summarizes the report in one phrase. An effective subject line states the subject or topic of the report and either states or clearly suggests the purpose of the report. It attracts the readers' attention and motivates them to respond.

An effective subject line has these characteristics (Figure 5.5):[1]

- It is specific, making both the topic and purpose (or document type) clear.
- It uses understandable and standard terminology, avoiding abbreviations, acronyms, and jargon.
- It is short (rule of thumb: ten words or fewer).
- It puts important nouns in initial positions.
- It minimizes use of generalized terms such as "report on," "investigation of," "study of."

A subject line should apply to its report and no other possible report. The topic should be specific and the purpose of the report clear. A subject line such as, "Request for Funds to Meet Increased Demand for Department of Mental Health Services," has these features. It applies to one specific report. The purpose is stated directly: it is a request for funds. The topic of the report also is stated: increased demand for Department of Mental Health services.

Similarly, a subject line such as, "Reactor 526 Baseplate Failure: Recommendation for Corrective Actions," has these features. The subject line was not "Investigation of Reactor 526 Baseplate Failure." Although it was an investigation report, the purpose of the report was to recommend corrective actions. The topic of the report was the baseplate failure. The use of a colon is an effective stylistic technique for combining topic and purpose in the subject line. In fact, as you can see, it is easy to put the topic before the colon ("Reactor 526 Baseplate Failure"), and the purpose after the colon ("Recommendation for Corrective Actions"). The colon thus provides a nice means of connection between the two appropriate elements of a subject line.

Features of a Good Subject Line

- It is specific, making both the topic and purpose (or document type) clear.
- It uses understandable and standard terminology, avoiding abbreviations, acronyms, and jargon.
- It is short (rule of thumb: ten words or fewer).
- It puts important nouns in initial positions.
- It minimizes use of generalized terms such as "study of," "report on," "investigation of."

Examples

Recommendation to Prosecute Robert Smith for Manslaughter: Report of the Fatal Accident on 15 January 1989.
Request for Funds to Meet Increased Demand for Department of Mental Health Services
Reactor 526 Baseplate Failure: Recommendation for Corrective Actions
Proposal to Introduce Line Heating into ASBC Production Procedures

Negative Examples

Operator Interruption of High Pressure Injection
Test of New 1500 lb. General Equipment Screw Jack for Use with SB6-LB6 Line
Possible Delay in Pilot Plant Scale Up
Engine Accessory Mountings

FIGURE 5.5 **Characteristics of an Effective Subject Line.**

Subject lines which are imprecise and do not signal the purpose of the report can fail to attract the readers' attention. For example, the two crucial Three Mile Island reports had the same subject line, "Operator Interruption of High Pressure Injection," although they had different topics and different purposes. The first report recommended changes in operating procedures to avoid operator interruption of high pressure injection. The second report rejected the recommendation and requested resolution of a technical issue. As you can see, the reports were completely different, yet they shared the same subject line—not a helpful cue for hurried readers, certainly. Indeed, neither the topic nor the purpose was implicit in the subject line for either report. When the same subject line is used both to make a recommendation and to reject it, you know you do not have a good subject line.

Other negative examples of subject lines have the same imprecision or ambiguity. The report on the "Test of New 1500 lb. General Equipment Screw Jack for Use with SB6-LB6 Line" should have been a recommendation to reject use of the jack. In much the same way, a report with the subject line "Proposed Changes to Personnel Evaluation Procedures" was completely misleading in that the report really discussed the effects of changes in procedures, not how the procedures were to be changed. The report on "Possible Delay in Pilot Plant Scale Up" was not to alert the readers to a possible delay; its actual purpose was to initiate action to avoid possible delay. A report on "Engine Accessory Mountings" could be a request for a test, a requisition of test specimens, a production status report, identification of a design problem, a recommended design change, and so on.

Your subject line should make your intended audiences pay attention to the report and prepare them to respond appropriately.

Audiences

The heading should identify the primary and secondary audiences for the report by role and department. These readers are identified in the "To" line on a memo and on "Distribution" lists of other persons to whom the report is sent. These three types of information are important if the communication process is to be efficient.

First, the audiences should be identified and classified. Otherwise, they might not receive the report. If they receive it but are not identified by name and singled out as important, they might not pay attention to it. This means you should not send a report to "Distribution" and just put names on a distribution list. A particularly interesting example is the heading of the Kelly report (Figure 3.1). Mr. Kelly addressed the report to "Distribution" and put seven names on the distribution list. (Unlike the Dunn memorandum, Figure 2.3, the names on the distribution list for the Kelly memorandum were the appropriate names. In this chapter we are not talking about identifying the appropriate audiences; we are talking about how you specify them in the heading after you have identified them.) Yet Mr. Kelly says, "I recommend the following guidelines be sent." Insofar as he intended his memorandum to be a recommendation report, then, Mr. Kelly should have sent the report to the person responsible for acting upon the recommendation, not to a distribution list.

If at all possible—that is, if your organization protocol does not prevent you—you should address your report to your primary audience. Also you should separate your important secondary audiences from the distribution list, which usually is for information purposes. Important secondary audiences can be put in the heading after the primary audience or they can be put in a separate group on a distribution list. Other secondary audiences and nominal audiences can be included on the distribution list. The distribution list also should be functional: the names should be in some order relevant to the purpose of the report, not listed alphabetically, or serially by department number, or worse, randomly. One effective way to handle this distinction between primary audiences and other audiences is simply to acknowledge the difference by making two distribution lists, one of them labeled "For Action" and one labeled "For Information." That makes explicit what the audience should perceive to be their responsibilities as readers.

Distribution lists have an additional function. Some persons on the distribution list are not audiences who will use the report. They are persons—usually your department colleagues—who worked with you on the investigation, who reviewed your report, who even helped you to write the report, or who are supervisors and managers signing off on the report. These persons are on the distribution list so that the real audiences know who besides the writer is familiar with the investigation, with the report, or with the decisions or actions discussed in the report.

In addition to identifying primary and secondary audiences, you should identify the role and department of each person in the heading, including the distribution list (Figure 5.6). These roles enable all readers to grasp the organizational context of your report immediately. The roles and departments make the communication network clear to the reader. Along with the subject line, they also make the decision-making or implementation process clear. Most persons in an organization, especially managers, can get a pretty clear idea of the entire decision making process for a report just by seeing the roles and departments of everyone involved in the communication process.

Furthermore, a report exists over time. When a report is retrieved 6 months after it is written, some of the audiences may no longer be in the same job or even with the organization. The roles and departments still remain, however; if they are identified in the heading, the fact that the names have changed can be unimportant.

Source

The source of a report is identified in the "From" line on a memo or, in some formats, in a signature line at the end of the report. You should identify yourself by name, role, and department. For many of your audiences, your role is more important than your name. Once a report leaves your office, it reaches many readers who are not familiar with you or the other persons in your department. Your role or title rather than your name has meaning: B. L. Ross, Test Engineer, or Iris Kanter, Personnel Analyst.

The source of a report has another feature that most persons in organizations know and accept as standard operating procedure. The source of the report—the name in the "From" line—might not be that of the actual writer of the report. Instead, it might be a supervisor, project head, or department manager's name. Often, the source of a report operationally is a department rather than a person. That is, you often write a report for your department. That report is then issued under your manager's signature. The report is from

NBIn
NATIONAL BANK OF INDIANAPOLIS

MEMORANDUM

October 11, 1990

TO: Frank E. Stewart, Vice-President, NBIn Wisconsin
John S. Turner, Vice-President, NBIn Indiana
Lee A. Carlson, Vice-President, NBIn Ohio

FROM: Frank B. Angelo, Municipal Trader, NBIn Corporation

SUBJECT: August 1990 Investment Portfolio Activity

SUMMARY OF ACTIVITY

Current Activity:

- Completion of the 110 million leverage program approved at the May Corporate ALCO.
- Sale of 60 million in Asset-Backed Securities for reinvestment in Agency-issued CMO product.

Portfolio Review:

- The on-balance sheet portfolio surpassed $4.0 billion for the first time. MBS represents about 63% of portfolio securities. 76% of this is in CMO/REMIC product.

CURRENT ACTIVITY

August saw the completion of the 110 million leverage program initially approved at the May Corporate ALCO. The 110 million in purchases done in the three months following the May ALCO were primarily in Agency-issued CMO product, except for approximately $28 million in seasoned 15-year FNMA and FHLMC paper described in last month's activity memo. In total, the purchases averaged 9.65% yield, with an average life of about 3.6 years. The maturity sensitivity of the purchases is not expected to shorten to less than 2.0 years or extend much beyond 7.0 years.

The other major activity during the month centered around the sale of just more than $60 million in Asset-Backed Securities (ABS), with reinvestment in Agency-issued CMO product. The swap was done for three reasons. First, CMO spreads to AMB spreads were relatively attractive. Second, we were able to pick up yield, 5-10 basis points,

FIGURE 5.6 **Heading with Audiences Identified by Name, Role, and Location.**

Purchasing, Contracts, Production Engineering, Personnel, and is signed by the appropriate supervisor or manager, depending on its topic and purpose. When this is so, you usually put your name and role on the distribution list, often first. Given the topic of the report, your name on the distribution list signals whom a recipient of the report should contact for any follow up.

Reference Information

The reference information dates the report and identifies specific organizational items to establish the organizational context of a report. This information places a report in the communication sequence of which the report is a part. Usually a report has been preceded by previous communication, written or oral. The relevant communications are identified by subject line and date or by file number. A report often is related to a project or a contract, so these identification numbers are provided. A report also can refer to standards, specifications, or regulations, so these are specified by number and title. This reference information enables the reader to place the report in its specific organizational context.

In this sense, reference information for a report differs from bibliographic references. Bibliographic references identify the sources and background of information or ideas in a document. Students are familiar with bibliographic references to books and articles at the end of essays and journal articles. Bibliographic references are rarely made in organizational reports, especially informal reports. Most organizational reports are based on firsthand investigation, analysis, and data. Secondary sources usually are not used, unless they are "off the shelf" specifications or standards. References in an organizational report usually refer to prior communication, documentation, and various code numbers.

Typical types of reference information include:

Memoranda and reports
Letters
Minutes of meetings
Phone calls
Facsimile communications
E-mail and related electronic messages
Bulletins
Forms
Standard Procedures
Standards and Specifications
Regulations
Tests
Drawings
Parts Numbers
Contract Number

Project Number
Work Action or Number
File Number
Report Number

Your references usually will include several of these items.

In addition, your reference information should be detailed. A reference to a previous memorandum, for example, could be:

> Memo L. Hernandez—V. White, 4 August 1986: "Complete Wing Ballast Tank Drainage"

Although it is not practical to document all reference items so completely, do so as appropriate.

One last but important bit of information is the date. Don't forget to date your report. The date is important for more than information retrieval purposes. It places the report in the communication sequence. It tells the readers when others have received the report. This can have contract and legal implications as well. Without a date, reports can be useless, and of course without being connected to other communications, they can be confusing.

By the way, even in writing dates you have to be careful not to confuse readers. Do not write dates in this format: 12/5/90. As you can see, that date could be either December 5th, 1990, or May 12th, 1990. Which is it? You cannot be sure because there are two conventions for writing dates in the United States: the military and international convention (12 May 1990) and the popular convention (December 5, 1990). Because the abbreviated format gives no cues as to which convention is being used (unless the numbers go over 12), you should avoid the abbreviated format. We recommend the military format because it tends to be the format preferred by businesses which interact with the government and because it is much more widely used around the world than the traditional American format.

❑ 5.3 ❑
The Purpose Statement in Informal Reports

The purpose statement establishes the function of the report in its organizational context. As we explained in Chapters 2 and 3, the organizational context consists of the organizational problem, the objective of your investigation or activity, and the actions required of the persons receiving your report. The purpose statement makes this context immediately clear to your reader.

The purpose statement, therefore, does not introduce the report in the traditional sense of providing background to introduce the technical discussion. When you state the objective of your investigation, you do so to relate it to the organizational problem and the communication purpose of the report, not to lead into the discussion of the investigation.

The purpose statement is the first paragraph or section of your report:

> To: S. Gutherie
> Assistant General Manager
> From: J. Andrulis, Manager
> Production Engineering
> Subject: Proposal to Introduce Line Heating Technology into ASBC Production Procedures
>
> At the present time, the technique of cold-forming shell plating limits our productivity and flexibility. The fact that we rely on this technique limits our competitiveness in some bid situations. I therefore propose that ASBC introduce line heating technology in our facilities, and have scheduled a meeting on 16 July 1986 to discuss their proposal and its implementation. I believe this is an important matter, as line heating technology will enable us to stay competitive with other U.S. shipyards.
>
> Line heating technology will (1) increase our productivity, (2) facilitate more complex shell plate curvature and lower costs, and (3) improve production flow. Improved fabrication quality will result in the yard being more competitive. Implementation of line heating technology will require approximately one year, including planning.

As you can see, the purpose statement provides the organizational context necessary for the readers to understand the summary of important information which follows immediately.

In Chapter 3 we presented a 3-step outline of the purpose statement and a heuristic for formulating it. Those three elements—the organizational problem, the investigation, and the communication purpose—are the core of the purpose statement. Given your purpose and the information the readers need to understand the organizational context, your purpose statement could include these and other items as well:

1. The organizational problem, need, or conflict at issue as perceived in the organization.
   - The names and roles of persons or the departments involved in the issue.
   - The person or department requesting the investigation.

2. The investigation or analysis needed to resolve the issue.
   - The assignment, specifically what the writer or the writer's department was asked to do or undertook to do.
   - The specific questions or tasks addressed in the investigation, and perhaps the hypothesis or proposed solution.

3. The communication purpose of the report.

The purpose statement of the report on Line Heating Technology states the problem in two sentences and the communication purpose in the third and fourth sentences. The positive examples of purpose statements in Chapter 3, from memoranda, letters, and formal reports, illustrate how the statements vary according to the type of document you are writing. Their functions, however, are the same: to establish the purpose of the report in its organizational context. For the purpose statement, choose only the information necessary to do so.

❑ 5.4 ❑ The Summary in Informal Reports

The summary of your report presents the most important information in the report. The summary enables readers to respond to the report without having to read further. As we discussed in Chapter 4, the summary has a secondary function of providing a framework for readers who are going to read some or all of the rest of the report.

When your management audiences finish reading the summary of your report, they should have enough information so that they do not have to read further in the report. If they are to make a decision or to take action, the summary should provide the information they need to do so. If they are just being informed, as executives often are, they should have the salient information that keeps them familiar with the broad outline of the organization's activities. Other readers need just enough information to know what actions are going to occur in order to determine if and how their responsibilities and departments will be affected.

To accomplish these functions, the summary presents four types of information: significant results, conclusions, recommendations, and implications for the organization (Figure 5.7). These should be understandable in the context introduced in the preceding purpose statement. Most of the summary information consists of selected generalizations.

Selected Generalizations

The summary presents selected generalizations that contain the most important information in the report. For many professionals, this information consists of high-level generalizations: conclusions and recommendations. The summary paragraph in the Field Density Tests report presents only conclusions and recommendations:

> Our testing indicates that the soil of the Delhi Road site is naturally in a highly densified condition [conclusion]. Virtually no settlement will occur at this site

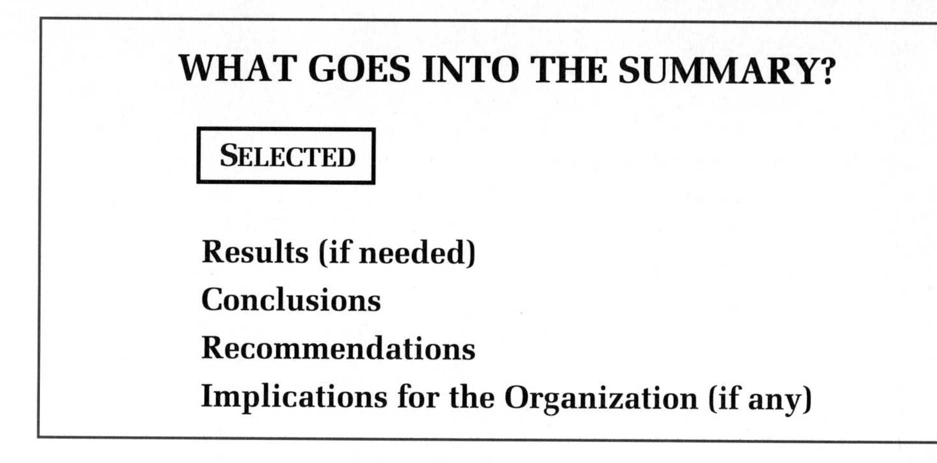

FIGURE 5.7 The Contents of the Summary.

> under the proposed loading [conclusion]. Therefore, we consider this site an excellent location for the placement of vibratory machinery [implicit recommendation].

In this summary an intermediate conclusion (the soil is in a highly densified condition) supports a final conclusion (no settlement will occur), which in turn supports the implicit recommendation. In the Front End Sheet Metal report, two final conclusions (standardized F.E.S.M. mounting is satisfactory and A/C condenser mounting is satisfactory) support the recommendation to approve use of both.

You must be selective if you are to write efficient summaries. The sample report on Line Heating Technology (Figure 5.8) illustrates the concept of selectivity. The discussion of that report itself consists mostly of generalizations—and rather high-level generalizations at that. These are efficiently summarized in the first two sentences of the summary, which is the second paragraph in the report (the first paragraph is the purpose statement). Being selective requires you to present only the most important information in high-level generalities. This is information that you judge necessary to convince your primary audiences to accept your recommendation, if it is a recommendation report, assuming that they accept your analysis without question. In the discussion of a report where you present your analysis, a recommendation is based on a conclusion, which might be based on intermediate conclusions, which in turn are based on bottom-line results, intermediate results of calculations, and even selected data. In the summary, you can stop with the conclusion when you assume that your reader will not question your analysis.

To present only the most important information means that you are not trying to summarize the entire discussion: you are selective in content as well. The summary of the Line Heating Technology report does not sum-

Atlantic Shipbuilding
MEMORANDUM

File No. 12-5T
Date 3 July 1986

To: S. Guthrie
Assistant General Manager

From: J. Andrulis, Manager J. Andrulis
Production Engineering

Subject: Proposal to Introduce Line Heating Technology into ASBC Production Procedures

At the present time, the technique of cold-forming shell plating limits our productivity and flexibility. The fact that we rely on this technique limits our competitiveness in some bid situations. I therefore propose that ASBC introduce line heating technology in our facilities, and have scheduled a meeting on 16 July 1986 to discuss this proposal and its implementation. I believe this is an important matter, as line heating technology will enable us to stay competitive with other U.S. shipyards.

Line heating technology will (1) increase our productivity, (2) facilitate more complex shell plate curvature and lower costs, and (3) improve production flow. Improved fabrication quality will result in the yard being more competitive. Implementation of line heating technology will require approximately one year, including planning.

<u>Limitations of cold forming technique</u>. Our facilities for cold-forming shell plating do not suit the needs of current designs. These facilities can perform adequate shaping only up to 3/4 inch thick plate and are not capable of giving plates longitudinal curvature. The specifications of our cold-forming facilities and optimal specifications for contract TY236-83 are listed in Appendix A.

<u>Line heating technology</u>. Line heating is the process of inducing curvature in structural shapes and plates by controlled heating and cooling. The process is also scientifically applied for fairing structural intersections and removing distortion due to thermally induced stress.

<u>Advantages of line heating for forming curved shell plating</u>. By utilizing line heating technology, we can eliminate the restrictions we now face in the forming of shell plating. The advantages of line heating are:

1) Increase in productivity, as seen in other U.S. shipyards (Appendix B).

2) Ability to accurately form shell plates with compound curvature (Appendix C itemizes parameters and dimensional control for plates on Hitachi Hull Y-14-2), which results in a cost savings.

FIGURE 5.8 Recommendation Report Illustrates Selectivity of the Summary.

3) Ability to accurately fit curved parts to curved shell with minimal force, which requires fewer manhours than cold forcing.

4) Accommodation of larger plate sizes, thereby minimizing butts and seams.

5) Optimization of existing facilities through improved scheduling and elimination of bottlenecks.

6) Facilitation of subsequent assembly by eliminating distortion prior to forwarding an interim product to the next level of production.

7) Ability to accurately and productively fair structural intersections.

8) Enhanced worker safety.

Our statistical accuracy control system would be modified to assure the normal process capability, including standard range and tolerance limits.

In conclusion, line heating shapes material more accurately with less effort and provides more production flexibility.

Implementation of line heating. Introduction of line heating would be best accomplished by contracting with a Japanese firm to provide technical information and training. Our own shipyard would be responsible for the adaptation of existing equipment and the construction (and purchase) of new equipment needed for this method. Finally, ABS approval will be needed before this technique is adopted in actual production. The implementation of line heating will require approximately one year.

Upon approval of this proposal I will prepare a detailed procedure and cost estimate for implementation of line heating technology. Final cost estimates can be made after technical training with the Japanese firm. Qualitative benefits then can be calculated for selected proposals and contracts, based on the percentages of shell plates which could be formed in whole or part by line heating.

I have scheduled a meeting in the Administrative Conference Room for 0900 16 July 1986 to discuss this proposal and implementation plans.

Appendices: A-Yard Specification and optimal specification, Contract TY236-83
B-Effect on Productivity of Line Heating Technology
C-Compound Shell Curvature for Hitachi Hull Y-14-2

cc: S. Gutherie, Asst. General Manger
G. Landon, Prod. Engineering
C. Sherman, Welding Engineering
A. Vaslo, Chief Naval Architect
V. White, Quality Assurance
F. Domino, Construction
W. Loman, Purchasing

marize information from the sections on "Limitations of cold forming technique" and "Line heating technology." Furthermore, some of the intermediate conclusions in "Advantages of line heating" (for example, "enhanced worker safety") are not summarized or aggregated in a more comprehensive term in the summary. In addition, the final sentence of the summary, "implementation of line heating technology will require approximately one year," just repeats one item of information in "Implementation of line heating." It does not generalize about that item, and it ignores all of the other information in that section. Again, you select information that is important to your primary audiences and to your purpose. You ignore the secondary information and conclusions in your discussion in order to emphasize your basic conclusions.

Finally, being selective requires you to use as few words as possible. The summary should be proportionate to the discussion in some noticeable ratio. This summary has 43 words; the discussion following has 432 words. The ratio therefore is 1:10. If the ratio of your summary to the discussion is 1:2 or 1:3, then you are not being selective for the reader who only wants the "bottom line" and has no time to read more than a page.

Some reports have additional types of information. In a benefit/cost report, bottom-line results might be necessary to support the conclusions so that the reader doesn't have to search the discussion to verify the conclusion. An experimental report might have a statement of the methodology or procedure in the summary for the same reason. If you feel the need to include additional information in the summary, make sure that this is information the readers absolutely need in order to respond to the report.

Significant Details in the Summary

Although the summary can consist only of generalizations, such as the summary of the Field Density Tests report, often it contains significant details such as results as well. This information usually specifies implications for the organization. (Occasionally this particular information is not repeated in the discussion of a short report, such as a 1-page report. In longer reports it usually is.)

Generalities in the summary often are supported by results, either qualitative or quantitative. The summary of a progress report said, "Safety also has been good," a statement that by itself would be a vague generalization. The following sentence said, "We had no disabling injuries in the department during the reporting period." This statement of results, "no disabling injuries," specifies what the writer meant by "good."

A summary can contain considerable particular information as well, as in the following example from the Request for Adjudication report (Figure 5.4):

> Modifications to the motor, controllers, and resistor bank drew a preliminary cost estimate by the Contractor of $42,221 for the five ships. The final estimate is $279,866. The Owner feels that costs related to a change in gear ratio, struc-

tural changes to bulkheads, modification of gears and coupling, and related material costs are not chargeable to the Change Order and should be considered as a development of the contract. The Owner furthermore questions the costs due to disruption in man-hours (678). The Owner also observes an escalation in costs for successive vessels not provided for in the contract. The final estimate by the Contractor, therefore, is excessive, and should be made consistent with the preliminary estimate and with the specific changes requested in Change Order No. 16.

The writer's conclusion is based on intermediate conclusions as well as specific cost and man-hour data. The specific data considerably strengthen the summary, especially the quantification of the preliminary cost estimate ($42,221) and final estimate ($279,866). But note that these are selectively presented, significant details; they are not all of the details. The key word is "significant."

The summary of the letter on the clutch to disengage the turbine starting pump has specific performance data on the pump as well as a cost figure. The front end sheet metal report summary has a part number (P/N). Other types of specific information often found in summaries include dates, schedules, meeting times, etc. These are significant because they specify organizational implications. The date when you were assigned or finished a project may not be relevant. The date when parts will arrive or bids will close could be crucial. As with the purpose statement, in the summary you select information that you consider organizationally important for your report to accomplish its purpose.

The summary, then, can contain both generalizations and particulars. It can include other types of information in addition to conclusions and recommendations. A summary can include some but usually not all of these various types of information, depending on the length and complexity of the report:

Conclusions
Recommendations

- Objective or hypothesis of the investigation, if this is not in the purpose statement.
- Quantification or particularization of the problem, if this is not in the purpose statement.
- Criteria or standards used to evaluate the results.
- The methodology, experimental procedure, or test procedure.
- Significant results.
- Implications for the organization
- Subsequent actions required or to be taken
- Costs
- Benefits
- Personnel implications

- Space implications
- Dates, schedules, times.

Some of this additional information, whether or not it restates or summarizes information in the discussion, is necessary for the report to fulfill its organizational function.

The order of the items in the summary usually follows the order of the information in the discussion. In fact, in reports which make multiple conclusions or multiple recommendations, often the discussion is structured as a series of analyses leading to those conclusions or recommendations in the same order in which they appear in the summary. A summary that follows the arrangement in the discussion is useful for the reader who uses the summary as a framework for reading the discussion or searching selectively for information in the discussion. In the same way, a progress report might enumerate five or six projects and summarize the progress on each in the overview. In the discussion, the same sequence would be followed to add more detail. But for many readers only a few of the projects might be worth further reading; hence, the structural tie between the summary and the discussion would assist readers in reading selectively.

The order of the items in a summary, however, does not have to follow the order of the information in the summary. The primary function of the summary is to provide important information for readers so that they do not have to read the report. In terms of your purpose and your primary audiences' interests, therefore, you might rearrange information in the discussion. For example, you might start off a summary with the recommendation, which comes at the end of the discussion. Or, you might use a most-to-least important order, as in the sample Minutes of a Meeting (Figure 3.4), where information in the summary is ordered differently than the information in the discussion, which was ordered according to the agenda of the meeting.

Our discussion of particulars in the summary should not lead you to be unselective. Remember, the summary does not summarize the entire report: it summarizes the significant information in the report. When some of the preceding items of information listed, such as methodology, are not significant in terms of the function of the report, omit them. Your summary must be selective if it is to receive the attention it deserves. (If you overwrite the summary, you risk hiding important information in unimportant information. Hence the need for selectivity.)

❑ 5.5 ❑ The Overview in the Formal Report and Technical Report

The principles for writing the overview apply to formal reports and technical reports as well as to informal memoranda and letters. However, these reports have functions which are somewhat different from those of informal organizational reports. Consequently, the heading, purpose statement, and summary information are adapted accordingly.

The Cover, Title Page, and Table of Contents

Formal reports and technical reports superficially differ from the informal reports we have been discussing because they are bound with a cover and have a title page. The cover and title page contain the heading information for those reports and serve the same functions as the heading in informal reports.

The cover and title page for a formal report, such as a consulting report, are similar (Figure 5.9). Both have the title, the primary audience, and the source. Both have reference information, with the title page having more reference information than the cover. In technical reports, the cover and title page also are similar in appearance and content (Figure 5.10). Some technical reports have a documentation page (Figure 5.11) in addition to or in place of a title page. A documentation page is a standard form similar to those of memoranda headings, but with more items of information. It usually includes an abstract (discussed in a following section).

The title of a formal or technical report is the same in terms of its function as the subject line of an informal report; however, it is presented a little differently. It often is separated into a title and subtitle, the first element of which presents the topic, the second of which presents the purpose, as in the following example:

CHARLEVOIX, MICHIGAN WATER
SUPPLY SYSTEM IMPROVEMENTS

Recommendations to Meet
Projected Water Demands

The only differences between this formal report title and the subject line of an informal report are that the title and subtitle are not introduced, as a subject line is, with the word "subject" and that the title and subtitle are not separated with a colon. Otherwise there are no differences and the same guidelines can apply. This title states the subject and the subtitle states the purpose of the report. The complete title is understandable to all readers; it is short; it has no unnecessary generalized terms.

These reports do have an item that an informal report does not have: a table of contents. The table of contents enables readers to use the report efficiently. It provides an overview of the report content and structure as well as an index of main headings (and subheadings) for selective reading.

The table of contents lists the heading and gives beginning page references for each major section of the report (Figure 4.13). These headings should include those for the appendices; the appendix should be subdivided into major sections (appendices), each with its own title. If there are sufficient

CSF LTD.

2200 FULLER ROAD · ANN ARBOR, MICHIGAN 48105 · 313/761-1846

CORPORATE HEADQUARTERS

CENTRALIZATION OF MEDICAL

RECORDS TRANSCRIPTION SERVICES

Sunnybrook Medical Centre
Toronto, Ontario

CENTRALIZATION OF MEDICAL

RECORDS TRANSCRIPTION SERVICES:

Reducing Staff and Increasing Efficiency

Sunnybrook Medical Centre
Toronto, Ontario

Submitted by: CSF Ltd.
Project Engineer: G. F. Braidwood
Project Number: AA-SB-18
Date Submitted: 20 September 19XX

Submitted to: Mr. J. R. Hamilton, Assistant Executive Director, General Services
Mr. M. M. McCandless, Head, Medical Records
Ms. E. Johnson, R.R.L., Medical Records Supervisor, Coding and Indexing
Ms. J. Evans, Medical Records Supervisor, Stenographic Services

FIGURE 5.9 Cover and Title Page from a Formal Report by a Hospital Management Consulting Firm.

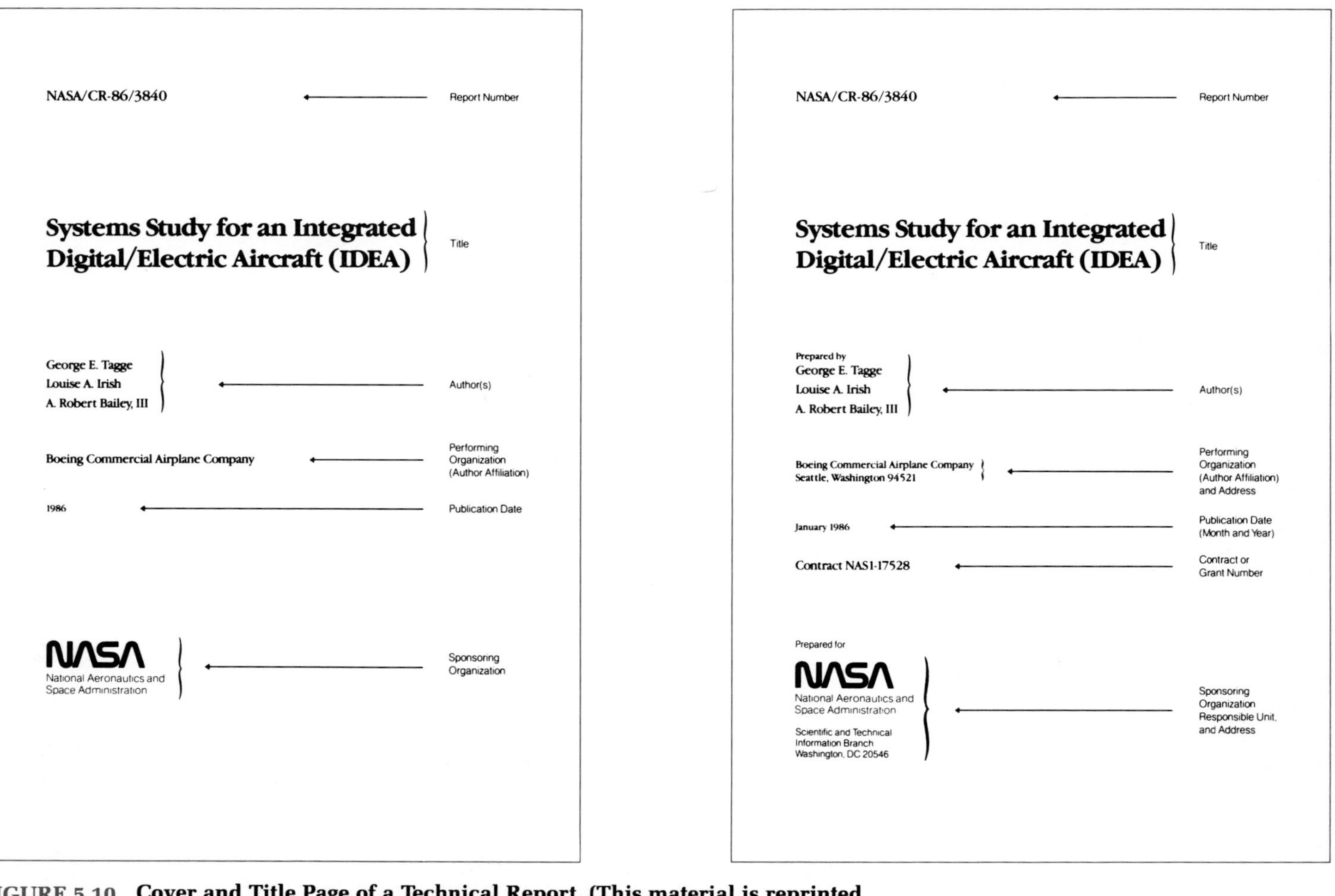

FIGURE 5.10 Cover and Title Page of a Technical Report. (This material is reprinted with permission from American National Standards Institute *Systems Study for an Integrated Digital/Electric Aircraft*, copyright 1983 by the American National Standards Institute. Copies of this standard may be purchased from the American National Standards Institute at 1430 Broadway, New York, N.Y. 10018.)

REPORT DOCUMENTATION PAGE	1. REPORT NO. NASA/CR-86/3840	2.	3. Recipient's Accession No.

4. Title and Subtitle
Systems Study for an Integrated Digital/Electric Aircraft (IDEA).

5. Report Date
January 1986

6.

7. Author(s)
George E. Tagge; Louise A. Irish; E. Robert Bailey III

8. Performing Organization Rept. No.
BOEING TR-85-107

9. Performing Organization Name and Address
Boeing Commercial Airplane Company
Seattle, WA 94521

10. Project/Task/Work Unit No.

11. Contract(C) or Grant(G) No.
(C) NAS1-17528
(G)

12. Sponsoring Organization Name and Address
National Aeronautics and Space Administration
Scientific and Technical Information Branch
Washington, DC 20546

13. Type of Report & Period Covered
Contractor Report (Final)

14.

15. Supplementary Notes
George E. Tagge (206) 943-7401
NASA Technical Monitor: Dr. Robert E. Cozgrove (804) 865-2904

16. Abstract (Limit: 200 words)

This document presents the results of the Integrated Digital/Electric Aircraft (IDEA) Study. Airplanes with advanced systems were defined and evaluated as a means of identifying potential high-payoff research tasks. A baseline airplane, typical of the 1990's airplane with advanced active controls, propulsion, aoerdynamics, and structures technology, was defined for comparison. Trade studies led to the definition of an IDEA airplane with extensive digital systems and electric secondary power distribution. This airplane showed a reduction of 3 percent in fuel consumption and 1.8 percent relative to the baseline configuration. An alternate configuration, an advanced technology turboprop, was also evaluated and showed greater improvement supported by digital-electric systems. Recommended research programs were defined for high-risk, high-payoff areas appropriate for implementation under NASA leadership.

17. Document Analysis a. Descriptors

b. Identifiers/Open-Ended Terms
Digital systems
Electric secondary power
Data bus
Electric actuation

c. COSATI Field/Group

18. Availability Statement	19. Security Class (This Report)	21. No. of Pages
National Technical Information Service 5285 Port Royal Road Springfield, VA 22161	Unclassified	32
	20. Security Class (This Page)	22. Price
	Unclassified	

(See ANSI–Z39.18) OPTIONAL FORM 272 (4–77)

☆ GPO : 1983 O - 381-526 (8393)

FIGURE 5.11 Documentation Page from a Technical Report. (This material is reproduced with permission from American National Standard *Systems Study for an Integrated Digital/Electric Aircraft,* copyright 1983 by the American National Standards Institute. Copies of this standard may be purchased from the American National Standards Institute at 1430 Broadway, New York, N.Y. 10018.)

items to warrant separate listings, lists of figures, tables, symbols, or glossaries of technical terms are sometimes included. These usually are on separate pages immediately following the table of contents and are the first items on the table of contents. (The ANSI standard for *Scientific and Technical Reports* notes that if the table of contents is short and the lists are short, the lists of figures and tables can be below the table of contents on the same page. It also notes that if there are few figures or tables, they can be combined in one list.[2])

The headings and subheadings in the table of contents should repeat those in the text word for word. (The ANSI standard is even more specific: "The headings in the contents are numbered, worded, spelled, capitalized, and punctuated exactly as they are in the report.") However, as the ANSI standard notes, the table of contents does not have to list all levels of headings; two or three levels of headings are usually sufficient. Fourth- and fifth-level headings (perhaps even third-level headings in short formal reports) are unnecessary. A table of contents that lists every heading and subheading loses its utility as an overview to a report when it goes more than two pages; the pattern of the main headings becomes obscured by the particulars.

The Introduction and Summary or the Executive Summary

The introduction and summary of formal and technical reports are similar in function and content to those of an informal report. (In an "executive summary," the introduction and summary are combined and formatted as a separate prefatory section.) Because of the length of these reports, however, the introductions and summaries are longer: they have more explanation or detail. They also include more of the items of possible information for the purpose and summary (listed above) than do organizational reports.

You write the introduction to a formal or technical report as an extended purpose statement. If the report serves a research function rather than an organizational function, as do many technical reports as defined by ANSI, you replace the organizational problem definition with a research problem definition, as discussed in Chapter 3. If the report serves a management function, as do many consulting reports, you write a more detailed purpose statement than for an informal report. Instead of explaining the organizational problem or need in one or two sentences, you explain it in detail—perhaps a paragraph. Instead of just stating the objective of the investigation or analysis, you itemize the specific steps or tasks required to perform the investigation and the criteria required to analyze the results. Instead of making a single declarative statement of communication purpose, you itemize the sequence of actions that need to be taken or the subsequent considerations to be made.

In the summary you amplify the conclusions and recommendations so that the summary comprehends more of the information in the discussion. You present the essence of your argument or analysis, not just the conclusions without this framework of support for them. You introduce intermediate conclusions to support the primary conclusions, and you state the results that support the intermediate conclusions. In a formal report, you also will explain the implications for the organization in detail, such as benefits, costs, schedule, or staffing. In a technical report, you will explain the significance of the investigation—the new knowledge resulting from the research. Your summary is necessarily longer because a formal report or a technical report presents a more complex analysis (and therefore is considerably more lengthy) than does an informal report. To enable most of your readers to get the important information without having to read the body of the report requires you to explain your conclusions rather than just present them.

The overview of the consulting report on "Centralization of Medical Records Transcription Services" illustrates the overview of a formal report (Figure 5.12). The "Purpose of the Study" presents a 2-paragraph purpose statement. The first paragraph states the organizational problem; it is a more detailed explanation of the problem than would appear in an informal report. The second paragraph states the objective of the investigation and the purpose of the report. These are more detailed than those statements in an informal report.

The "Synopsis of Recommendations" presents specific recommendations in addition to the primary recommendation of "centralization of the medical transcription function." It states the benefits of centralization, financial as well as administrative, then itemizes ten "specific recommendations." These recommendations specify ten means of implementing centralization. Thus, they clarify the implications of centralization for the organization. This summary focuses on implementation because the report is management oriented; it does not summarize many of the results and conclusions in the body of the report. These detailed recommendations enable the hospital management to understand the implications of the conclusions and recommendations without having to read the body of the report.

Cover Letters of Transmittal

Many formal reports are accompanied by a cover letter of transmittal. The letter of transmittal should serve as an additional overview of the report. That is, the letter of transmittal should not merely introduce the report, essentially saying: "you requested this study; here is our report—read it." The letter of transmittal should summarize the key information in the report. Formal reports usually are addressed to high-level administrators or upper management. These persons might not even bother to read the overview of the report. The letter of transmittal is the primary means of communication with those persons. It should be substantive in the sense that it very succinctly presents important information from the report, although it can also be a polite device by which to present the report and to thank clients or sponsors for supporting the work represented by the report. In other words, letters of transmittal have a politeness function as well as a summary function.

The report on "Centralization of Medical Records Transcription Services" has a substantive letter of transmittal (Figure 5.13). It presents the communication purpose in the first paragraph, the issue and important results and conclusions of the analysis in the second paragraph, and statements of appreciation in the third paragraph. (Sometimes the letter of transmittal serves as the overview; that is, the report itself does not have a summary up front. Most of the time, however, as with the report on "Centralization of Medical Records," the letter of transmittal is in addition to the overview of the report itself. This seems the preferable alternative.)

A copy of a substantive letter of transmittal is bound with the report, usually after the title page (although occasionally before). The report on "Charlevoix, Michigan Water Supply System Improvements" (Figure 3.6) had the letter of transmittal bound after the title page; the "Medical Records" report had the letter of transmittal bound before the title page. A copy of the letter of transmittal is bound with the report for documentation purposes, as the unbound original can easily be separated from the report and not be available when the report is used or referenced.

❑ 5.6 ❑
Abstracts in Scientific and Technical Reports and Articles

Scientific and technical reports and articles are accompanied by an abstract.[3] This is a compressed summary of the document which serves two important functions. First, the abstract is an information retrieval device. It is put into databases and published separately for reference purposes. In an electronic search, readers will call up abstracts to determine whether or not they need to read the report or article itself. Second, an abstract is informative. Many readers scan abstracts to see what kinds of work are being published in an area rather than to locate a specific document or article. Even with a document itself, readers often will read only the abstract. With a scientific or technical article, the abstract serves the additional function of providing an overview for the reader. (The abstract, however, should be distinguished from the overview or executive summary of a report, which are management oriented.)

The functions of an abstract require it to be a mini-document in itself. An abstract is a true summary of a document, a linear reduction of a document rather than a selective summary of the document's most important information. For a research report or article with a method of investigation design, for example, the abstract would have statements of the problem or purpose of the investigation, the methodology or procedure, the results, and the interpretations or conclusions of the analysis, and the recommendations, if any. The *Publication Manual of the American Psychological Association* says the abstract of a report of an empirical study should describe the problem, the subjects, the experimental method, the findings, the conclusions, and

CENTRALIZATION OF MEDICAL
RECORDS TRANSCRIPTION SERVICES

THE PURPOSE OF THE STUDY

At the request of the hospital administration, CSF Ltd. has analyzed the medical transcription function of the Medical Records Department at Sunnybrook Medical Centre. The service, which presently consists of twenty-six (26) full-time Medical Dicta-Typists distributed throughout the hospital, appeared to be both costly and inefficient. Miscellaneous clerical tasks, which should be performed by the forty-seven (47) cost-shared secretaries, have by habit and tradition become the responsibility of the Medical Dicta-Typists. Making appointments for patients, collecting x-ray and laboratory reports, filing admission and discharge cards, xeroxing, and running errands, the Medical Dicta-Typists are being used in a way which significantly departs from stated policy.

The purpose of this investigation was to determine if centralization of the medical transcription function of the Medical Records Department would make the service more efficient and more economical than it presently is. The study documents present activities, determines workload frequencies of present activities, develops required staffing projections for alternative organizations (both centralized and decentralized), considers the qualitative advantages and disadvantages of centralization, and recommends operating policies and procedures to implement centralization.

The remainder of this report shows that by centralization it is possible to reduce staff and to improve the quality of the medical transcription function. The next section summarizes our major recommendations.

FIGURE 5.12 Overview from a Hospital Management Consulting Firm Report. (3 pages)

CENTRALIZATION OF MEDICAL
RECORDS TRANSCRIPTION SERVICES

SYNOPSIS OF RECOMMENDATIONS

On the basis of our investigation we recommend centralization of the medical transcription function of the Medical Records Department. Not only is this consistent with recommendations of the Canadian Council on Hospital Accreditation, but it would achieve an annual savings of approximately $82,000 by staffing reduction. It would also eliminate performance of non-medical records work by Medical Dicta-Typists, afford better supervision and control, more efficient scheduling, and more even distribution of workloads. In addition, centralization would free one area on each of the ten floors and provide hospital-wide knowledge of work backlogs.

Our specific recommendations are as follows:

1. Centralize the Medical Records transcription function in the Steno Pool located on D-1. The required staffing is 11.3 full-time-equivalent employees. This staffing level includes discharge summary and operating room note transcription, administrative typing at the current level, supervision, transportation, and ample allowances for fatigue and unavoidable delays. Implementation should be accomplished via attrition. When coupled with the three positions described in the next two recommendations, the resulting 14.3 positions represent a savings of 11.7 positions, or $82,080 annually.

2. Maintain the present staffing of 1 full-time position on the Ground floor assigned to the Medical Records Department (as distinguished in terms of location from the Steno-Pool).

3. Transfer 2 employees permanently to the Pharmacy and to Neuro-Psychology (C-8). This essentially is a budget transfer because 2 employees are already working in these areas.

4. Establish a regular messenger service within Medical Records to provide adequate pickup and delivery of charts, discs, and reports for all areas of the hospital.

5. Do *not* implement a centralized dictating system. Further analysis is needed before this concept can be recommended.

6. Use the Hospital Messenger Service and the Print Shop (under the Materials Management Department) for all photocopying work.

7. Implement an index card file to control inpatient charts and, after discharge, to monitor transcription backlogs and physician reviews and signatures.

8. Implement regular Management Information Reports from the Steno-Pool supervisor to the Medical Records Department Head.

9. Review the design and number of copies of the Discharge Summary form, presently a 6-part carbon set.

10. Implement a well-communicated and administered secretarial replacement function which features a "first-come, first-served" policy until the maximums of two per day (or four per day for summer months) are reached.

September 20, 19XX

Mr. J. R. Hamilton
Assistant Executive Director
General Services
Sunnybrook Medical Centre
2075 Bayview Avenue
Toronto, Ontario Canada

Dear Mr. Hamilton:

CSF is pleased to submit our analysis of the Medical Records transcription function at Sunnybrook Medical Centre.

The study primarily addressed the issue of centralizing this function in the first floor Steno-Pool. Centralization of the transcription function was deemed advisable as a result of the analysis. A staffing reduction from 26.0 full-time-equivalent positions to 14.3 positions is one advantage of the centralization. This reduction will result in an annual savings of over $82,000, including fringe benefits. It will also result in several non-economic advantages.

We would like to take this opportunity to acknowledge the excellent cooperation and assistance received throughout the study from Mr. McCandless, Ms. Evans, and Ms. Johnson of Medical Records. Their efforts significantly contributed to the success of this analysis and are sincerely appreciated by CSF Ltd.

We are looking forward to assisting the hospital in implementing the approved recommendations.

Respectfully submitted,

G.F. Braidwood

G.F. Braidwood,
Associate

GFB/dv

FIGURE 5.13 Substantive Letter of Transmittal from a Hospital Management Consulting Firm Report.

the implications or applications.[4] In contrast, the summary of an organizational report on an investigation could be restricted to statements of the conclusions and implications for the organization.

An abstract should be as informative as possible. The ANSI standard for *Writing Abstracts* says the abstract must be understandable independently from the report.[5] To be independent it must be substantive or "informative."[6] The abstract should present what is in the report, not just indicate what readers can find in the report if they want to read it. Therefore, when you write the abstract, resist the temptation to indicate or describe what is in your report. Instead, present the most economical statement possible of your report's contents. For example, do not say: "The net present values of irrigation agriculture are calculated for two cropping intensities." Instead say: "The net present value of irrigation agriculture is −$12,000,000 with a cropping intensity of 180% and −$23,000,000 with a cropping intensity of 120%."

The sample abstract on "Systems Study for an Integrated Digital/Electric Aircraft (IDEA)" (on the report documentation page, Figure 5.11) illustrates the form and content of an effective abstract. The seven sentences follow the order of this systems study, from the purpose of the study to the results and recommendations. Each sentence is informative, presenting substantive information. The reader does not have to refer to the report to understand the study, its methodology, and its results: "This airplane showed a reduction of 3 percent in fuel consumption and 1.8 percent relative to the baseline configuration." In an information system, this abstract functions as an independent communication.

Because of length constraints, you might not be able to make your abstract completely informative. Some information systems limit your abstract to as few as 50 words. The *Publication Manual of the American Psychological Association* says the abstract of an empirical study should be from 100 to 150 words. The documentation page limits the abstract to 200 words. Length limitations can require you to generalize in some statements without being substantive (for example, "recommended research programs were defined"). Our advice is to be as informative as possible. Be informative on the most important information, given the purpose of the investigation; this usually consists of bottom-line results and interpretations of results (conclusions).

The abstract has an additional feature to serve its information retrieval function: reference information such as retrieval code numbers, report numbers, and key words. The abstract on the Integrated Digital/Electric Aircraft has key words, called "Identifiers/Open-Ended Terms": digital systems, electric secondary power, data bus, electric actuation. These key words cross-reference the report by subject matter. Notice that these key words are substantive nouns; they enable the searcher to identify this as a report that has information on digital systems, electric secondary power, the data bus, and electric actuation of aircraft. Two nouns in the title are not key words: systems study, and aircraft. The report is not about systems study as a methodology; that is, it is not a report that provides new information on how to conduct a systems study. The report is about aircraft, but such a term would be far too general to be a key word for a report in a NASA information system because it applies to a generic catagory of NASA reports.

Because of database information search technology, these key words have special value: they—more than the title and the author—are the means by which documents can be found in information searches. We emphasize, however, that you must give careful thought to what the key words should be. As author you are in much the best position of anyone to second-guess searchers. How might someone try to find your article? Anticipate their terms for what you are talking about. Unfortunately, we know from experience in companies that authors often treat key wording as an unpleasant, necessary act; they go through the motions and put some "key" words in, but give little thought to it. The result is poor information access by readers. Indeed, we know of one company which has been automating its library of some 50,000 technical reports. In the process, they have examined the abstracts for these reports (over a 3-year period) and have found fully half of them inadequate and requiring rewriting. Even among the newly written abstracts, fully a fourth are inadequate, and often a major source of the inadequacy is in the key wording.

You should not write an abstract until you have finished your report; the abstract must represent the report that a reader can actually read, not one that you are planning to write. As we have said, this applies to the overview as well. After you have written the discussion for your report, even if it is a 1-page report, you should examine and revise your overview to represent the report that you actually wrote. Your initial draft served to help you design and write the discussion. Your final draft of the overview addresses the reader so that the report accomplishes its function.

NOTES

1. H. L. Chadbourne, "Titling Technical Reports for Optimum Use and Retrieval," West Coast Navy Laboratories, August, 1965.
2. *Scientific and Technical Reports: Organization, Preparation, and Production*, American National Standards Institute (ANSI), Z39.18-1987: 18, and below, 17.
3. Our concern here is for the abstract that you have to write as the author of a report or article. Bibliographic references contain abstracts written by persons other than the writers. They abstract reports either objectively or critically, depending on the function of the bibliography.
4. *Publication Manual of the American Psychological Association*, American Psychological Association, Washington, DC, 1983, rev. 1984: 24. Note that this outline is an example. Not ail abstracts necessarily follow the method of investigation. For example, *Directions for Abstractors*, published by the Chemical Abstracts Service, stipulates this order: 1. purpose and scope of work; 2. new compounds, materials, reactions, and techniques; 3. new applications; 4. results and conclusions. In this case, however, the abstractor is not the writer. As the writer, you make your abstract closely follow the order of your report: it is a linear reduction.

5. *American National Standard for Writing Abstracts*, American National Standards Institute (ANSI), Z39.14-1979: 9.
6. The ANSI standard classifies abstracts as "informative," "indicative," and "informative-indicative" (7). An informative abstract presents quantitative or qualitative information from the document; it is substantive. An indicative abstract only describes or indicates the type of document and its contents. An informative-indicative abstract limits informative statements to important information from the document and uses indicative statements for other types of information.

C ❑ H ❑ A ❑ P ❑ T ❑ E ❑ R 6

Planning and Writing the Discussion

When you have determined your purpose for writing and understand the essential points you want to make, you are ready both to develop a detailed plan for the discussion part of your report and to begin to write. However, planning the discussion of your report is more difficult than planning its overview. For the task of planning the overview, we could present some fairly specific guidelines to help you cover the necessary points in the heading, the purpose statement, and the summary. For the discussion we can suggest only some fundamental strategies by which you can select and arrange the report's major supporting information. The particular strategy which you choose for an individual report must depend upon the material, your purpose for writing, and the audience and situation in which your report exists.

In this chapter, then, we outline three high-level strategies for selecting and arranging information in a discussion. Also, we explain how to begin and end the discussion. We conclude with some suggestions on how to outline effectively.

❑ 6.1 ❑ "Hierarchy" Defined

Before we discuss organizing your discussion, we need to look at the concept of a report's "hierarchical" structure.

Think of a shipping container which holds several crates which in turn each contain several cartons.

Obviously, the shipping container itself has a design and specific dimensions; it is made out of some specific material; and we could talk about its large-scale architecture and material independently of what is contained within it. Yet, just as obviously, the crates contained inside the shipping container have their own specific designs, dimensions, and contents. They exist within the container, but they are also independent of it to a certain extent, and we could look at them and talk about them without paying much attention—if any—to the larger box which contains them. We could continue in the same way with the cartons inside the crates. They too can have distinctive designs, dimensions, and contents even though they exist down at a third level within other, larger crates and containers.

Structure within structure within structure. This is the concept of "hierarchy," and it is a concept important to our discussion of report structure and important to your understanding of the relationships among structures discussed in this chapter and the next three chapters of this book.

The discussion section of a report is not just a single-level, linear sequence of ideas. Rather, it is a combination of structures within structures. The report's discussion is made up of major sections or divisions ("sections" we call them). These sections, in turn, are made up of groups of related paragraphs ("units" we call them). Units are in turn made up of paragraphs.

Our approach to report design is to look at the largest levels of structure first, and then to switch the focus of our attention down to the smaller and smaller levels of structure. That is, we believe, the process by which designs are always developed—for buildings, for machines, for programs, for projects, and for reports. In that light we have already indicated that the overall report architecture is a 3-part architecture consisting of an overview, a discussion, and a documentation section. And we have already talked about the internal architecture of one of those elements, the overview. Now, in turning to the internal architecture of a discussion, we are still examining structure at a high level, but we are moving down within the hierarchy.

For example, consider the outline of a report by a purchasing agent who is proposing a new system for evaluating vendors supplying parts to her company (Figure 6.1). The outline of her report illustrates the hierarchical structure of the discussion.

The outline shows three levels of structure. The first level (the numbered units 1 through 6) is the primary structure of the report:

1. Need for a New Vendor Rating System
2. Objectives of the Vendor Rating System

3. Advantages of the Proposed Vendor Quality Index
4. Disadvantages of the Monthly Vendor Summary
5. Comparison of Vendor Quality Index Ratings with Monthly Vendor Summary Ratings
6. Implementation of the Vendor Quality Index Rating System

Independently of the fact that this report has both an overview and a documentation section as well as a discussion, there is structure within the discussion itself. That is, the writer chose to discuss six topics, not four or five. She chose to put the six units into a particular order for some particular reason. She began with "Need" and she ended with "Implementation." She

Level	Heading	Length
section	1. The Need for a New Vendor Rating System	***two paragraphs***
section	2. The Objectives of the Vendor Rating System	***two paragraphs***
section	3. The Advantages of the Proposed Vendor Quality Index	***one paragraph (overview)***
unit	3.1. The Vendor Quality Index	***one paragraph (overview)***
unit	3.1.1. Defect Severity Classification System	***two paragraphs***
unit	3.1.2. Computation of Vendor Quality Index Rating	***four paragraphs***
unit	3.2. Use of the Vendor Quality Index	***three paragraphs***
unit	3.3. Effectiveness of Vendor Quality Index	***one paragraph (overview)***
unit	3.3.1. Effect of Defect Severity on Vendor Rating	***two paragraphs***
unit	3.3.2. Efficiency of Defect Information Retrieval	***three paragraphs***
unit	3.3.3. Additional Cost of the Index	***one paragraph***
section	4. Disadvantages of the Monthly Vendor Summary	***one paragraph (overview)***
unit	4.1. Current Ratings by Monthly Vendor Summary	***two paragraphs***
unit	4.2. Failure to Rate Vendors Effectively	***one paragraph (overview)***
unit	4.2.1. Severity of Defect Unweighted	***two paragraphs***
unit	4.2.2. Inaccuracy of Vendor Ratings	***two paragraphs***
section	5. Comparison of Vendor Quality Index Ratings With Monthly Vendor Summary Ratings	***four paragraphs***
section	6. Implementation of the Vendor Quality Index Rating System	***three paragraphs***

FIGURE 6.1 A Sample Hierarchical Structure of a Report.

put "Advantages" before "Disadvantages." And she followed "Disadvantages" with an explicit comparison of her proposed new system for rating vendors with the existing system for rating them. For all of these choices of what to put in and what order to put them in, she operated from deliberate principles of selection and arrangement.

The same principles apply on the second level of structure (for example, in sections 3.1–3.3 or 4.1–4.2). These lower levels of structure have their own internal structure:

3.1. The Vendor Quality Index
3.2. Use of the Vendor Quality Index
3.3. Effectiveness of the Vendor Quality Index

Again, these three items are selected and arranged according to some principle or pattern.

Further, the outline contains some structural elements at a third level of structure, each with its own structure (for example, inside section 4.2):

4.2. Failure to Rate Vendors Effectively
    4.2.1. Severity of Defect Unweighed
    4.2.2. Inaccuracy of Vendor Ratings

These items, too, are selected and arranged according to some pattern.

Thus, the discussion is made up of three levels of structure. In fact, as might be apparent to you, there are actually fourth and fifth levels of structure not represented in the headings of the outline. For example, within the "Defect Severity Classification System" section (3.1.1) there are two paragraphs. These paragraphs have a relationship to one another; in addition each has its own internal architecture.

Each level of information has its own pattern or method of selecting and arranging information, and each part of each level has its own pattern. The arrangement of the six units on the first level of structure is not necessarily the same as the arrangement of the two units of information on the second level, and each of those units has its own structural pattern. From your report as a whole to each individual paragraph, you are selecting and arranging information according to a method appropriate for that unit of structure as a separate unit. The final result is a hierarchical structure of all of these levels of structure.

This chapter and the following three chapters are based on the concept of hierarchical structure. In this chapter we discuss three means of selecting and arranging information at the first level of your discussion. In the next two chapters we present a number of alternative strategies for selecting and arranging information in the sections and units of your discussion. And then in the following chapter we discuss strategies for developing the paragraph-level structures that form the basic building blocks of your report. (After that, we move on to sentence-level issues and to other aspects of report design, such as format.)

❑ 6.2 ❑ Three Organizational Strategies

Let's assume you are a staff member in the Executive Personnel Department in the statewide headquarters of an insurance company. The Manager of Branch Offices has called a meeting to select a new manager for the Brighton branch office. The purpose of the meeting is to select a person to be recommended to the Vice President for appointment as Manager of the Brighton office.

At the meeting with you are several persons from corporate headquarters and several managers from important branch offices. Because you represent Executive Personnel and are the junior person present, you are to take notes and write the recommendation for the Manager of Branch Offices to approve and sign (note that you will write the report for her signature.)

The meeting is scheduled for 8:30, and by 8:50 almost everyone has arrived. Chairing the meeting, the Manager of Branch Offices begins by reviewing the situation. Then she outlines the qualifications for the position to be filled. Others suggest additional qualifications, not all of which seem to be relevant. Then she presents the list of possible candidates, primarily based on seniority with the company. Several additional names are added to the list at the suggestion of several branch managers. And then the meeting gets interesting.

There is an open-ended, round table discussion of the qualifications of the various candidates for the position. Comments about each candidate or several candidates together bounce back and forth: positive, negative, irrelevant. At times, several persons speak at once, and several separate conversations are going on simultaneously. You had planned to follow the discussion with a simple matrix which listed the candidates and matched them against the qualifications, but you quickly find that your plan can't cope with the reality of the meeting, and your notes become more and more chaotic as the meeting goes along. The chair finally stops the discussion about ten minutes after you think it has ceased to be productive. It is 10:30, past time for a coffee break.

At 11:05, most people are back in the conference room, and the feasibility of hiring from outside the company is brought up, then dropped. The manager briefly summarizes the evaluations of the various candidates and presents two or three as the obvious top candidates. As these probably were everyone's choices at 8:50 in the morning, there is no dissent. Arguments pro and con are again voiced, and at 11:45 the chair announces that the group appears to have a consensus choice. Everyone breathes a sigh of relief and starts picking up papers and getting ready to leave for lunch. One person voices a serious dissent and wants to reopen the discussion in favor of going outside the company for a candidate. Mumbles, glares, and glances at watches quickly silence him, and the chair reaffirms the recommendation of the group.

At 12:06 you are the only person left in the room. Everyone else will have a nice lunch, secure in the knowledge that a good choice has been made and that you will handle the chore of writing up a report of the morning's decision.

Now you have to write the recommendation for the Manager to present to the Vice President, with copies to all of the persons present at the meeting.

That will take careful diplomacy, but also it will take careful selection and arrangement. You have an overwhelming amount of information to work with. Seven persons have been talking without interruption, at times several at once, for more than two and one-half hours. You have dossiers on all of the candidates considered. You have official personnel forms and position descriptions. Where do you start?

You are faced with a composing process typical of those which all writers face after they have finished an investigation or analysis. First, you must segment the mass of information into coherent parts. Then you must select from among the parts those which are appropriate to include in terms of your communication purpose, the readers, and the situation. Finally, you must arrange those parts in some logical order appropriate to your purposes. When you have finished this outlining process (Figure 6.2), you have a basic outline for your discussion that you can flesh out and start writing. Of course, the process is dynamic. At any given point in the process, you might have to reconsider prior choices (as indicated by the feedback loops on Figure 6.2.). For example, arranging the parts might force you to consider information you initially had discarded—or to discard information you initially had included.

To segment, select, and arrange information, however, assumes some structural approach or strategy. How you approach your outlining process determines how you segment, select, and arrange the material to be included in the report. This becomes apparent when you consider the approaches you could take to write a report recommending the candidate selected at the meeting.

In our view, you have at least three workable, but quite different, ways of arranging the information: according to the process of selecting the candidate, according to the candidates and their qualifications, or according to the recommendation which is being made. Of course, you could also just tell the story of the meeting, using a straight chronological order, but we reject that as a valid organization: it would be inefficient for all readers and would risk obscuring the consensus of the meeting by emphasizing the dissent and pro-and-con discussion. Whatever its success in reaching a good decision, the meeting was a chaotic event. No point in reproducing that chaos in a report.

The Process of Selecting the Candidate

Look for a logical, not chronological, order. Behind the apparent chaos of conversation and commentary, the meeting actually did have a useful organization: it was organized according to a logical problem-solving process:

Problem
Criteria: Qualifications for the Position
Qualifications of the Candidates
Evaluation of the Candidates Individually
Comparative Evaluation of the Candidates
Conclusion and Recommendation: Choice of Candidate

You could cut through the chaos of the information to impose an organization based upon a logical process of problem solving. Organizing the details in your notes according to the logic of this procedure would give you fairly clear guidelines for segmenting, selecting, and arranging the information that you finally include in the report.

The Candidates and/or the Necessary Qualifications

The qualifications of the candidates for the position and the necessary qualifications for the position were the topics or subject of much of the meeting. If we assume four candidates (A, B, C, and D) and four qualifications (1, 2, 3, and 4), you could organize the information according to either the candidates or their qualifications. In either case you would have to begin with a discussion of the problem and end with a presentation of the conclusion and recommendation:

By Candidate for the Position

Problem
Candidate A (strongest)
  Qualifications 1 through 4 (+ −)
Candidate B
  Qualifications 1 through 4 (+ −)
Candidate C
  Qualifications 1 through 4 (+ −)
Candidate D (weakest)
  Qualifications 1 through 4 (+ −)
Conclusion and Recommendation

By Qualification for the Position

Problem
Qualification 1 (most important)
  Candidate A
  Candidate B
  Candidate C
  Candidate D
Qualification 2
  Candidate A
  Candidate B
  Candidate C
  Candidate D
Qualification 3
  Candidate A, etc.
Qualification 4 (least important)
  Candidate A, etc.
Conclusion and Recommendation

As you can see, the topics of the meeting provide another method of segmenting, selecting, and arranging the information in your notes. (Note that you do have to impose an order on either the candidates or the qualifications; in either case, the order would be order of decreasing importance, not order of occurrence.)

The Purpose of the Report

The purpose of the report that you are writing to the Vice President, and for your Manager's signature, is to recommend a candidate for the position

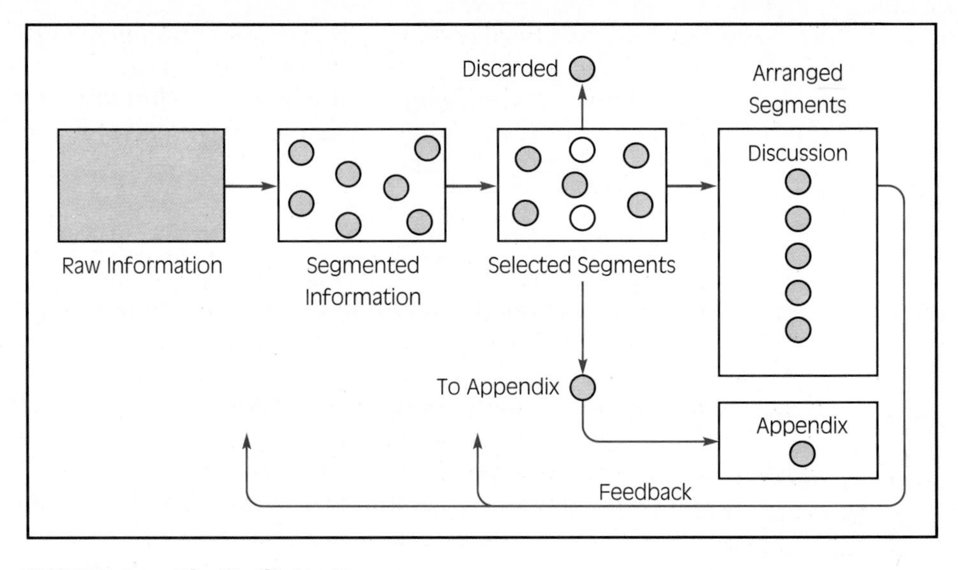

FIGURE 6.2 The Outlining Process.

of Branch Manager of the Brighton Branch. That purpose provides yet another means of organizing the information:

Problem
Criteria: Qualifications for Position
Recommendation of Candidate A
Reasons Candidate A is Recommended
  Qualification 1 (strongest reason)
  Qualification 2, 3, etc.
Other Candidates Considered and Why They Were Rejected
  Candidate B (closest contender)
  Candidate C, D, etc.

As you can see, your recommendation and the reasons for that recommendation yield yet another method of segmenting, selecting, and arranging information.

Which is Best?

As we have suggested, you could organize the report in any of these three ways (or perhaps other ways as well). Which is best? Which should you choose?

The answer is that the choice depends upon your purpose for writing as well as upon your intended readers and the situation.

You might choose the third organization, a persuasive pattern, if you saw your purpose as simply to report the person chosen for the job. Other considerations, however, might make another pattern more appropriate. For example, you might want to convey the idea that the selection process was systematic and that the person conducting the meeting did an excellent job of managing an objective decision-making procedure. If so, an organization based upon the process or method of choosing a candidate might be a good organization to consider. Or you might want to stress the candidates and their qualifications, so you could choose to organize by candidates. Or you might want to stress the relationship of the choice to the criteria, so you could choose to organize by criteria. The point is that you choose a pattern that will accomplish what you want to accomplish in relation to the readers you anticipate and the situation in which your report functions.

As this example suggests, you have to adopt some organizational strategy before you can outline a discussion. The three strategies we have introduced are rather common in scientific and technical documents because they are derived essentially from the basic stages of the communication process. (1) Your communication activities begin when you undertake an investigation in response to some problem or need. The process of investigation provides a strategy for outlining. In the preceding example, the outline by the problem-solving process of the meeting is organization according to the method of investigation. (2) The subject of your investigation provides the content of your communication, so the subject matter provides a second strategy for outlining. In the example, the outlines by candidate or qualification are organization by subject matter. (3) The purpose of your communication yields a third strategy for outlining: you organize the report to persuade the audiences to accept your conclusions or recommendations. In the example, the outline recommending a candidate is organization by persuasive purpose.

Thus, you have at least three basic organizational strategies for thinking about and outlining your discussion:

1. Organization based on your method of investigation.
2. Organization based on your subject matter.
3. Organization based on your persuasive purpose.

In our discussion of these strategies, we start with organization based on your persuasive purpose. We do so because the purpose of many of your organizational communications will be to present your professional judgment persuasively. In addition, this outline directly addresses the needs of the audiences for most of your reports.

Organization Based Upon Persuasive Purpose

Much of your communication as a professional serves to support decision-making processes within an organization. Your role is to present and support your judgment, which usually is stated as a recommendation or as conclusions and recommendations. When you write reports of this sort—no matter what their topic or length—you have essentially a persuasive purpose and you can structure these reports as arguments which present conclusions plus support.

A basic persuasive pattern can have this 5-step outline:

1. Statement of the Conclusion or Recommendation (or of the Problem and Solution).

2. Statement of the Criteria for Evaluating the Evidence (if needed).

3. Support for the Conclusion or Recommendation (or solution), arranged in descending sequence of significance or as positive/negative evidence.

4. Anticipation and Refutation of Alternatives.

5. Restatement of Conclusions, Recommendations, and Implications.

Do not hesitate to adapt this outline (or any other outline which we present) to the specific needs of any communication situation. However, this outline might provide a good place to start thinking about how to structure and outline your discussion.

You first state the conclusion you are going to support (of course after having introduced the topic of discussion). This could be a conclusion of an analysis, a recommendation for a change, or a solution to a problem. It could also include a summary of the important points of your argument, which will serve as a forecast of the following discussion.

Next, you state the criteria for evaluating the evidence, if you have not already done so in an introduction to the discussion. When the criteria are readily apparent or objectively stated as specifications about which there is no disagreement, they can be mentioned in an introduction. When the criteria are not explicit or immediately evident to your readers, they need to be established in a separate step in your outline. For example, if scheduling of computer facilities is an important criterion in the selection of a computer system for an office function such as inventory control, then you might want to include this item in your outline. You certainly should do so if you are going to recommend a system that can be dedicated to inventory control but at a higher initial cost than some of the alternative configurations you will reject in step 4.

The heart of your argument, of course, is step 3, where you present your support for the conclusion, recommendation, or solution. You arrange your support in descending sequence of significance or as positive/negative evidence. Present your most important support first, then the second most important, and finally the third.

After this positive support, then discuss the negative arguments that might be made against your position. You must either reject a negative argument or accept it and argue a trade-off in favor of your positive arguments. For example, you might accept the fact that a procedure which you recommend is going to be more costly than an alternative. But you point out that cost considerations are offset by such other factors as employee safety or product quality.

Next you have to anticipate and refute any alternative conclusions, recommendations, or solutions that could be formed given the problem, the alternatives, or the evidence. If you are recommending a brand of personal computer for office work stations, then you probably have to argue against one or several alternative brands after you present the support for your recommendation. You have to do so because your audiences know there are many brands of personal computers and probably have their own favorites. Most decisions in organizations involve trade-offs, which means you usually will have to consider limitations or negative arguments in step 3 or discuss alternatives in step 4.

If you have introduced negative arguments or alternatives, you usually restate your conclusions, recommendations, or solution. You want to end your report on a positive note, not on some minor point having to do with a rejected alternative. Also, you need to discuss any organizational implications of your solution, as we discuss in Section 6.4, Ending the Discussion, so you provide an ending which brings the document back to its main point.

When you write steps 2, 3, and 4 of the outline for a persuasive report, you have to keep your audiences in mind. You discuss criteria that you know are important to your readers, especially to your management readers. You discuss your support in terms of reader knowledge, expectations, and values. You discuss limitations, negative evidence, and alternatives according to your understanding of what you need to cover to convince your readers by using logical evidence. Thus, your discussion is not an exhaustive exercise of systematically covering all of the arguments that could be made for and against a conclusion. Nor is your discussion limited to the criteria you used in your own analysis. Instead, it is a discussion designed according to your audience analysis—their values, their needs.

The report by the purchasing agent proposing a new vendor evaluation system (Figure 6.1) illustrates a persuasive pattern. The discussion of the report has this basic outline:

1. The Need for a New Vendor Rating System

2. The Objectives of the Vendor Rating System

3. The Advantages of the Proposed Vendor Quality Index

4. The Disadvantages of the Monthly Vendor Summary

5. Comparison of Vendor Quality Index Ratings with Monthly Vendor Summary Ratings

6. Implementation of the Vendor Quality Index Rating System

The first section of the report discusses the problem and states the recommended solution. The problem was that the current system of evaluating parts suppliers did not enable the company to monitor suppliers whose parts were excessively prone to defects. The writer proposes a new vendor rating system.

The second section of the report establishes the criteria for evaluating a vendor rating system in the form of stating the goals or objectives of such a system. The question she answers is, "how would you know a good system when you saw it?"

The third section presents the arguments in favor of the recommended system, including a discussion of a limitation ("additional costs").

The fourth section argues against the alternative, the present system.

The fifth section summarizes by comparing and contrasting the ratings of the two systems, arguing the trade-off in favor of the proposed system to restate the conclusion.

The sixth and final section discusses implementation of the new system (which is a further argument in favor of the proposed solution because, as the discussion makes clear, the proposed system is feasible to implement).

A report of a reactor baseplate failure also has a persuasive pattern. The writer had to convince management of his conclusions about the cause of the failure and of his recommended solution because of the potential loss of $2.7 million dollars should the failure occur again. The 5-page discussion of his report has the following outline:

1. Description of the incident.

2. Weld slag and flux caused the 416 reactor to fail.
   2.1. Weld slag and flux blocked the valve.
   2.2. Weld slag and flux were in the system.

3. Water was in the system but did not cause the failure.

4. Why the failure occurred in the 416 reactor and not the 415, 417, and 418 reactors.

5. The valve was small enough to be plugged in throttled position.

As you can see, the writer often used sentences rather than phrases for his headings (except the first heading), an unusual but effective practice that he used to make his argument clear.

The discussion opens with a description (with an attached figure) of the incident, which is a statement of the problem, although it included background narrative detail to introduce the analysis. The item, "Weld slag and flux caused the 416 reactor to fail," is both a statement of the conclusion (in one paragraph, 2.1) and discussion of three points of support for the conclusion, arranged from direct to inferential evidence (in two pages, item 2.2). After his support for his conclusion, the writer argues against an alternative possibility (item 3) and an anticipated objection to his analysis (item 4). The writer concludes with a further argument (item 5) that he treated as a restatement of his initial conclusion.

Organization Based on Subject Matter

Organization based on subject matter is another frequent pattern for organizing a report, especially an informative report. After all, you are writing about some thing or some idea. Often you can segment and arrange material according to logical "joints" in the subject itself. Many design reports, for example, are arranged according to subject matter. The design, whether it be of a building, a boat, or a health delivery system, is broken down into its constituent parts, which then are explained one by one in an order based upon the system itself.

The basic pattern of organization based on subject matter has this 4-step outline:

1. Overview: Presentation of the Whole, with emphasis on function or significance.

2. Forecast: Subdivision of the Whole into system-based parts or other logical subdivisions.

3. Presentation of the Parts in a functional, system-based order or other logical order.

4. Discussion of Implications, Future Work, Applications, etc.

Whereas the persuasive pattern is arranged as an argument in support of a recommendation, organization by subject matter is information oriented. The recommendation, if there is one, usually comes at the end. The subject matter is the focus of the discussion. Although organization by subject matter enables you to segment and arrange your discussion very clearly, it provides few guidelines for selecting information. You can go into considerable detail as you explain each part of the system; deciding how much detail is appropriate requires matching your design to your purposes and the needs of your readers.

A report illustrating subject matter organization is *Water Resource Management and Gambia River Basin Development*, a socioeconomic and environmental impact study prepared for the United States Agency for International Development (USAID).[1]

1. The Need for Water Resource Management
2. Development Actions and Impact Networks
3. Land and Water Management
4. Agricultural Management
5. Fishery and Aquatic Resource Management
6. Urban and Regional Planning
7. Management Actions to Enhance Water Resource Development
8. Benefits and Costs of Development Projects
9. Further Considerations

The report presents the impacts of the construction of several dams on the entire ecosystem of the Gambia River Basin, including the social and economic infrastructures of the human populations. The overview of the ecosystem and the division into subsystems is in section 2. These subsystems are then discussed one by one (sections 3 through 6). The recommendations are in section 7, and implications are in sections 8 and 9. The report, therefore, closely follows the four-step outline of a report organization based on subject matter.

The Dow Chemical progress report (Figure 6.3) has a simple organization by subject matter:

Brief
Productivity
Safety
The 5th Shift (Expanded Day Program)
Conclusion

The "Brief" is the overview of the report and of the pattern. It also serves as a forecast of the subdivisions of the report. The three headings, "Productivity," "Safety," and "The 5th Shift," are the discussion, along with the "Conclusion," which is the ending. The order of the three parts of the discussion (item 3 on the outline for organization by subject matter) follows the standard outline for progress reports in that organization, which always lead off with "Productivity."

Organization Based on Method of Investigation

The activities of many professionals consist of investigations, tests, experiments, and field studies, especially in research and development organizational environments. When the purpose of the reports of these activities is primarily informational rather than decision making, you often use the familiar method of organization based upon the investigation to outline your report.

Organization by method of investigation focuses the reader's attention on your problem-solving process, the intellectual process by which you arrived at your conclusions. By following this process, you can lead your readers to understand and accept your conclusions because they recreate in their minds the logic of your analysis. They are able to do so because the logic behind this intellectual problem-solving process is an accepted standard in the professional community. Therefore, you outline your reports according to this standard.

When you perform an investigation, you follow a problem-solving process that can be divided into eight stages:

1. You or your organization encounters a problem or need requiring an investigation.

2. You define the specific objectives of the investigation or questions to be answered.

3. You determine how to answer the questions or perform the investigation—the equipment and procedure (means and methodology).

4. You perform the investigation according to a specified series of steps.

5. You collect and tabulate the results.

6. You analyze the results and draw conclusions.

7. You formulate recommendations or implications for further research or activities.

8. You write a report or article to present your conclusions and, if any, recommendations.

Most students are familiar with this problem-solving methodology because it is taught in many of their laboratory and professional courses. Professionals

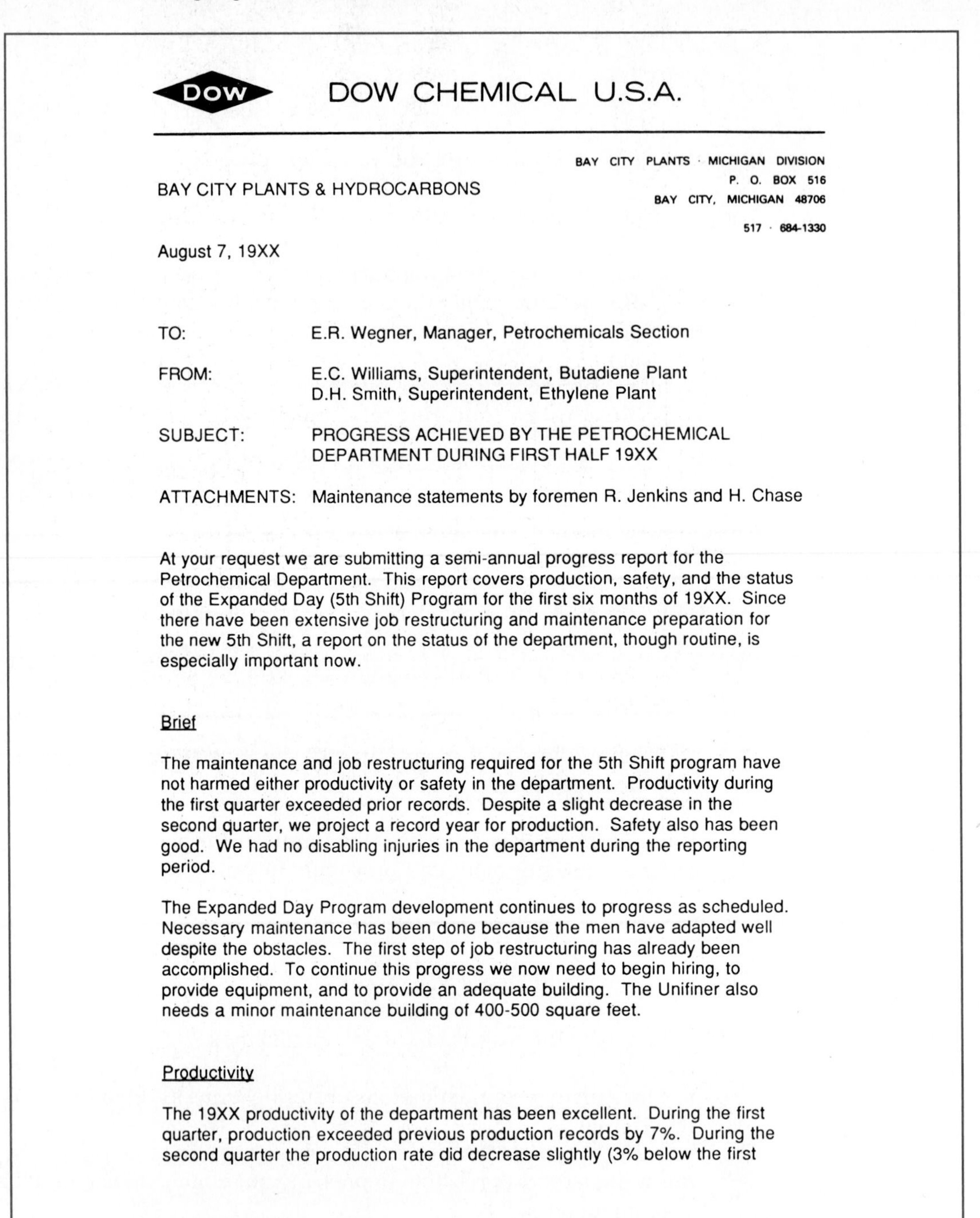

DOW CHEMICAL U.S.A.

BAY CITY PLANTS & HYDROCARBONS

BAY CITY PLANTS · MICHIGAN DIVISION
P. O. BOX 516
BAY CITY, MICHIGAN 48706

517 · 684-1330

August 7, 19XX

TO: E.R. Wegner, Manager, Petrochemicals Section

FROM: E.C. Williams, Superintendent, Butadiene Plant
D.H. Smith, Superintendent, Ethylene Plant

SUBJECT: PROGRESS ACHIEVED BY THE PETROCHEMICAL DEPARTMENT DURING FIRST HALF 19XX

ATTACHMENTS: Maintenance statements by foremen R. Jenkins and H. Chase

At your request we are submitting a semi-annual progress report for the Petrochemical Department. This report covers production, safety, and the status of the Expanded Day (5th Shift) Program for the first six months of 19XX. Since there have been extensive job restructuring and maintenance preparation for the new 5th Shift, a report on the status of the department, though routine, is especially important now.

Brief

The maintenance and job restructuring required for the 5th Shift program have not harmed either productivity or safety in the department. Productivity during the first quarter exceeded prior records. Despite a slight decrease in the second quarter, we project a record year for production. Safety also has been good. We had no disabling injuries in the department during the reporting period.

The Expanded Day Program development continues to progress as scheduled. Necessary maintenance has been done because the men have adapted well despite the obstacles. The first step of job restructuring has already been accomplished. To continue this progress we now need to begin hiring, to provide equipment, and to provide an adequate building. The Unifiner also needs a minor maintenance building of 400-500 square feet.

Productivity

The 19XX productivity of the department has been excellent. During the first quarter, production exceeded previous production records by 7%. During the second quarter the production rate did decrease slightly (3% below the first

FIGURE 6.3 **Progress Report Organized by Subject Matter. (3 pages)**

quarter). The decrease was caused by steam plant boiler outages, a power failure, an exchanger leak, and warmer weather. Despite this downturn, if we continue our present rates of production, 19XX will be a record year.

Safety

The department continues to work safely. There have been no disabling injuries (DI's) and very few near-miss incidents during the reporting period. We attribute this good safety record to our on-going drive to produce a safer work climate. For example, more platforms have been installed, sight glasses on equipment have been eliminated, and housekeeping has improved. Safety discussions and instruction, of course, continue. As our new people are becoming more experienced, we can expect even better safety performance.

The 5th Shift (Expanded Day Program)

Preparing for the 5th shift by maintenance and by job restructuring is progressing well. We now need to begin hiring and to expand facilities.

What we have done so far

We have good reason to be proud of the maintenance work being done to prepare for the Fifth Shift. The foremen and production supervisors deserve a lot of credit for their enthusiasm and good work. It is especially gratifying to witness the same thing from the operating technicians. Their willingness to learn and show us what can be done if they are allowed and trusted to perform well is encouraging. (Please read the attached statements by two of the shift foremen regarding the maintenance performed by their crews.)

The maintenance items performed were necessary to the smooth operation of the plant, and they have been accomplished in the face of reduced manpower due to (1) job restructuring, (2) the Midland strike, (3) vacation, and (4) unusual sickness.

A second and very important part of the program is the job restructuring. The first step was accomplished two months ago. The furnace-dehydrator duties were combined into one job (and the pay level upgraded) and the ethylene loading, depentanizer and furnace assistance duties were added to the transfer and lab tech duties. Again, the operating people responded well and the results have been excellent. Not only have our people learned their own duties, but many are, on their own and under the direction of the shift foremen, learning other jobs as well. Again, a good deal of credit goes to the shift foremen and those people who are taking up the challenge and doing something about it.

We feel strongly that we should promote training and equipment, and provide supervisory encouragement so our people can show what they can do. The plant will run better, human relationships will be improved, and vacations etc. can be more easily accommodated.

Here is where we need some help

Our people have shown that they are very capable and have responded over and above what we had proposed. To continue the progress and to make the 5th shift a solid success we need to do the following:

1. Begin Hiring: By September 1, 19XX we need 6 people. (A request for hiring dated 17 July 19XX has already been submitted to the Personnel Department.)

2. Provide Equipment: We have provided small hand tools already. In addition, we need a drill press, a pipe threader, and two trucks.

3. Provide an Adequate Building: We have consulted an outside architect regarding the building proposed for Area 4. The architect's report indicates 3000 square feet would be required for shop area and offices. In addition, a minor maintenance building for the Unifiner (approximately 400-500 square feet) is necessary. The present Diphenyl building would do, but it is being used for research purposes.

Meeting these needs is crucial to the success of the Expanded Day Program.

Conclusion

As you can see, preparation for the Expanded Day Program is moving ahead on schedule. At the same time the department has been able to maintain high production and safety records. We therefore feel our progress during the first six months of 19XX has been excellent.

E.C. Williams

E.C. Williams, Superintendent,
Butadiene Plant

D.H. Smith

D.H. Smith, Superintendent
Ethylene Plant

are very comfortable with the methodology, even if they don't use it as an outline for most of their reports, because it is the accepted means of organizing research reports and scientific journal articles.

Here is a basic pattern of organization based on the method of investigation:

1. Statement of Objectives of the Investigation (including Findings if this is the overview of the report).

2. Background of the Investigation: Previous Work, Theory, Specifications, etc.

3. Procedure of the Investigation: Equipment, Methods.

4. Presentation and Analysis of Results.

5. Summary of Conclusions and Recommendations, if any, and Implications, Future Investigation, etc.

Usually, you have to adapt this outline according to the context of your investigation (such as that of a research lab, test facility, or police department), or the type of investigation (such as experiment, analytic investigation, or field test.)

The *Publication Manual of the American Psychological Association* (APA), the standard for scientific articles in the social sciences, presents guidelines only for organizing a manuscript by the problem-solving or method of investigation pattern:[2]

1. Introduction
   - Introduce the problem
   - Develop the background
   - State the purpose and rationale

2. Method
   - Subjects
   - Apparatus
   - Procedure

3. Results

4. Discussion (evaluation and interpretation of results, conclusions, and implications)

The *Manual* is concerned with documents which deal primarily with experimental studies. As you can see from the outline, the Introduction is an extended problem statement (as we discussed in Chapter 3 for the introduction to a scientific and technical article). The Method "describes in detail how the study was conducted" so that the reader can evaluate the results. The Results "summarizes" the main results and findings and presents the data needed to justify the conclusions, which are derived in the Discussion.

The ANSI standard for the organization, preparation, and production of scientific and technical reports also has guidelines only for organizing a report by the problem-solving process:[3]

1. Introduction (subject, purpose, and scope of the investigation and forecast of the structure of the report).

2. Methods, Assumptions, and Procedures.

3. Results and Discussion (these can be in separate sections).

4. Conclusions (findings and implications).

5. Recommendations (additional areas for study, alternate design approaches, production decisions, etc.).

The ANSI standard applies to several types of reports in addition to experimental studies, but the outline is very similar to the APA outline except for the last item, Recommendations. The Recommendations section is appropriate for reports of studies such as "field trials, specific design problems, feasibility studies, and market appraisals." The ANSI standard, however, mentions that "the organization of a report depends on its subject as well as on its purpose," so your organization might vary from the suggested outline.

Segmenting and arranging by method of investigation does not mean that you present a chronological step-by-step description of your actual investigation. The actual chronology of events of any investigation (step 3 of our generic outline) is not the same as the methodology, and is the least important stage intellectually. It certainly should not yield the structure of the report. Your intellectual process begins before you enter the laboratory and ends after you have your results. The method of investigation models your intellectual problem-solving process, not your physical activities.

A good example of how the intellectual problem-solving process (rather than the actual process of the investigation) determines the structure of the report was given to us by a staff member of the National Highway Traffic Safety Administration. He described how he reported his study—as opposed to how he actually did his study. The study was "A Human Factor Analysis of Most Responsible Drivers in Fatal Accidents." Here is his explanation:

> The study was originally designed as an exploratory investigation of the human factors present in fatal motor vehicle accidents. We decided to key upon the driver most responsible for the accident and gather a standard set of data on

him via an interviewing technique. After data collection, we were going to try to relate some of the findings to the causes of the accident and see if these drivers differed in any respects from other types of drivers. The objectives were general and exploratory in nature.

We then designed the data forms, gathered the data, and began analyzing the data. We created certain groupings and categories on a logical basis. For example, we separated drivers who were killed (Type I) from drivers who survived (Type II) and from drivers who killed pedestrians (Type III). When analyses showed differences between these groups, this aroused our curiosity. We were finding something we did not expect. Type II drivers were differing significantly from the other two groups in very important respects. Therefore, we formulated the null hypothesis and then went on to show that it could be rejected with respect to these variables. Thus the basic thesis of our report was formed.

In summary, our problem-solving process went this way:

(1) We formulated some general objectives due to a need for information.
(2) We designed the methodology for data collection.
(3) We collected data, and began tabulation.
(4) We began to group the data in logical categories.
(5) We did exploratory factor analysis on the data groups.
(6) We discovered something that we did not expect.
(7) We formulated the null hypothesis knowing that we would reject it.
(8) We narrowed the study to this objective.
(9) We completed the study.
(10) We wrote the paper.

These ten stages were the sequence of the investigation. However, we wrote the report as if the investigation had been conducted this way:

Data need
Hypothesis
Design
Method
Results
Discussion/Recommendations

His actual report outline closely follows the outline in the APA *Manual*. This structure clarifies his intellectual problem-solving process, whereas a chronological narrative of the actual investigation would have obscured it.

So far we have discussed the problem-solving process in terms of the experimental method of investigation, which is its most common form. For reports other than research reports, however, you might find other types of problem-solving processes more appropriate than the method of investigation. You base your organization on the form of the logic you used to solve the problem.

The consultant firm report on Charlevoix, Michigan Water Supply System Improvements has this outline:

Introduction

1. Water Supply Requirements

2. The Present Water System and Recommended Improvements

3. Cost Estimates

This three-part organization (in items 1 and 2) follows a simple problem-solving formula:

Needs − Capacity = Necessary Improvements

"Cost Estimates" (item 3), of course, is an item of implications for the organization that you often include at the end of any report, regardless of its method of organization.

Organization by method of investigation is one which you often can use. Whether or not you should use it depends on your purpose. We encounter this method of organization in many research laboratory reports whose purpose is to document research results. The report on the reactor baseplate failure was written in a research laboratory. Because of its purpose, however, it was arranged according to a persuasive pattern, not according to the writer's method of investigation of the failure. The writer determined that selecting and arranging information in terms of his persuasive purpose would be more appropriate for his management audiences than organization in terms of his method of investigation.

❑ 6.3 ❑ Beginning the Discussion

For short reports, the beginning strategies which we discussed in Chapter 5, Planning the Overview, will suffice to begin the discussion. For longer reports, however, in addition to an overview of the sort we have already discussed, an introduction to the discussion itself (a "beginning") is needed to provide further introduction to the discussion and to the details that follow in the body of the report (the middle), which culminates in a final section or conclusion to the report (end). In other words, some reports—mostly long and complex ones—need discussions with their own beginning, middle, and end. That makes the discussion able to stand on its own, but does not do away with the need for the overview. You can think of this type of report as consisting of two self-sufficient elements, one on a high level of generalization (the overview) and one on a low level of generalization (the discussion).

With short reports, an additional beginning is usually unnecessary. The overview for these reports provides an adequate beginning. For many reports, especially for long and complex ones, you may need to frame the discussion with a beginning of its own. If so, we suggest you consider three strategies of approach for writing the introduction to the discussion:

- Amplify the purpose statement. Either expand on the organizational problem or present additional details of the technical investigation.
- Forecast the structure of the discussion. If the pattern for the particulars in the discussion does not suggest a forecast and if the summary of conclusions and recommendations does not forecast the structure of the discussion, then the beginning of the discussion often outlines the details to follow. Such a forecast explains the reasoning behind the arrangement and selectivity of the details.
- Explain criteria and unfamiliar concepts. As with several other patterns, before an explanation of the details you often have to present background material so that the reader can understand the explanations that follow. That is, you do not want to interrupt the explanation to introduce background information, so introduce it before you get underway with the details of the discussion.

These three strategies for beginning the discussion often are used in combination.

Amplify the Purpose Statement

Because of the importance of having a concise and direct purpose statement in the overview, you often need to amplify the purpose statement at the beginning of the discussion before you get into the particulars. You expand on the statement of the organizational problem or of the investigation.

The report on standardization of front end sheet metal discusses the organizational problem at the beginning of the discussion in the "Background" section:

> Sport-Utility and Conventional vehicles are now being built in the same plant. Standardization in location of F.E.S.M. [front end sheet metal] mounts has thus become more important to simplify plant handling. This standardization should also result in a cost reduction on some models. For these reasons, F.E.S.M. mount standardization was tested as outlined in the following "Discussion."

A report on request for adjudication of a cost estimate expands on the investigation to introduce the discussion. In the purpose statement, the writer just summarizes the investigation:

> The Owner considers this final estimate unreasonable, and in accordance with standard contractual procedure asks MARAD to establish a fair and reasonable cost.

To introduce the discussion, the writer states the specific questions that must be answered to determine if the final estimate is unreasonable:

In examining the reasonableness of the final cost estimate submitted by the Contractor, the Owner thinks the following questions must be addressed:

1. Most important, do the charges relate specifically to the particular modifications required by the Change Order, or should the charges be considered as a development of the contract?
2. Are the charges realistic or inflated on the basis of typical construction procedures?
3. Are the charges provided for in the contract?

The disparity between the Contractor's preliminary and final cost estimates suggests that the final estimate must be scrutinized.

To answer these questions, the modifications to the original contract specifications required by Change Order No. 16 must be noted. Then specific items in the Contractor's final estimate can be examined in light of the original contract and modifications. Other charges in the estimate can be examined separately.

The last paragraph in this introduction is a forecast of the structure of the discussion.

Forecast the Structure of the Discussion

A forecast of the structure of the discussion usually is necessary in long or complicated reports. It serves to divide the discussion into manageable parts and to predict their sequence. Here is an example of such a forecast:

An electrocardiogram (ECG) records electric signals from the heart muscle as it beats. A normal ECG is taken while the patient is lying perfectly still because electric signals from the motion of other muscles can interfere with the signals from the heart. However, doctors have found that the shape of the ECG sometimes changes if it is taken while the patient is exercising because the heart activity increases. By comparing the ECG taken at rest with the one taken during exercise, doctors can obtain additional information about the condition of the patient's heart.

The problem is that the muscle activity during exercise produces electric signals which interfere with the electric signals from the heart. The muscle activity makes the exercise ECG difficult to read. To improve the exercise ECG signal, we investigated methods of eliminating the interference from muscle activity. The method had to be conducive to further research.

This report examines the four questions addressed in the technical investigation.

1. What is the nature of the ECG signal, and why is it useful to study the electrocardiogram during exercise?

2. Why was filtering, and not signal averaging, chosen as a means of eliminating the interference from the exercise ECG?
3. Why is filtering not the solution to eliminating the interference?
4. What are some alternative solutions?

With this forecast as a guide, the reader knows what the sections of the discussion are and what their sequence will be.

Explain Criteria and Unfamiliar Concepts

Factual background or potentially unfamiliar concepts may need explanation. This is particularly true in reports which have a complex "story" behind them or which involve recent technology and new concepts. Sometimes the discussion can profitably begin with a section labeled "Background," in which details of the situation behind the work are presented. (But watch the temptation to overdo this; a lot of reports need no background but get a dump of it anyway.) At other times a discussion might profitably begin with an extended definition. Here is an example:

> **1.2 Volume Control vs. Weight Control**
> Modern destroyers are "volume-controlled," so the significant design features are those relating to utilization of volume. Formerly, destroyers were "weight-controlled," and the naval architectural characteristics such as buoyancy and stability were those chosen to carry the weight of hull, machinery, ship systems, people, and payload. The resulting enclosed space was ample to contain the total volume demanded by these elements. Modern destroyers are volume-controlled in that the volume required by these elements is greater than that associated with an equivalent weight-controlled design. Thus the hull must be larger and heavier; this produces secondary increases in machinery, fuel, and system weights, all of which result in a total displacement increase above a weight-controlled design.

This section briefly explains the new concept of volume-controlled design. Without the explanation, the subsequent discussion might be unclear to many readers.

❑ 6.4 ❑
Ending the Discussion

Although short reports may be able to end without anything more elaborate than a restatement of the main point and an offer for further contact, sometimes you may need a more developed ending for the discussion. If so, we suggest that you consider three strategies for ending a report:

- Address the organizational implications of the report. The direct organizational context and problem your report addresses may not lead you to discuss some important organizational implications of your conclusions and recommendations in the discussion itself. Therefore, especially for management, you often will find it appropriate to end by introducing additional organizational implications of what you have concluded and recommended. Typical topics for ending in this way include the following: actions to be taken or that are being taken; personnel; schedules; facilities; and cost implications.
- Introduce additional technical considerations. You sometimes have to limit the technical content of your discussion to the specific technical investigation, design, or analysis. When you do so, you might omit technical considerations that could be relevant to some of your readers. If you present a preliminary process design, for example, you might not have considered control systems.
- Summarize the main points of the discussion. If your discussion has been long and detailed, and if your discussion hasn't ended with a summary, at times you can end your report by summarizing the main points of the discussion.

Address the Organizational Implications

In "working" organizational reports, as distinct from scientific and technical reports and articles, perhaps the most important way to end the report is by addressing the organizational implications of your conclusions and recommendations. This is because those reports are action-oriented rather than informational. Most of the readers, especially managers, are concerned for the organizational implications of accepting or implementing your recommendations.

Here are the last sentences or paragraphs of reports (which we have shown you either in whole or in part) that address organizational implications:

> Upon approval of this proposal I will prepare a detailed procedure and cost estimate for implementation of line heating technology. Final cost estimates can be made after technical training with the Japanese firm. Qualitative benefits then can be calculated for selected proposals and contracts, based on the percentages of shell plates which could be formed in whole or part by line heating.
>
> I have scheduled a meeting in the Administrative Conference Room for 0900 16 July 1986 to discuss this proposal and implementation plans. [The ending of the line heating technology report, Figure 5.8]
>
> With issuance of this letter, TB-3056 is considered closed. [The ending of the test report on front end sheet metal mounts, Figure 3.3]

With your approval we will proceed with the second phase of our work with Waterco. This phase will have them prepare detailed equipment specifications for all new equipment required to upgrade the plant. [The ending of the report on modification of the boiler feed water treating unit, Figure 7.9]

Introduce Additional Technical Considerations

Many of your professional activities are part of ongoing processes and investigations. That means that your ending could pose specific questions for subsequent investigation. This will be the case if your investigation is part of a larger investigation, if your research suggests additional avenues of research, or if your design allows for further modification. Here is an example from a report on an experimental device used to measure the effects of mercury poisoning on the central nervous system.

RECOMMENDATIONS FOR FURTHER INVESTIGATION

There are four major recommendations for further research based on this study.

1. It appears as if the reliability of the measurement system is strongly related to the S/N ratio and possibly to learning effects. Therefore, added effort should be devoted to attempting to find better ways to condition the input signal. An increase in S/N ratio may serve to make the system more reliable.
2. Some sort of skill in which people differ greatly may be involved in performing the tracking task with the designed machine. This hypothesis should be tested. If the reliability does vary significantly from person to person, an analysis should be made to determine what this skill is so that operators may be picked accordingly.
3. The effect of better quality components should be investigated. This has a cost trade-off and should be considered in relation to its effect on the entire system's worth.
4. The variability of the mode in a test-retest stuation must be evaluated to determine the reliability of the test-retest scheme as a detection device and the acceptable deviation at the man-machine level.

Summarize the Main Points of the Discussion

For some reports, it is sufficient merely to summarize:

As you can see, preparation for the Expanded Day Program is moving ahead on schedule. At the same time the department has been able to maintain high production and safety records. We therefore feel our progress during the first six months of 19XX has been excellent. [The ending of the Petrochemical Department progress report, Figure 6.3]

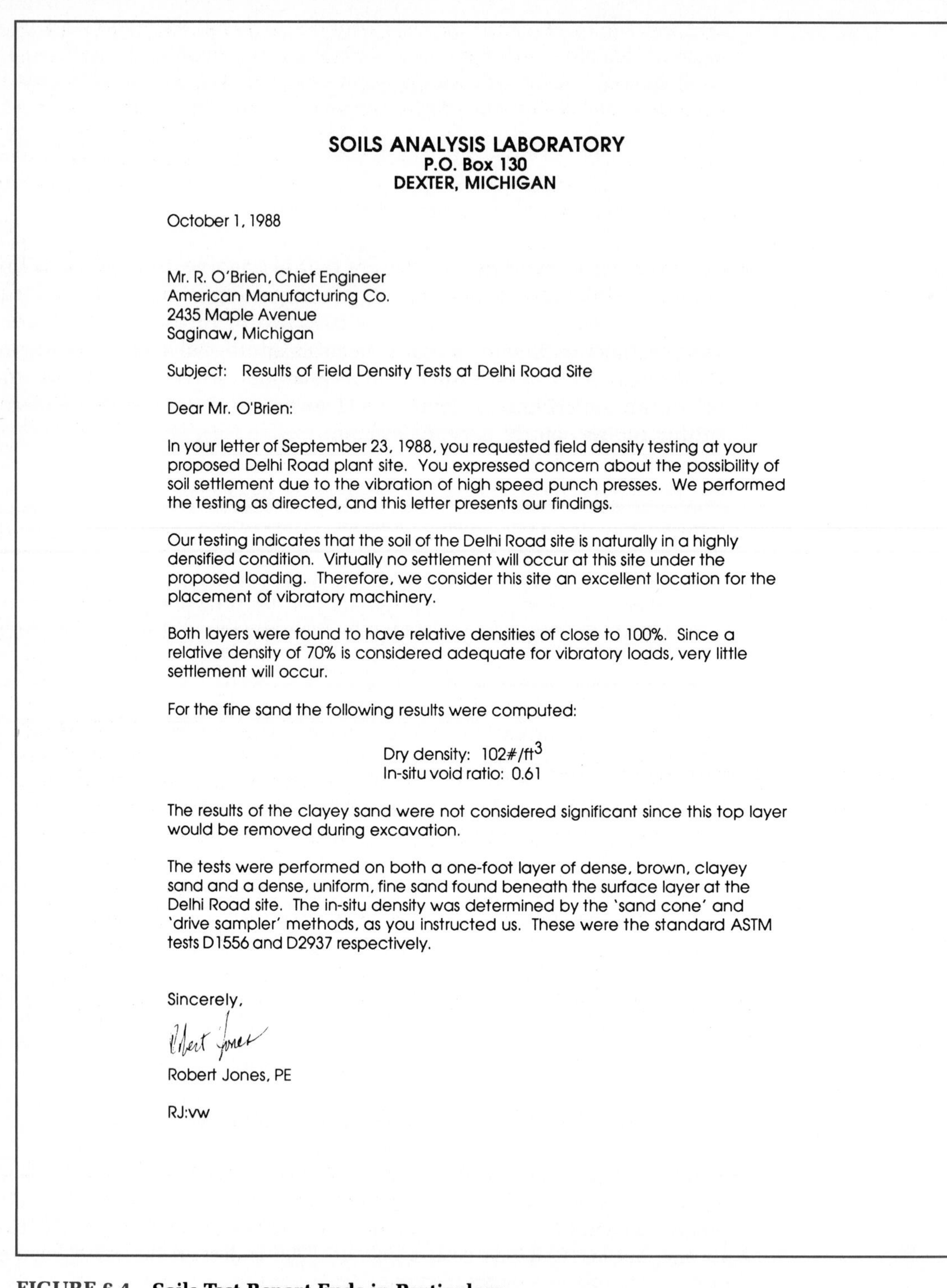

SOILS ANALYSIS LABORATORY
P.O. Box 130
DEXTER, MICHIGAN

October 1, 1988

Mr. R. O'Brien, Chief Engineer
American Manufacturing Co.
2435 Maple Avenue
Saginaw, Michigan

Subject: Results of Field Density Tests at Delhi Road Site

Dear Mr. O'Brien:

In your letter of September 23, 1988, you requested field density testing at your proposed Delhi Road plant site. You expressed concern about the possibility of soil settlement due to the vibration of high speed punch presses. We performed the testing as directed, and this letter presents our findings.

Our testing indicates that the soil of the Delhi Road site is naturally in a highly densified condition. Virtually no settlement will occur at this site under the proposed loading. Therefore, we consider this site an excellent location for the placement of vibratory machinery.

Both layers were found to have relative densities of close to 100%. Since a relative density of 70% is considered adequate for vibratory loads, very little settlement will occur.

For the fine sand the following results were computed:

Dry density: 102#/ft^3
In-situ void ratio: 0.61

The results of the clayey sand were not considered significant since this top layer would be removed during excavation.

The tests were performed on both a one-foot layer of dense, brown, clayey sand and a dense, uniform, fine sand found beneath the surface layer at the Delhi Road site. The in-situ density was determined by the 'sand cone' and 'drive sampler' methods, as you instructed us. These were the standard ASTM tests D1556 and D2937 respectively.

Sincerely,

Robert Jones, PE

RJ:vw

FIGURE 6.4 Soils Test Report Ends in Particulars.

> It is unfortunate, to say the least, that an Owner in making an economic decision must rely on the engineering expertise of the shipyard to submit a reasonably accurate preliminary estimate only to find the final estimate to be in error by almost 600%. That the final estimate is in error seems clear for reasons explained above. The Owner therefore requests your review and adjudication of this estimate. Enclosed for your use are all of the references, as well as the Contractor's back-up material for this charge. [The ending of the request for adjudication report, Figure 5.4]

As we said, however, you do not necessarily have to have an ending. The field density test report (Figure 6.4) ends on particulars:

> The tests were performed on both one-foot layer of dense, brown, clayey sand and a dense, uniform, fine sand found beneath the surface layer at the Delhi Road site. The in-situ density was determined by the 'sand cone' and 'drive sampler' methods. These were the standard ASTM tests D1556 and D2937 respectively.

❑ 6.5 ❑ Implementing Your Plan: Developing an Outline

The planning strategies which we have discussed so far should give you a helpful way to think about the structure of your reports. They can also give you starting points for building an outline from which to actually write. But many writers make the mistake of bypassing both the planning and the outlining, starting to write a first draft before they have thought carefully about what they have to say. A better procedure is to think carefully before beginning to write and then to begin the actual writing process with an outline, not with a first draft of the document. Again, however, many writers have trouble, even if they do outline, because few of them know how to go about outlining effectively. In fact, many writers do not understand the value of an outline and do not like to outline simply because they do not know how to do it.

Perhaps this simple 8-step procedure for outlining will help. This procedure is best implemented by a form that you can set up on the word processor (Figure 6.5). Merely translate the procedure into a simple form to be filled out.

Step 1. State your thesis or purpose in a single sentence.

You should always be able to use one sentence to express the entire thesis or purpose of a document, no matter how complex or long the document will be. This statement is the conclusion at your highest level of generalization. It is the most succinct expression of the "main message" of your document, and will be implicit in your subject line or title and subtitle. If the idea is

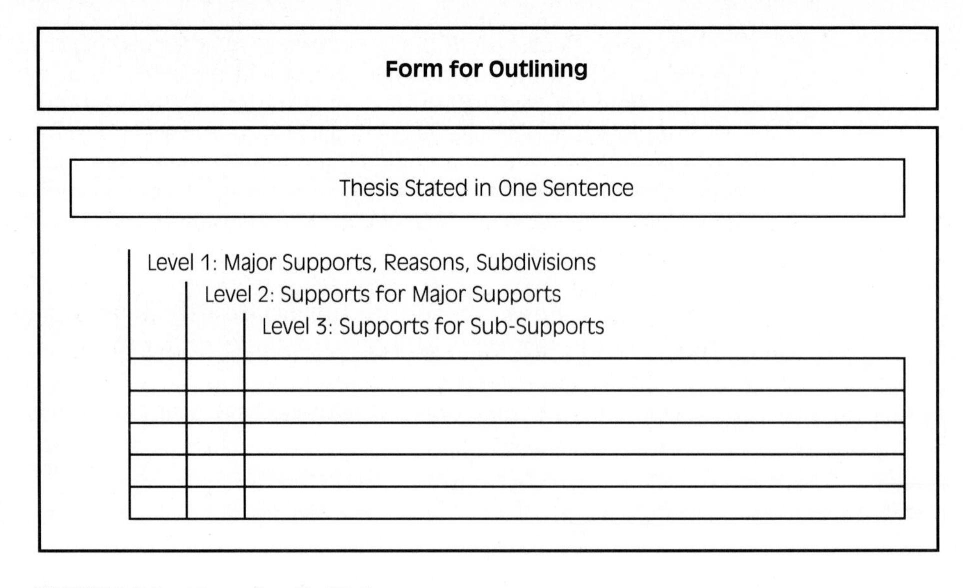

FIGURE 6.5 Form for Outlining.

not explicitly clear to you as a writer, it is probably not going to be clear in the document itself. Moreover, if the idea is not clear to you, it cannot exercise any controlling influence over your planning process of segmenting, selecting, and arranging the information in your document. In other words, it is important to know the main message before beginning to develop the plan for the document which will support that message.

An example of a thesis in a single sentence (for the report outlined in Figure 6.1) is:

> A Vendor Quality Index would be more effective than the Monthly Vendor Summary used to evaluate vendors.

This sentence is a high-level generalization which comprehends that entire report. The very act of formulating your thesis or purpose in a single sentence will force you to clarify your purpose and your topic.

Step 2. State the primary topics of the sections of the discussion.

Just as you should state your main idea, you should be able to make single-sentence statements of every support upon which that main thesis of your document will rest. These too will be high-level generalizations, but less high than the statement of the main thesis. In a sense, they are the "steel girders" which will ultimately support the structure of the thesis when your document is actually written or "built." They will be the topic sentences or core ideas of the major sections of the document. They are your second level of generalization. (Our organizational strategies will provide guidelines to help you decide when you have stated all of the topics needed to develop and support the thesis.)

State each supporting idea in one complete sentence, just as you stated the thesis in one sentence. Doing so again will force you to clarify your ideas and the direction of your thinking.

As you state your supporting ideas, be sure to keep these sentences consistent in terms and form. Grammatical parallelism is a good idea. Later, when these statements are distributed throughout the document—separated by many pages—you might find it difficult to see that these sentences do not match in terminology or form if you are careless at this point. Accordingly, it is good practice to be rigorously consistent as you state the major topics supporting and developing your thesis.

Step 3. Order the primary topics of the discussion in a logical sequence.

After you have listed the primary topics of the discussion, examine their order and revise it if necessary. When writers grope along trying to discover an idea at the same time that they are writing it, they often ignore the issue of logical order. Accidental order, not planned order, is the result. The document will ultimately have an order, but there is no guarantee that the thought process that has led to it will be either efficient or logical. Accordingly, it is much better to begin by examining the skeletal content of the document—looking at the "steelwork"—before all of the details of the document begin to cover up and obscure that steel.

Simply examine your thesis statement and establish a basic organizational strategy appropriate for developing it. Then look at the order of the topics you listed in Step 2. If it is not appropriate for explaining and supporting your thesis, reorder them. Later on, when the document is actually written, moving information around will be much harder than now, when your entire "document" consists of only four or five sentences.

By clarifying and revising the order of these statements, you will be creating additional meaning that might otherwise remain fuzzy. There is meaning in arrangement. If you can clarify the logic of your arrangement, you have clarified your thinking and the global coherence of your writing will be improved.

Step 4. Treating each primary topic as a subthesis, move to the second level of the outline and state the topics of the subsections or units of each section of the discussion.

After you have completed Step 3 of this procedure, you will have a skeletal document consisting of statements of your thesis or main idea and the major topics supporting it—two levels of generalization. You will know that the basic structure of the idea is both complete and logically ordered. In a sense, the basic steelwork will be in place, and you can then move to filling in that steelwork with the subordinate structural supports that will be required in the actual writing of your document.

Just as you did in Step 2, state your support for the major topic of each section of the document; this time you do so on a third level of generalization. (These are second-level topics, each of which is the topic of a subsection or unit of the report.) Make sure these second-level topics add up to a complete and logical pattern of support or development of the major topic. (The outlines for the basic organizational strategies discussed in this chapter and the outlines for organizing information in sections of a document, discussed in the next two chapters, provide guidelines to help you in this step.)

As you state your supports for the topic of each section, be sure to complete the statements of all of the points of support for the topic of a section of your document before moving to another major topic and section. That is, solve one problem at a time, just as you did in Step 2 when you were outlining the topics of support for the thesis you developed in Step 1.

Again, remember to keep your terms consistent and to make the grammatical forms of the sentences parallel if you can.

Step 5. Order the topics for each subsection.

Here you do exactly as you did in Step 3, but this time you work on ordering topics on your third level of generalization. Examine each subthesis and its topics of support in turn. Decide whether the order is both complete and logical. If it is, move on. If it is not, revise the order or make whatever additions or deletions are necessary. (Our suggestions for organizing various types of information in sections and units of your document, presented in the next two chapters, provide guidelines at this level of outlining.)

Step 6. If the document is to be long or complex, repeat Steps 4 and 5 on the next level of generalization. (If it is to be short, go to Step 7.)

For most relatively short documents, a three-level outline (thesis plus two levels of support) is sufficient. However, for longer documents, you may need to move to a fourth, or even fifth, level of generalization. If so, merely repeat the procedures you used for Steps 4 and 5, stopping only when you are certain that you have expressed all of the core ideas that will support the details of the document when it is actually written.

Step 7. Eliminate unpaired headings, at least on the higher levels of generalization.

When you reexamine your outline, you may find that you have "divided" a topic (on any level) into "one part" (a "2.1" without a "2.2," for example). Usually, this is a symptom of having gone too far in moving down the levels of generalization. In fact, often it is a symptom of including examples in your outline. Admittedly, on the lower levels of the outline, this will cause no problem; but on the higher levels it usually indicates that your thinking has become muddled as you have moved among levels of generalization. Either the unpaired heading does not belong there, the higher level heading is somehow at fault, or something is missing. If you find unpaired headings, look again and consider what the source of the difficulty may be.

Step 8. Examine the outline for omissions, logical inconsistencies, inconsistent terminology, or unparallel grammatical forms.

Now you have a completed outline. Almost. Just one more thing to do. That is, before you start fleshing out the outline, look to be sure that you have been consistent in terms, logical in your order, and parallel in your grammatical form. Later, when all the steel has been covered up by concrete, wall coverings, and trim, you will have a very hard time in seeing whether you were effective in erecting the steel in the first place. Consistent terms, logical connections, and parallel forms are the bolts that join the girders of your structure.

This 8-step procedure for outlining is a practical way by which you can establish a coherent hierarchical structure for your report. As we discussed at the beginning of this chapter, the report has a hierarchical structure of different units and types of information. Effective outlining enables you to make each unit of structure coherent in itself as well as interconnected with other units to form a coherent whole.

Figure 6.6 shows the steps by which the outline of an insurance company legal brief was built from a single sentence into a fully developed sentence outline. As you can see, with this outline as your guide, you could probably write the brief yourself.

A Comment about Outlining on a Word Processor Many writers still cling to the idea that they should write on paper and then copy the text on the word processor. They also think of the outline as a "waste product" of writing—one that perhaps serves a purpose, but which must be discarded when the document is actually written. From our point of view, these writers are not taking advantage of the special power of the word processor for developing an outline and then helping the writer to turn the outline into a finished text.

Example Outline

(Step 1)

THESIS:

The requested judgment for repairs, insurance, and taxes should not be awarded.

FIGURE 6.6 Steps by Which the Outline of an Insurance Company Legal Brief Was Built. (4 pages)

Example Outline

(Steps 2 and 3)

THESIS:

The requested judgment for repairs, insurance, and taxes should not be awarded.

1. The mortgagee took possession of the building without notice and prior to the agreed upon date.
2. The mortgagee made repairs without giving the mortgagor the opportunity to make necessary repairs.
3. The mortgagee took out insurance and paid premiums without demand for or prior agreement for reimbursement.
4. The mortgage paid back taxes upon the property without demand for or prior agreement for reimbursement.

Example Outline

(Steps 4 and 5)

THESIS:

The requested judgment for repairs, insurance and taxes should not be awarded.

1. The mortgage took possession of the building without notice and prior to the agreed upon date.

   1.1. The mortgagee took possession of the property without informing the mortgagor.

   1.2. The mortgagee entered and occupied the building five weeks prior to the agreed upon date.

2. The mortgagee made repairs without giving the mortgagor the opportunity to make necessary repairs.

   2.1. The mortgage spent $1673.35 on electrical repairs within one week of entering the building but without informing the mortgagor of the repairs.

   2.2. The mortgagee spent $3216.35 on plumbing and heating repairs within three weeks of entering the building but again without informing the mortgagor of the repairs.

3. The mortgagee took out insurance and paid premiums without demand for or prior agreement for reimbursement.

   3.1. The mortgagee arranged for insurance on the building by July 10, 1982—five weeks prior to the agreed upon date of possession—and without prior agreement of reimbursement.

   3.2. The mortgagee paid a second premium upon his newly-arranged insurance within six days of entering the building but again without prior agreement for reimbursement by the mortgagor.

4. The mortgagee paid back taxes upon the property without demand for or prior agreement for reimbursement.

Example Outline

(Steps 6, 7, 8)

THESIS:

The requested judgment for repairs, insurance, and taxes should not be awarded.

1. The mortgage took possession of the building without notice and prior to the agreed upon date.

    1.1. The mortgagee took possession of the property without informing the mortgagor.

    1.2. The mortgagee entered and occupied the building five weeks prior to the agreed upon date.

2. The mortgagee made repairs without giving the mortgagor the opportunity to make the necessary repairs.

    2.1. The mortgagee spent $1673.35 on electrical repairs within one week of entering the building but without informing the mortgagor of the repairs.

    2.2. The mortgagee spent $3216.35 on plumbing and heating repairs within three weeks of entering the building but again without informing the mortgagor of the repairs.

3. The mortgagee took out insurance and paid premiums without demand for or prior agreement for reimbursement.

    3.1. The mortgagee arranged for insurance on the building by July 10, 1982—five weeks prior to the agreed upon date of possession—and without prior agreement of reimbursement.

        1. The premium invoice was mailed by the insurer to the mortgagee on 12 July, 1982.

        2. The mortgagee made no attempt to contact the mortgagor during the period between arranging the insurance and paying the first premium.

        3. There is nothing in the written agreement between the parties that requires the mortgagor to repay the mortgagee for insurance coverage on the building.

    3.2. The mortgagee paid a second premium upon his newly-arranged insurance within six days of entering the building but again without prior agreement for reimbursement by the mortgagor.

4. The mortgagee paid back taxes upon the property without demand for or prior agreement for reimbursement.

With its quickly set tabs and its special capability for cutting, pasting, and quickly revising text, the word processor is an ideal instrument for brainstorming the document before you write it. Moreover, it is ideally suited to working on one level at a time without losing track of what is ahead of or behind you. Wherever you insert text, the rest of the text is merely pushed ahead, and you can always scroll back and forth quickly to make modifications in your plan and to see the parts in relation to the whole.

After you have an outline that you are comfortable with, you can revise the outline on the screen to become part of your document. The primary topics of the sections of your discussion become main headings. The topics of the subsections become subheadings. And the numbering system can often be kept in the report exactly as it appeared in the outline.

Naturally, both outlining and writing are dynamic processes in which you will plan, revise plan, write, revise plan, write, revise plan many times. However, the word processor is a powerful tool for allowing you to state your topics, see them on the screen, and develop them in text. Even if you are skeptical about on-screen writing, our advice is to give it a fair try. It may take some time to learn to use the tool, and it may take some time before your keyboard skills are adequate to the task, but the payoffs can be substantial.

Conclusion

As a professional, you will have many details at your fingertips, far more than you can put into a report. What you need, then, is a planning strategy which will help you to segment, select, and arrange your material before you then turn that planning into a detailed outline. The strategies presented in this chapter should give you some useful ways of going about these tasks.

Once they are done, you are ready to write.

Notes

1. *Water Resource Management and Gambia River Basin Development*, Center for Research on Economic Development, University of Michigan, Ann Arbor, MI, 1985.
2. *Publication Manual of the American Psychological Association*, American Psychological Association, Washington, DC, 1983, rev. 1984: 24–28.
3. *Scientific and Technical Reports: Organization, Preparation, and Production*, American National Standards Institute (ANSI), Z39.18-1987: 18–19.

CHAPTER 7

Outlining and Writing Persuasive Sections

In Chapter 6 we presented three high-level strategies to help you to think about how to select and arrange the materials within the discussion of a report. These three were organization by persuasive purpose, organization by subject matter, and organization by method of investigation.

In this chapter, and the next, we want to push our discussion of structure to a somewhat more detailed level. That is, we want to discuss additional, more detailed strategies or patterns for selecting and arranging material within the discussion of your reports. In long reports, these patterns can help you to outline and write the individual major sections and units which combine to make up your overall discussion. In short reports, the patterns which we discuss in this chapter (and in the next) can help you to outline and write the whole report discussion.

In this chapter we present four persuasive patterns for selecting and arranging information. The four are alike in one central respect: they all make an assertion and then defend it. In other words, these patterns are explicitly persuasive. Granted, the nature of the assertion and support differs from pattern to pattern, but behind each of these four patterns is an explicit attempt to convince somebody of something.

Of course, you could argue that all technical documents are in a sense persuasive. A progress report, for example, implicitly asserts that progress has been made; a test report implicitly argues that the conclusions logically follow from the results and that the procedures followed were sensible ones, well executed. Indeed, all technical reports implicitly attempt to show reasonable people working reasonably. Yet in the case of the four patterns we discuss here, the persuasive element is explicit: the writer wants to convince a reader or group of readers of a conclusion by presenting to them the logical evidence upon which that conclusion rests. In an organizational context, the conclusion leads to a decision or action.

The four persuasive patterns which we discuss are these:

Persuasion
Problem/Solution
Cause/Effect
Comparison and Contrast

For each, we will explain both when the pattern might be used and how it might be designed.

In the next chapter we present four informative patterns for selecting and arranging information, again explaining both when you might want to use these patterns and how you might design them:

Analysis
Process or Instructions
Description
Investigation

Obviously, the specific organizational pattern which you choose to follow in any individual section of a long report (or within the whole discussion section of a short report) is a function of a number of considerations about audience, purpose, context, and material. Therefore, we can present only generic patterns for you to use as points of departure in your detailed planning. They are not rules; they are merely suggestions to help you deal with the planning and design decisions you must make in order to write the discussion section of a report, and we encourage you to use these patterns with an appropriate sense of freedom and flexibility. Nevertheless, we can give you some suggestions which we have found to work very well as report planning devices when we have shared them with professionals in business, industry, and government.

Before we discuss these four persuasive patterns, however, we need to introduce some basic principles which apply to all patterns, persuasive and informative.

❑ 7.1 ❑
Four Basic Principles of Patterns

The methods of organization which we discuss in this chapter and the next chapter have four common principles. First, patterns must both create and fulfill readers' expectations. Second, patterns should move appropriately from general to particular information. Third, they should present selected information. And fourth, they need clear content and format cues to help hurried readers to recognize the pattern of development quickly and easily. Although of course these four principles apply to an entire report, they are particularly relevant to the design of the sections of a report.

Patterns Create and Fulfill Expectations

Everything you say "from word one" in a technical document will create expectations on the part of the reader. To a large extent, the clarity of the document will depend upon your creating the appropriate expectations and then fulfilling them.[1]

For example, consider the opening paragraph of a section of a report shown in Figure 7.1, a section called "Secondary Treatment."

> **3.2. Secondary Treatment**
> Secondary treatment is a continuous biological process followed by a physical separation. Bacterial flocs formed in the biotreater are settled out in the clarifier and further removed by a rapid sand filter.

Look at the structural expectations which just these few words establish on the part of readers, even if you read them in isolation, without reading any of the surrounding text:

1. You can be almost completely certain there was a prior segment called "Primary Treatment." (You don't know whether there is a later segment called "Tertiary Treatment," but there certainly could be.)

2. You know that the prior segment was numbered 3.1., and that if there is a following tertiary section, it will be numbered 3.3.

3. You know that the probable purpose of the prior segment was to explain how the first stage of a process is performed.

4. You know that the probable purpose of this section is to explain how a second stage of the process, actually a subprocess, is performed. (The heading term is "treatment," which suggests process. In addition,

3.2. Secondary Treatment

Secondary treatment is a continuous biological process followed by a physical separation. Bacterial flocs formed in the biotreater are settled out in the clarifier and further removed by a rapid sand filter.

3.2.1. Biotreatment

Wastewater from the equalization basin enters the biotreater where it is completely mixed and aerated. Aerobic biological growths recycled from the clarifier are mixed with the wastewater. Insoluble organics are removed by physical enmeshment or absorption into the biological flocs. Soluble organics are removed through direct oxidation by the biological growths. Residence time in the biotreater is approximately eight hours. The water is then pumped into the clarifier where it is settled.

3.2.2. Clarification

Effluent from the biotreater enters the clarifier where the biomass is settled out and returned to the biotreater. Approximately 5% of the solids, however, are dewatered by a rotary vacuum filter and then incinerated. This removal is performed to help maintain a constant level of biological activity in the biotreater. Overflow from the clarifier is passed through a rapid sand filter for further solids removal.

3.2.3. Rapid Sand Filtration

Overflow from the clarifier passes through the rapid sand filter where the level of . . . .

FIGURE 7.1 A Pattern Creates Expectations.

the first sentence uses the word "process" to confirm the hint provided by the heading and to strengthen your expectation that the section will indeed explain process.)

5. You suspect that either of two possible paths is likely to be followed in further text within section 3.2. Sentence 1 gives you a hint that a 2-part structure might follow: "Secondary treatment is a continuous *biological process* followed by *physical separation*. (Possible part 1 for further discussion: biological process; possible part 2: physical separation.) The following text might very reasonably take its cue from that structural forecast; however, there is a second possibility. The second sentence says that "Bacterial flocs formed in the *biotreater* are settled out in the *clarifier* and further removed by a *rapid sand filter*." This sentence gives a hint of a 3-part structure that the rest of the section might follow. Possible part 1 for further discussion: biotreater; possible part 2: clarifier; possible part 3: rapid sand filter.

You can't be certain at this point which of the two possible paths is likely to be followed, but probably you do expect the writer to take either one or the other of those two paths. Anything else—although not impossible—would be a bit of a surprise, an unexpected departure from the generic "pattern" which you have already diagnosed as being the apparent pattern of this section of the document. In other words, you think you are reading a process description, and having read hundreds of them over a lifetime of reading, you pretty well know what is coming, even though you don't know the particular details of *this* process.

As you read the text which follows the opening in the example, you will be looking, unconsciously of course, for confirmation of the pattern you expect. If you find this confirmation, your reading will be assisted. If you don't find it, you will have to puzzle out the relationship between your earlier expectations and what the writer is actually doing.

We think that the opening paragraph rather clearly creates the expectation that the following paragraphs are going to explain the process of secondary treatment. Further, we find that most readers are led to expect that the process is going to be broken down into three stages: the biotreatment stage, the clarification stage, and the rapid sand filtration stage. In addition, we find that most readers quite reasonably expect that the order of the explanation will follow the same as the order forecast: biotreater, clarifier, rapid sand filter (as it does in the actual document). And finally, we find that some readers correctly anticipate that the sub-subheadings in further text will be "Biotreatment," "Clarification," and "Rapid Sand Filtration," not "Biotreater," "Clarifier," and "Rapid Sand Filter." These readers recognize that the emphasis of the first-level heading, "Secondary Treatment," is upon the process, not upon the hardware of the process. Accordingly, they guess that the later sub-subheadings are likely to be more process-oriented ("Biotreatment") than hardware-oriented ("Biotreater"), and in fact that is the case in the actual text.

We know what readers' expectations are in this instance because we have tested it numerous times. We project the paragraph on a screen and ask students to tell us what the probable function of the unit is and how it is likely to be developed. Most can correctly do so, and a few always correctly predict that the probable sub-subheadings will be "Biotreatment," "Clarification," and "Rapid Sand Filtration." When we show them the rest of the section, Figure 7.1, most readers find that their predictions of the structure of the section have been confirmed.

As this example suggests, a particular text creates expectations with just a few words of text. Because the pattern behind a particular piece of text is not unique to that text or its particular subject matter, we recognize generalized, generic patterns which we have read hundreds of times before in similar texts. In that sense, our expectations for a particular text are partly based upon the cues it gives us and partly based upon mental models we all carry around with us of what apparently similar texts generally do and should do.

Patterns Move from General to Particular

As we discussed when we explained the basic structure of technical documents, general-to-particular design is an important principle of technical document structure. At all levels of the document, always give the "bottom line" first; then explain how you got there. This is a standard principle of technical writing presented in virtually every company guide or style standard for report writers as well as in most technical writing textbooks. More importantly, it is a principle evident in the practices of well-written journal articles, reports, manuals, proposals, letters, memoranda, test reports, feasibility studies, design reports, minutes of meetings, progress reports, and so on. In short, "general to particular" is a basic principle of technical writing—perhaps it is even "the" most basic of technical writing principles.

You should apply this principle at all levels of structure in the report:

- whole documents: put the "Overview," "Executive Summary," or "Abstract" first (as in the sample reports illustrated by the various figures in this text).
- sections and units of the discussion: put the "core paragraph" first (as in the section on secondary treatment).
- paragraphs: put the topic sentence first (more about this in Chapter 9).

Putting the general information first creates the readers' expectations. Also, it states the conclusion or the core idea clearly so that readers can't miss that idea. Finally, it provides the reader with a conceptual framework that permits him or her to read subsequent text purposefully or to skim (or even skip) the particulars at any level of the document without fear of missing important general information.

Patterns Present Selected Information

In addition to providing an outline for the presentation of information, patterns—to varying degrees—also provide guides for selecting information. In implementing any pattern, you must pay careful attention to selectivity. That is, you must decide what elements within the details of the pattern need fleshing out.

You have two criteria for selectivity. First, you need to keep your communication purpose clearly in mind: you must decide what information is needed in order to accomplish your purpose. Tangential information and information suggested by your personal interest in your subject matter are types of information that can be nonfunctional. Second, you must be aware of the needs of your readers. You have to decide what information they need in order to understand your conclusions. Your tendency will be to overestimate the types and amount of information most readers need.

As an example of appropriate selectivity, consider the "Advantages" section of the "Proposal to Introduce Line Heating Technology into ASBC Production Procedures":

> *Advantages of line heating for forming curved shell plating.* By utilizing line heating technology, we can eliminate the restrictions we now face in the forming of shell plating. The advantages of line heating are:
>
> 1) Increase in productivity, as seen in other U.S. Shipyards (Appendix B).
> 2) Ability to accurately form shell plates with compound curvature (Appendix C itemizes parameter and dimensional control for plates on Hitachi Hull Y-14-2), which results in a cost savings.
> 3) Ability to accurately fit curved parts to curved shell with minimal force, which requires fewer manhours than cold forcing.
>
> . . . . . .

As you can see, the writer had considerable information at his disposal, some of which he put into appendices. In this section, however, the writer focuses on conclusions without supporting detail, as in item 3. He judged that his management audiences would be interested only in the conclusions. They would not need the details on "increase in productivity," and they would not challenge his assertion about accurate fit with "minimal force." He further judged that his audiences would not need him to specify an "accurate fit" in terms of tolerance (millimeters). The writer's purpose is to marshal the conclusions and support for his argument, not to explain how the conclusions were derived.

As the brief excerpt makes clear, a report on line heating technology with different audiences and a different purpose could have considerable detail. Increase in productivity could be quantified, accuracy could be specified, man-hours could be quantified, and so on. Yet, this writer did not go into these details because his purpose was to persuade a management audience, not to instruct an audience of line personnel.

Patterns Have Clear Cues of Content and Format to Help Hurried Readers

Whatever the pattern of a document, section, or unit, you should use clear cues of structure to signal that pattern to the reader. In technical documents, which often must be read quickly and which must be completely clear, these cues of structure are great helps for the reader.

We discuss these cues in greater detail in Chapter 12, but it is useful to introduce them here. You will see these cues in the various examples in this and following chapters on patterns and paragraphs.

Structural cues can include all of the following: white space, formatted lists, headings and subheadings, numbering, and variations in typography. To these cues might be added language cues of structure (sometimes called "metalanguage," language about language). Language cues are explicit statements of structure and transitional words that either tell the reader what pattern you are using at that point in the document or connect different types of information within a pattern. For example, if you say, "for example," the reader knows what is coming and how it relates to what went before. Similarly, if you follow with the word *similarly*, that word signals another example or parallel idea. And if you say, "First, we discuss four general observations about patterns: A, B, C, and D," readers can predict what will follow and what the order will be because you have explicitly told them what to expect. In addition, with the language cue "First," you have told readers that the "four general observations" are the first part of a larger section of the document.

The section on Secondary Treatment (Figure 7.1) embodies three format cues: numbering, subheads, and white space. The numbers, 3.2.1, 3.2.2, and 3.2.3, identify the sequence of the three stages of the process. The subheads, Biotreatment, Clarification, and Rapid Sand Filtration, identify these three stages. The indented white space signals that the three stages of the process are particulars subordinate to the main idea in the core paragraph. The main head has format cues, 3.2. and Secondary Treatment, that connect this section to the prior section of the document.

Well used, format cues and metalanguage cues of structure are extremely helpful for the reader at the section and unit level as well as at the whole document level. (If you can use them effectively, they also are very helpful to you as you compose your document.) A cautionary note, that we explain in Chapter 12: overuse of cues can obscure your organization rather than reveal it.

❑ 7.2 ❑ The Persuasion Pattern

The basic persuasive pattern which we discussed in Chapter 6 is often used in sections of a report as well as in an entire report. But the basic pattern has many variations based upon differences in the specific nature of the persuasive purpose. We reintroduce the basic persuasive pattern in this chapter and discuss it first. Then we follow with a discussion of three variations of the pattern. For each we provide an outline which you can adapt for your own specific purpose in any part of a report or for entire reports, one page or several pages in length.

The persuasive pattern is a good pattern to use when your purpose is to justify a conclusion or recommendation, to present the case that can be made in support of a judgment.

As appropriate, the pattern consists of the following:

- Conclusion: Statement of conclusion.
- Support: Justification for the conclusion arranged in a descending order of importance.
- Alternatives: Anticipation and refutation of objections or alternatives, if needed.
- Summary: Restatement and explanation of conclusion, if the section is long enough to need restatement.

Begin with a succinct statement of your conclusion, the point you are going to support in the details that follow. This should be a brief core paragraph consisting of the conclusion and perhaps several sentences summarizing your supports. These will serve as a forecast of the pattern as well as a capsule summary.

Next, present your justification or positive support for the conclusion in a descending order of importance. Each item of support usually needs to be explained in some detail, so the support might consist of several paragraphs with one item of support plus explanation in each paragraph. In short persuasive sections the support sometimes can be presented in one paragraph.

The concept of descending order of importance is based upon both your own interpretation of the support and your anticipation of the readers' priorities. It is not given. For example, do cost considerations come before time considerations? Only your audience analysis and your own sense of priorities can answer that; but if you think about it consciously, the organizational context usually provides criteria by which to determine an appropriate order of importance.

Descending order of importance also suggests when to stop itemizing positive support for your conclusion. Often, a few items of support are sufficient, as minor items would be relatively unimportant and diminish the significance of your important items of support. A manager in industry was quite definite about this point; she said she tells her engineers to limit their support of conclusions to three items. Although we are uncomfortable with such a prescriptive approach, the concept which it suggests is a reasonable one. What the manager really wants is for her staff to present necessary and sufficient support to accomplish their tasks. She does not want them to waste her time and theirs merely because they know other minor details that might be pitched into the pile of supports. She wants them to do what they should, not what they can.

Finally, anticipate and refute any objections or alternative arguments implicit in your analysis or which your readers might raise. At times, you will address either or both of two types of refutations. First, you might have to refute any plausible objections to your conclusion, either against one of your arguments of support (such as negative evidence) or a negative point that might be raised (such as cost when it is not a positive item of support). Second, you may have to refute any alternative conclusions that reasonably might be made by your readers.

Be sure to take into account that "refutation" does not necessarily mean to argue that the objections are unsound, unreasonable, or invalid. Credible objections can be raised to many conclusions. However, professionals must always evaluate trade-offs in order to formulate conclusions. This means that you often must accept negative evidence and arguments. To refute them merely means that you must argue against their importance or weight relative to your positive support.

Also remember that your purpose is to support your conclusion by presenting positive support. You can never win an argument by showing that someone else has lost it. Accordingly, keep your emphasis on the positive supports for your conclusion and don't devote more space than necessary to anticipating and refuting the objections which reasonable readers might raise to your conclusions or to discussing alternatives which you considered and rejected. Do what is necessary, of course, but watch the natural tendency to bludgeon those who hold different views from your own or to overwrite the sections of a document dealing with alternatives. Be selective and keep your eye on your purpose.

If the section is long or if you introduce negative arguments, you can appropriately conclude your section with a summary or at least a restatement of your conclusion. At times, you need to conclude by explaining how your positive support outweighs the objections or alternatives. Again, you need to analyze your audience to determine when you have to do so, or even if you need to summarize.

As an example of an effectively arranged persuasive pattern, the section on "Lack of evidence of rock salt deposits" (Figure 7.2) is worth close examination. This is a very important persuasive section from a long geological report on the siting of the Enrico Fermi nuclear plant now in operation near Monroe, Michigan. No salt can be present in the strata beneath a nuclear plant, as any groundwater might dissolve the salt and weaken the soil foundation supporting the plant. Thus, the consultant had to argue his conclusion rather than present the results of an investigation.

The first paragraph (8) directly states the conclusion that no rock salt is present in the formation beneath the proposed site. Then the writer summarizes his argument, forecasting the structure of the rest of the paragraph cluster. There are three supports, and the writer names them before taking them up one at a time in the subsequent paragraphs.

The first item (8.1) presents the most important support for the conclusion: visual inspection of the core borings. This is direct, empirical evidence.

The second item (8.2) presents secondary support, argument by authority. The writer summarizes the published literature, which also concludes that no salt is present at the Fermi site. Secondary support for the conclusion is needed because the primary support, based on empirical investigation, is limited to "several borings" which, because they are expensive, cannot be made at every point throughout the entire area of the plant site. Inherently, selecting some sites for borings leaves other sites untested. Hence, the published literature is backup evidence. (By itself, however, the published literature would be weak evidence because it covers the entire geographical area of Michigan and Ontario, not just the particular construction site.)

The third item (8.3) on the surface seems to be a third and weakest item of positive support—an examination of well-logs and drillers' reports. Obviously, it deals with unscientific literature, which is less persuasive than either of the two types of evidence presented earlier. Yet it is still evidence of a sort. Notice, however, that the writer shifts from the positive support of his conclusion to a refutation of an alternative conclusion. That is, 2 (of over 100) well-logs and drillers' reports do mention salt somewhere in the area of the Fermi site. The writer does not focus on the 98 reports which support his conclusion that there is no salt present; rather he focuses upon rejecting the 2 that cast doubt on his conclusion. Although 98% seems strong support, when it comes to siting a nuclear plant, 98% is not sufficient.

The pattern concludes with a restatement of the conclusion with a summary of the argument (8.4). Perhaps this last paragraph should not have been numbered as 8.4. It is not a fourth subdivision, parallel to the three elements of support. Rather, it is a concluding paragraph, parallel in its level of generality to the unnumbered introductory paragraph of section 8. Nonetheless, this example is an excellent illustration of a basic persuasive pattern.

While the section on rock salt deposits comes from a long formal report, the persuasive pattern is common in short reports as well. The police report on a traffic accident (Figure 7.3) consists of persuasive sections within a persuasively arranged report. The section on "Establishing Gross Negligence" is a straightforward persuasive section arguing that the driver "was grossly negligent."

The opening paragraph (3) directly states the conclusion. In addition, it presents the criteria on which the argument is based, which is an item required in many persuasive patterns. Finally, it summarizes the argument to forecast the pattern that follows.

The three items of support that follow (3.1, 3.2, and 3.3) are arranged in descending order of importance, as the writer states in his presentation. For the first item, he states that "driving while intoxicated satisfies all the criteria for gross negligence as it shows a willful, wanton and reckless disregard of

8. Lack of evidence of rock salt deposits

Our study indicates that there is no salt present in the Silurian Salina Formation which underlies the Enrico Fermi site. This conclusion is supported by visual inspection of cores taken from the upper part of the Salina Formation at the site, by published reports on the Salina in Michigan and Ontario, and by well-logs and drillers' reports from the area.

8.1. First, on the basis of visual inspections of cores from several deep borings made for Detroit Edison both at the Fermi site and the Monroe site, I can report that no salt was encountered in that portion of the Salina penetrated (approximately 210 feet at Fermi and 150 feet at Monroe). As the Salina is approximately 500 feet thick in the Monroe area I can therefore say that the upper 30 to 40 per cent of the Salina is free of salt.

8.2. Second, the published literature on the Salina in Michigan and Ontario indicates that no salt is present at the Fermi site. Statements from four of the most recent such reports verify this finding.

8.2.1. K.K. Landes, in the text accompanying the U.S. Geological Survey Oil and Gas Investigations Preliminary Map 40 (1945, The Salina and Bass Islands Rocks in the Michigan Basin) says of the Salina salt beds, "The F and lower salts disappear in southern Wayne County. The D salt is not present in the Macomb County well, and it disappears with the higher and lower salts in southern Wayne County. The B salt, like the D and F salt, ends abruptly in southern Wayne County."

8.2.2. C.S. Evans, in a paper entitled, "Underground Hunting in the Silurian of Southwestern Ontario" (1950, Proc. Geol. Assoc. of Canada, Vol. 3, pp. 55-85), includes a map showing the southern limit of salt a mile or so south of Amherstburg, a town on the east bank of the Detroit River seven miles north of the Fermi site.

8.2.3. B.V. Sanford (1965, Salina salt beds, Southwestern Ontario, Geol. Surv. Can.: Paper 65-9, 7 pp.) states that the B salt has the widest distribution of any of the saline units. Both his figure showing the distribution of the B salt and the figure showing total salt thickness within the Salina and its distribution show that there is no salt within the Salina Formation in Ontario south of a point one mile south of Amherstburg.

8.2.4. In an American Chemical Society Monograph (1960, Monograph 145, edited by D.W. Kaufman), K.K. Landes discusses the salt deposits of the United States (Chap. 5). In the section on Michigan (pages 71-74) he states that the bedded salt deposits of the state are in either the Salina (Silurian) or Lucas (Detroit River Group-Devonian) formations. The Devonian salt occurs only in the northern part of the Southern Peninsula. All salt mining is done from the Salina and in the Detroit-Windsor area. His Figure 1, on page 72, showing the aggregate thickness of Silurian salt deposits in northeastern United States and Ontario shows the southern limit of Silurian salt to be north of the Wayne-Monroe County line; in fact, according to the figure there is no salt present anywhere in the section beneath any of Monroe County. The natural brines (distinct from the bedded salt deposits discussed above that are either mined or artificially brined) utilized within Michigan are from the Parma, Marshall, Berea, and Dundee formations, all of which occur only to the north and northwest of the Fermi site and are higher stratigraphically than the rocks that underlie the Fermi site.

FIGURE 7.2 Section of a Geological Report with Persuasive Pattern.

8.3. Third, well-logs and drillers' reports do not indicate that salt may be present at the Fermi site. A careful reading of over 100 well-logs for the Monroe County area turned up only two in which salt was even mentioned, and in neither case is the evidence persuasive that salt might be present at the Fermi site. One report mentioned salt near Milan, where one might expect salt to be present. The other mentioned salt near Lambertville; however, the latter report, not backed up by samples reviewed by a geologist, is perhaps questionable.

8.3.1. The first mention of salt in the logs reviewed was at Milan, Michigan, first reported by A.C. Lane in the Michigan Geological Survey Annual Reports for 1901 and 1903. The log for this well, based on samples, showed 5 feet of rock salt near the base of the Monroe group (which includes the Salina) at depths between 1540 and 1545 feet. The Milan area is about 30 miles west-northwest of the Fermi site; the possibility of salt occurring in the section should increase in that direction, which is towards the center of the Michigan Basin. Although twelve detailed logs (some of which are also based on samples) from the townships between Milan and the Fermi site were carefully looked at, no further mention of salt could be found.

8.3.2. The other mention of salt noted in the many logs reviewed was in that of a well located some two and a half miles north of Lambertville, about 24 miles southwest of the Fermi site. The well-log in question was based simply on the driller's log rather than on samples evaluated by a geologist; it noted "dolomite and salt" in a 15 foot interval near the base of the Salina, between depths of 525 and 540 feet. Mounted samples from a well some two miles away and logs of eleven other wells between the one near Lambertville and the Fermi site again give no further indication that salt is present in the section.

8.4. In sum, the evidence indicates no salt underlies the Fermi site. The evidence consists of close visual inspection of the cores taken from the upper part of the Salina for Detroit Edison, four recent reports based on a detailed study of carefully selected well-logs and cuttings, and all the older published reports each based on only a few rough drillers' logs.

3. Establishing Gross Negligence

Smith was grossly negligent in operating his vehicle at the time of the accident. Establishing gross negligence is critical to successful prosecution for manslaughter. Gross negligence is shown by "a willful, wanton and reckless disregard of the consequences." The Defendant's merely being careless is not enough to prove the charge. I will establish that Smith's behavior was willful, wanton, and reckless in three ways. First, Smith was intoxicated while operating his car. Second, Smith was driving at an excessive rate of speed. Third, Smith's vehicle crossed the centerline and traveled into the path of oncoming traffic, causing the accident.

3.1 Smith was Intoxicated

Smith was intoxicated while operating his car. Driving while intoxicated satisfies all the criteria for gross negligence, as it shows a willful, wanton, and reckless disregard of consequences. Drunk driving is not an accident. Smith's intoxicated state can be demonstrated in two ways.

3.1.1 Smith's blood alcohol level (BAL) was established at .18% through analysis of a sample (Appendix C, Crime Lab Report.) This level is significant in that it is .10% higher than allowable levels for driving. Studies have shown that driving ability is affected at levels as low as .04% BAL. The alcohol's effect would be even more pronounced on a 17-year-old like Smith who had limited drinking experience.

3.1.2 Smith admits drinking prior to the accident in a statement given to officers, "I had about two drinks of his whiskey" (see Appendix A). This statement shows Smith's willfulness in driving after he had been drinking. Additionally, testimony from the State Crime Lab witness will show that Smith had much more than "two drinks of whiskey."

3.2 Smith was Speeding

Smith was driving at an excessive rate of speed. Speeding, as drunk driving, is the result of purposeful behavior. Speeding shows a willful, wanton, and reckless disregard for the safety of others. While it is not possible to establish Smith's exact speed, it is possible to show that it was excessive through two facts: the length of skid marks found at the scene and the extreme amount of damage involved.

3.2.1 Smith's car skidded 251 feet prior to striking the Jones vehicle. Even with his brakes locked, Smith was unable to slow his vehicle enough, over this distance, to prevent the accident. That Smith was unable to stop over such a distance, as indicated by the unbroken skid marks found at the scene, indicates that he was traveling at an excessive rate of speed. Testimony from Officers Olsen and Ward will establish the skid length found.

FIGURE 7.3 Section of Police Report with Persuasive Pattern.

3.2.2 Physical damage in the accident was extreme. Even after having skidded 251 feed, Smith's car was torn in half by the force of the impact. Officers found the passenger compartment of Smith's car resting 30 feet from the engine compartment. Such extreme damage, in light of Smith's futile attempts to stop his car, establishes that speed far in excess of that deemed prudent was involved. A scale drawing of the scene and 35 mm color photos are available to supplement the testimony of officers.

3.3 Smith Caused the Accident

Smith's vehicle, not Jones', crossed the centerline on E. Roosevelt Ave., causing the accident. This could be attributed to Smith's intoxication and speeding. While not of itself enough to establish gross negligence, this critical loss of control was the sine qua non of the accident.

3.3.1 Joan Jones stated that she was lawfully in the eastbound lane when Smith's vehicle crossed the centerline to strike hers. She can testify as well to her unsuccessful attempt to avoid the Smith car by steering to the right (see Appendix B). Her testimony will be supported by the observations of officers concerning vehicle placement and skidmarks. As above, photos and a scale diagram are available to enhance the testimony.

Smith's gross negligence, then, is seen in his actions: driving while intoxicated, driving at an excessive rate of speed, and (whether accidentally or not) crossing the centerline of the road to collide with an oncoming car. The fact that three separate negligent acts are involved demonstrates even more forcefully my contention that more than mere carelessness is involved. This is a case of gross negligence. I will now show that Smith's negligent actions led to Glisson's death.

I

THE UNDERCOVER AGENT DOES NOT APPEAR TO BE A LADY OF MATURE APPEARANCE AND DEMEANOR.

A primary issue in this case is, did the Police Officers retain Ms. Conrad because she looked much older than 19 years of age? If Ms. Conrad's appearance was that of a mature woman, then the argument would follow that the Police had encouraged, set up, or induced the Defendants to commit a crime that they would not ordinarily have done.

Based on the facts of this case and the personal observation of this Court, Ms. Conrad does not appear to be a lady of mature demeanor and appearance. The Court bases this decision on the following facts:

(1) The fact that 22 of the 27 establishments requested Ms. Conrad to furnish identification.

(2) Witnesses at the Entrapment Hearing testified they estimated Ms. Conrad's age to be 22 or 23 years of age.

(3) The Court's observation of Ms. Conrad's appearance and demeanor in the Courtroom.

It is the Court's opinion that the appearance and demeanor of Ms. Conrad required that her identification be checked prior to the time she was furnished with alcoholic beverages, as was done by 22 of the 27 establishments. Even if the Defendants' guess was that Ms. Conrad was 23 years of age, they should have checked her identification because of the close proximity of 23 to the legal age to purchase alcohol.

FIGURE 7.4 Section of Court Opinion with Persuasive Pattern.

consequences." The evidence is strong that the driver was intoxicated. For the second item, he formulates the same conclusion, although it seems to be not quite as important as the first item. The writer has to argue that "it is possible to show that it (speed) was excessive." For the third item, he observes that it is "not of itself enough to establish gross negligence," so the item is the least important.

The writer concludes with a summary that, in fact, provides a further argument: "The fact that three separate negligent acts are involved demonstrates even more forcefully my contention that more than mere carelessness is involved." (The final sentence is a transition sentence to the next section of the report, a nice technique.)

The section from a court opinion (Figure 7.4) follows the same persuasive pattern. The first two paragraphs present the conclusion as well as the criteria on which the conclusion is based. Then the writer presents the evidence in support of the conclusion, apparently in decreasing order of importance. The writer concludes with a restatement that interprets the conclusion in terms of the basic issue.

These examples illustrate a feature of sections of reports: they are independently structured as self-contained patterns.

❑ 7.3 ❑ The Problem/Solution Pattern

The problem/solution pattern is a good one to use when your purpose is to demonstrate that you have solved a problem or when you have a solution to propose to a problem. Much of your professional activity consists of troubleshooting and problem solving. Thus, you will often find yourself writing documents or parts of documents that appropriately use a problem/solution pattern.

As appropriate, the pattern consists of the following:

- Problem: Statement of the problem or question at issue.
- Solution: Explanation of the solution or answer.
- Criteria: Statement of the criteria for a satisfactory solution. (Sometimes this is combined with the explanation of the problem.)
- Details of Implementation. Explanation of the solution, point by point in a descending order of importance or in the order of the necessary steps to be taken to implement the solution.
- Limitations: If necessary, the explanation of any limitations of the solution.

- Alternatives Considered: If necessary, explanation of alternative solutions considered and rejected (includes reasons for rejection).
- Restatement of solution (if limitations or alternatives are considered).

Begin with a general statement of the problem or the question at issue. This problem statement might differ from the explanation of the organizational problem which we discussed in Chapter 3. Here, we are discussing a technical problem or question, such as how to meet a proposed EPA regulation on lowering sulfur content or whether a proposed residential complex meets zoning requirements. State the problem as clearly and precisely as you can, and then explain any of the particulars necessary in order to understand the problem.

Next, state the solution to the problem or the answer to the question. This provides a summary as well as a forecast of the details that follow. In a short section, the statement of the problem and the statement of the solution often are combined in one paragraph which serves as an overview of the section.

For many problems, before you explain the solution, you need to establish the criteria for a solution. In some situations, this is a simple statement of specifications. In other situations, this is a matter of explaining and establishing the appropriate criteria for a solution. If your readers might ask, "how do you know when you have a good solution?" tell them.

Often, the criteria for a solution are explicit in the statement of problem. For example, in one report dealing with problems in the architectural design of a retirement center, the problem was that the design did not meet a specification which required that all levels of the center had to be exposed to the sun for a minimum of 30% of the day. In this case, the statement of the criterion is combined with the statement of the problem.

At other times, you will have several criteria which need to be enumerated and which are much less precise than the specification illustrated above. For example, "durability" and "cost effectiveness" are criteria for much technical work, but what do the terms mean? When is something judged to be "durable"? When is it "cost effective"? Answering those questions may take some text, but if the concepts are important to the following discussion, the text is necessary.

Then, explain the implementation of the solution in detail. This usually consists of breaking the solution down into its component parts or stages, and explaining each in some appropriate logical sequence. An appropriate order might be found in decreasing order of importance. Or an order might be implicit in a physical object, if the solution is a thing. Or an order might be found within the stages of a process, if the solution is a process or procedure.

In many situations, an explanation of the solution is sufficient and can conclude the pattern. However, it is a good idea to at least consider discussing two other types of information: limitations to the solution which you propose, and alternative solutions that were considered but rejected for credible reasons. If these types of information seem appropriate, wait until after you have demonstrated the feasibility or appropriateness of your proposed solution, then take up the limitations and the alternatives which you considered and rejected, discussing each in turn (probably you could take them in either order, but decide that upon the basis of your sense of the audience's needs.)

If you discuss limitations or alternatives, of course, you probably need to return to a restatement of your solution in order to end the section. You want to get back to the main idea, not leave the reader hanging on a limitation or a rejected alternative.

As an example of an effectively arranged problem/solution pattern, examine the section on "Impact of Proposed EPA Regulation" (Figure 7.5). The first paragraph succinctly states the problem and the primary criterion, the first item on the problem/solution pattern outline. Then, the solution is presented. The solution (a brief analytic section) consists of three changes that should be made to implement the solution. Notice that the three changes (the second and third are in the second paragraph) are presented in parallel sentence constructions. The first change is the primary or most important change.

Then the writer has a section on "Limitations" and a section on "Alternatives." The result is a straightforward problem/solution pattern.

The section on Development Office Plans for FY 1989/90 (Figure 7.6) presents an entire discussion arranged in a problem/solution pattern (only the first part of the discussion is shown in the figure). After presenting the objective of the plan and some background in the first paragraph, the writer presents the problem and solution in the second paragraph. Then the writer outlines the rest of the discussion. The promotional activities and fund-raising activities are the actual solution to the problem. The budget and staff as well as implementation schedule are means of achieving the solution, which seems a logical replacement for the "Limitations" section of the pattern. An "Alternatives Considered" section was not necessary for this particular report.

❑ 7.4 ❑ The Cause/Effect or Effect/Cause Pattern

The cause/effect or effect/cause pattern is a good one to use when your purpose is to answer the question "what caused X?" or "what will happen as a result of X?" In other words, this pattern either looks back into the past or projects into the future, in either case presenting a conclusion which has the highest degree of probability.

As appropriate, the pattern consists of the following:

- Issue: Statement of the issue—cause? or effect?—and statement of the conclusion.

Impact of Proposed EPA Regulation

The proposed EPA regulation requires that we lower the sulfur level in the product fuel oil from 0.2 percent to 0.05 percent. A solution to this problem needs to be flexible enough to allow for a sulfur level in the product stream of 0.2 percent if the proposed regulation is not enacted.

Solution. In order to reduce the sulfur level, the following revisions should be made to the current design (Figure 1). First, the reactor temperature should be increased from 700 degrees to 712 degrees in order to achieve the desired level of desulfurization. The product specification can then be altered by simply reducing the outlet temperature of the fired heater. In addition to the flexibility of this solution, no additional capital investment is involved since the reactor sizing is not altered.

Second, the supply of makeup hydrogen should be increased from 365 SCF/bbl to 391 SCF/bbl. This is necessary because hydrogen is consumed in the catalytic desulfurization process. Finally, the supply of DEA to the stripper should be increased from 105,000 lb/hr to 112,000 lb/hr. This is necessary since higher levels of hydrogen sulfide will need to be removed by the stripper.

Limitations. This solution has two drawbacks. First, higher heat duties will be required of the heat exchangers and air coolers downstream from the reactor. This will raise our capital investment and operating costs slightly, due to the increase in size of this equipment. Second, the reactor temperature will have to be carefully monitored to ensure that the maximum temperature does not exceed 825 degrees in order to prevent catalyst deactivation. The current control scheme should provide sufficient protection against this problem.

Alternatives. One alternative to increasing the reactor temperature is to increase the reactor size. This method has two severe limitations. First, in order to achieve the desired level of desulfurization, the reactor size would have to be increased by 50 percent. This would increase our capital investment significantly. Second, this option does not leave us much flexibility if the proposed regulation is not adopted.

NOTICE OF MEETING TO FINALIZE MODIFICATIONS

Considering the possibility of complications due to these modifications, a meeting has been set up between all parties involved on March 23 at 11:00 AM in the conference room.

Dist: Manager, Texas City Refinery
Refinery, Operating Representative, TCR
Manager, Planning
Manager, Marketing
Environmental Affairs Coordinator
Manager, Safety and Regulations

FIGURE 7.5 Section of a Refinery Report with Problem/Solution Pattern.

DEVELOPMENT OFFICE PLANS FOR F.Y. 1989/90

The Development Office was established 5 years ago. Cuts were made in the Development Office staff and budget for 1989-1990 to reduce costs. However, during this current F.Y. it is projected that the percentage of income vs. expenses will increase slightly from last year. It is also projected that for this school year the number of enrollment inquiries will also increase from previous years.

The problem is that the projected goals for increased contribution income and increased enrollment are greater for 1990-1991 than in previous years. To accomplish these goals the Development Office will have to expand its activities and staff.

This report examines two areas which address means of achieving contribution and enrollment income goals.

1. Promotional activities including publications and advertising
2. Fund raising activities including the development of a volunteer structure for implementation

The needs of the Development Office budget and staff and the implementation schedule then are explained.

1. <u>Promotional activities including publications and advertising</u>

   This office's responsibility is to publish and distribute quality publications that will effectively meet the recruitment and promotional goals of the school. This office also understands that this responsibility must be carried out within a limited budget. The publication plans described below will provide the best means for achieving these goals within a limited budget.

   1.1. School catalogs
   The Upper and Lower Schools each will have their own catalog. These catalogs will be distributed to ....

FIGURE 7.6 **First Part of a Discussion of a School System Administration Report with Problem/Solution Pattern.**

- Causes (Effects): Support for the conclusion; that is, analysis of the causes or effects, arranged in a descending order of probability or importance.
- Alternative Causes (Effects): Anticipation and refutation of alternative possibilities.
- Summary: Restatement of the conclusion, if the section is long enough to warrant restatement or if alternatives have been rejected.

Begin with a statement of the issue—is it one of cause or of effect?—and a statement of your conclusion. You often expand on this by summarizing or forecasting your argument: state that you base this conclusion on x number of points, and then itemize the points. At times, you will find it appropriate to explain the methodology upon which your argument is based. For example, you explain that your prediction of the effects of a price increase is based on interviews with the regional marketing managers. (You might have to introduce a methodology item into your outline, although usually you can use an attachment or appendix to present any necessary details.)

Then, present your analysis of the causes or effects. This is the listing of the causes or effects, with explanations of the connection between each and your conclusion. The order of these should be descending order of probability or importance. For example, some causes are *necessary* (effect could not happen without them); other causes are *sufficient* (effect could happen as a result of the cause, but there are other possible causes); and still other causes are *contributory* (effect could not happen as result of the cause alone). Start with the necessary causes, not with the contributory causes. Similarly, if there are causes which you have observed and which others have observed, start with those you observed: your own direct observation is inherently stronger evidence than the testimony of other observers.

Because you are arguing cause and effect, you next need to account for alternative causes or effects: you need to anticipate and refute alternative possibilities. After you support your conclusion, therefore, you need to address possible alternatives and to dismiss them insofar as is possible. Answer any reservations or questions that your readers might have and demonstrate that your thought process in identifying the causes or effects was comprehensive and reasonable, that you did not merely look for the first answer that seemed to suffice, ignoring possible alternatives. Defend your conclusion by documenting your thought process and by anticipating the thoughts of your readers.

Again, when you introduce alternative explanations, you probably should go back to your conclusion and perhaps a summary of your argument to conclude.

For an example of an effectively arranged effect/cause pattern, examine the section on "Failure Analysis of Impeller Shaft P/N 53127X" (Figure 7.7). This section is the entire discussion of a memorandum report, except for a figure and some calculations.

Failure Analysis of Impeller Shaft: P/N 53127X

An investigation was conducted to determine the cause of failure of the steel impeller shaft (P/N 53127X) in the 13PV Turbine Pump after only one month of service. Examination of the failed shaft and analysis of the stress situation both indicate that the shaft failed in fatigue. An inspection of the fracture surfaces reveals the beach marks and final rupture area that are characteristic of fatigue failures. Analysis also indicates that the fatigue strength of the shaft is marginal, with the situation worsened by the apparent use of a chipped cutting tool in the cutting of the threads where the failure occurred.

1. The appearance of beach marks on a smooth fatigue zone (see Fig. 1) and a coarse, crystalline final rupture zone clearly indicate that this is a fatigue failure. The fatigue crack initiated near point 'I' (Fig. 1) and progressed across the shaft to final rupture at area 'R'. The changing angular orientation of the beach marks as they progress to the final failure zone indicate that there was a rotating bending load causing the failure. This is consistent with the loading known to have been acting on the shaft.

2. An analysis of the fatigue strength of the steel impeller shaft with respect to its nominal loading shows that its endurance strength is only 3% above the expected minimum that would be required for infinite life. This is a very marginal value, especially when viewed in light of the high variability seen in the fatigue strengths of steel parts, even when tested under rigidly controlled conditions.

3. An inspection of the threads where the failure occurred reveals that they had been machine-cut (as opposed to rolled, which would have improved the stress situation by approximately 25%). Furthermore, microexamination seems to indicate that the cutting tool used to cut the threads was chipped, thereby decreasing the thread root radius and increasing the stress raiser effect of the threads.

A calculation of the total life of the shaft shows that the shaft failed after 20 million cycles, which is on the high side of the distribution of fatigue cycles at failure in steel parts. However, considering the marginal endurance strength and the possibility of occasional overloads beyond the nominal loading level, this fact is not inconsistent with fatigue failure experience. Therefore, it is my conclusion that the 13PV Turbine Pump impeller shaft (P/N 53127X) failed in fatigue because of an insufficient endurance strength, which was due in part to the stress raiser effect of an improperly manufactured threaded section.

FIGURE 7.7 Section of a Failure Analysis Report with Effect/Cause Pattern.

The section opens with a summary which states the effect and then the cause of the effect. Three types of evidence are summarized to support the conclusion that the shaft failed in fatigue. These also serve as a forecast of the explanation which follows.

The first paragraph on beach marks presents the most important evidence for fatigue failure. The next two paragraphs present additional but secondary evidence, which in themselves would not be conclusive evidence.

The final paragraph anticipates and refutes an objection that failure occurred "after 20 million cycles, which is on the high side of the distribution of fatigue cycles at failure in steel parts." This argument provides a nice opening for the writer to conclude by restating his conclusion.

The section on "Water Delivery System Efficiency" (Figure 7.8) illustrates a combined effect/cause and cause/effect pattern. The first paragraph states that inefficient water delivery systems are the effect of poor perimeter construction and other causes. These inefficiencies in turn are the cause of system deterioration and other effects. The rest of the pattern then consists of an effect/cause paragraph and a cause/effect unit which explains these conclusions.

❑ 7.5 ❑ The Comparison and Contrast Pattern

The comparison and contrast pattern is a good one to use when your purpose is to demonstrate the differences and similarities between two or several things or concepts. Depending upon your specific purpose, the pattern can vary in focus. It can be

predominantly comparative (shows likenesses);
predominantly contrastive (shows differences);
mixed comparison and contrast (shows both similarities and differences, but emphasizes similarities because it treats similarities first); or
mixed contrast and comparison (shows both differences and similarities, but emphasizes differences because it treats differences first).

Although occasionally you will write either a predominantly comparative or predominantly contrastive pattern, most of the time you will use mixed pattern because you will be arguing trade-offs among alternatives. That is, most of the time you will be arguing trade-offs among such items as schedules, facilities, vendors, products, processes, costs, plans, or procedures, among others.

As appropriate, the pattern consists of the following:

4.6.1 Water Delivery System Efficiency

Inefficient water delivery systems in irrigated perimeters throughout the Gambia River Basin are a major constraint to expansion of irrigated agriculture. The major causes of inefficient water delivery systems are poor perimeter design and inappropriate equipment combined with insufficient maintenance of irrigation systems. These inefficiencies contribute to system deterioration, low yields and cropping intensities, and possible perimeter abandonment.

Perimeter design, which includes the choice of an appropriate water delivery system and its construction and maintenance, is a major factor in water delivery system efficiency and cost. Incomplete feasibility studies and construction result in poor identification of different soil types within perimeters, drainage problems, and unforeseen water losses due to seepage and evaporation.

Inappropriate equipment choices and lack of funding for equipment repair and replacement, nonstandardized equipment, and lack of spare parts result in serious maintenance problems. This problem is exacerbated by the lack of maintenance personnel. Poor system maintenance, especially in the post-investment period, creates water delivery irregularities that have a direct and negative impact on system costs, yields, and farmer participation.

In general, these chronic problems with inefficient water management within perimeters have been a major reason for the low use rate and rapid deterioration of existing perimeters in the Gambia River Basin. Variations in soil conditions and contour within perimeters have resulted in problems of water distribution between plots. Unanticipated seepage from unlined canals and large water losses due to evaporation have resulted in high water extraction rates and pumping costs. The use of irrigation perimeters in the rainy season has been limited by failure to incorporate adequate drainage and flood protection systems.

Equipment failure further has weakened water security. Each irrigation project operating in the Gambia River Basin has introduced a different selection of equipment and machinery. The maintenance of this equipment has become a serious problem in the post investment period due to a lack of critical local capability to maintain equipment. Adequate maintenance of machinery and equipment involved in irrigation presupposes the existence of trained mechanics, adequate supplies of spare parts, and adequate funding for repairs and replacement of equipment. This constraint becomes more binding as the technical complexity of equipment increases.

Perimeter size is also a factor in determining water delivery system efficiency. In theory, large perimeters (200-1000 ha) can achieve economies of scale in water delivery; however, they require more sophisticated levels of technology, management, and maintenance. SAED's experience with irrigation suggests that small perimeters (15-20 ha), which are labor intensive and based on simple technology, achieve higher yields and cropping intensities than do large scale perimeters. The experience in The Gambia with small perimeters has, nonetheless, been poor.

The variation in agroeconomic conditions throughout the Gambia River Basin and the poor performance of irrigation to date suggest the need for further experimentation with perimeter design before large scale expansion of irrigation can be considered.

FIGURE 7.8 Section of Development Agriculture Report with Effect/Cause and Cause/Effect Pattern.

Conclusion: Statement of the conclusion and a forecast of the points of comparison and/or contrast.
Support:
For Comparison *or* Contrast: present a point-by-point comparison or contrast in a descending order of importance.
For Comparison *and* Contrast, emphasizing similarities: present a point-by-point comparison of similarities in descending order of importance, then the points of contrast or difference in a descending order.
For Contrast *and* Comparison, emphasizing differences: present a point-by-point contrast of the differences in descending order of importance, then the points of comparison or similarity in a descending order.
Summary: Restatement of the conclusion, perhaps with a synopsis of the positive support if the section is long enough to merit restatement.

Begin with a statement of your conclusion and a forecast of the points of comparison and contrast. This provides an overview of the section:

> From environmental and site development considerations, Site Area M is a more advantageous site for the Jennings power plant than the previously investigated Site Area N. Our judgment is based primarily on population, ecological, land use, location, and transportation factors.

In a longer section, the forecast would include substantive conclusions on each point, such as, "Site Area M has a lower population density, would not affect wildlife or aquatic ecology," and so forth.

If comparison or contrast, then present your argument point by point in a descending order of importance. In the siting example, the population criterion is the most important, while transportation is the least important. If a criterion's importance is not self-evident, then you need to explain the significance of the item as well as explain it in detail. Population might be a zoning criterion weighted significantly by a regulatory agency. Transportation might be an organizational criterion that is less important in this situation because both sites are accessible, although Site M is more accessible than is Site N. (In a contrast pattern, any minor comparisons are combined in a paragraph following the listing of points of contrast; the same is done in a comparison pattern.)

If comparison *and* contrast or contrast *and* comparison—when there are both important similarities and differences—then you have to adopt a 2-level persuasive pattern. If you argue similarities, then you itemize points of comparison first. You itemize points of contrast second, which is analogous to the refutation of alternative items in the persuasive pattern. Again, in each group of items you go in a descending order of importance.

Finally, restate your conclusion and, especially if you have used a 2-level pattern, summarize your positive argument.

You can vary the comparison and contrast pattern according to your information or your readers. The most important variation is the pattern in which the comparison and contrast is done by criteria rather than by positive and negative argument:

Statement of Conclusion
Population Considerations
  Site Area M
  Site Area N
Ecological Considerations
  Site Area M
  Site Area N
. . .
Transportation Considerations
  Site Area M
  Site Area N

By the headings, you can see that this pattern is weaker because the focus is not on the subject of the conclusion, Site Area M; instead, it is on the criteria. However, given the problem you are addressing and, perhaps, the values of your primary audience, you might adopt this variation on occasion.

Another variation of the comparison and contrast pattern includes a separate item on criteria. This would come between the Conclusion item and the Support item on the outline. As with other persuasive patterns, you sometimes have to explain the sources of your criteria or to justify the criteria you use to compare and contrast your subjects.

As an example of an effectively arranged comparison and contrast pattern, examine the section on the extra demineralizer train (Figure 7.9). The opening paragraph states the conclusion and summarizes the two basic points of support for the conclusion. These are points of contrast between the two options.

The next paragraph explains the first point of contrast, cost, illustrated by the cost table. This is followed by an explanation of the second point of contrast, the operating differences between the two options.

Then the writer concludes with a brief paragraph of comparison between the two options.

The section on "Funding" (Figure 7.10) also illustrates a comparison and contrast pattern. It opens with the conclusion that funding should not be an issue even though the three divisions of the Human and Social Services Departments are funded differently. The first two divisions are compared in terms of source of funding. Then the contrastive source of funding of the third division is discussed, but the same conclusion is argued: regardless of source, funding should not be an issue.

THE ADVANTAGE OF THE EXTRA DEMINERALIZER TRAIN OPTION

Adding a new demineralizer train is preferable to adding an electrodialysis unit. A new train has a lower first cost and is simpler to operate in combination with the existing equipment. It might be possible to add an ED unit later if the operating cost savings would justify the incremental investment.

Based on our needs for high quality, high reliability water supply, Waterco defined two water treatment options: adding a new demineralization train or adding an electrodialysis unit.

OPTION	INVESTMENT	CAPITAL COST DIFF.*	OPERATING COST DIFF.
Add new demineralizer train	$3.7 MM	$800 M in favor of new demineralizer train	
Add 400 gpm electrodialysis (ED) unit upstream of existing demineralizers	$4.4 MM		$400 M/yr. in favor of electrodialysis

*Based on G.A. Stevens estimate Feb. 11, 1983. Costs of major materials supplied by Waterco.

The capital cost differential is $800M in favor of the extra demineralizer train.

In addition, the ED unit has two drawbacks:

The ED unit is much more susceptible to damage from upsets in the clarator performance. Even though we would expect upsets to be infrequent, the resultant carryover of clarator sludge would quickly exhaust cartridge filters on the ED unit. If the filter cartridges were not changed quickly enough, ED membranes would be fouled. Cleaning them takes time and could reduce system capacity.

The ED unit requires different operating and maintenance procedures than do the existing demineralizers. The makes operation of the water treating plant more complicated with ED units compared to adding another demineralizer train identical to the existing equipment.

These drawbacks offset the operating cost differential in favor of the ED unit. However, American Petrochemical has a similar sized ED unit in service at its Chocolate Bayou plant. They are pleased with its performance so far although the unit has only been in service about a year.

Either option would be acceptable from a process standpoint, according to Waterco. The two options are equivalent at a 30 percent rate of return on capital.

FIGURE 7.9 Section of a Recommendation Report with Comparison and Contrast Pattern.

FUNDING

The three divisions, Social Services, Mental Health, and Public Health, are funded very differently. However, even though the divisions are funded very differently, funding should not be an issue for either the townships or the cities in any of the three divisions.

Social Services. Ottawa County residents do not pay anything for Social Service Programs. Social Service Programs are funded 100% by the state and federal governments (ref. c). Therefore, funding for Social Service Programs should not be an issue for either the cities or the townships.

Mental Health. Ottawa County residents pay only the small amount required for Mental Health Programs. Mental Health Programs are financed about 90% by the state of Michigan, 7-8% by local and county taxes, and 2-3% by fees and collections (ref. d). The state government funds most of the money for Mental Health Programs so that the programs will be uniform across the state of Michigan. Because of this, all Mental Health recipients across the state receive the same treatment. Ottawa County's 10% contribution in fiscal year 1987/88 was only $780,000. Since Mental Health Programs are required by the federal government and local funding is minimal, funding should not be an issue for Mental Health Programs for either the cities or the townships.

Public Health. In contrast to Social Service and Mental Health Programs, Public Health Programs are funded mainly by local means. In fiscal year 1986/87, the state and federal governments supplied 28%, fees and collections supplied 9.7%, and local funding supplied 62.3% of the total $2,583,461. (ref. e) In fiscal year 1988/89 the percentages should be relatively the same as the percentages in 1986/87. The state and federal governments will contribute 22%, local funds will contribute 60%, and fees and collections will contribute 18% of the total 3.5 million dollars.

Even though Public Health Programs are funded mainly by local means, funding should not be an issue for either the cities or the townships since usage by the cities and the townships is about equal. Anybody in the county, regardless of income and asset level, is able to utilize the programs. Also, many people in both the townships and the cities utilize programs without even realizing it.

Two Public Health Programs which people do not realize that they are utilizing are the public school programs and the environmental service programs. Children from both the townships and the cities derive benefit from the public school health programs. Also, everybody in the community benefits from environmental services. (Environmental services include such services as food inspection of restaurants, sewage inspection, and well-water testing.) However, most people take the public school programs and the environmental service programs for granted; they do not realize that they are benefiting from Public Health Services since they are not actually going into the service offices. Since everyone benefits from at least some Public Health Programs, funding should not be an issue for either the cities or the townships.

FIGURE 7.10 Section of a Social Agency Report with Comparison and Contrast Pattern.

Why Bother with These Outlines?

We have used these outlines as planning devices with literally hundreds of professionals in industry, business, and government. In the process, we have seen numerous instances of how the act of consciously thinking the outlines through helps writers to sort out what needs to be covered and to suggest possible orders for the various points. Sometimes they help writers to realize that their first impulses for organizing a report are not altogether successful. Here is a vignette which illustrates why we think you may find these outlines helpful.

Figure 7.11 shows a report written by one of our former students during a summer internship at an automobile company and subsequently handed in for one of our course assignments. (Her internship occurred between the junior year and senior year; the report was handed in a few weeks after the start of school in the fall of her senior year.) The writer received her BSE degree five years ago. She took a job with the same company with which she had interned, enrolling part time as a graduate student working on her MSE degree. She later received her master's degree, and is currently employed as a vehicle development engineer.

On first look, the report illustrated in Figure 7.11, which proposes a solution to a problem, looks pretty sound to most readers. (We have tested this reaction by distributing the report and asking numerous readers both in school and in industry to evaluate it.)

We find that readers tend to like the following features:

- The subject line is clear on the purpose of the report as well as on its topic.
- Paragraph 1 defines the problem clearly and economically.
- Paragraph 2 gives a clear statement of the purpose of the memo and summarizes the recommendation, focusing on important cost-related information.
- Paragraphs 3 and 4 give selective information about the design and the logic behind it, pulling important points from the attached three sheets of documentation. Readers like the selectivity.
- Paragraph 5 (and its numbered subparagraphs) presents a very clear synopsis of what would have to be done to implement the proposed solution—three actions. (Readers especially like this feature of the report. They also like the self-sufficient nature of the attached sheets which provide the particulars.)
- Paragraph 6 ends with a return to the purpose, points to the attached sheets, and provides a polite offer for follow up. Most readers regard this as a succinct but effective ending.
- The prose is fairly direct and simple, and the format makes the report easy to read quickly.

In general, these points summarize our first reactions to the report too.

Yet, look a little more closely, this time using the outline for a problem solution pattern as your guide.

Do you know what the problem is? We think you do: Old design for key allowed its storage in one place; however, new design for key requires its storage in another place. Must find new storage place.

Do you know what the criteria for a satisfactory design are? We think you do not, although hints of criteria are scattered throughout the text of the report. Among them are the following:

Cost (tooling cost and piece cost are mentioned, obvious issues)
Safety (coating is intended to protect against sharp edges)
Durability (coating was designed to prevent rust)
Security (after all, theft prevention is the objective, but this is not stated)
Accessibility (is mentioned)
Appropriateness (changing a tire requires using a jack; hence, putting the key near the jack makes sense—this is mentioned)
Compatibility (must fit with other aspects of the car's design—not clearly mentioned)

As you can see, certainly there are criteria for a successful design, but they are not all explicitly stated and grouped, as they might be. Rather, some of them are implicit in the text, fairly clearly hinted at, whereas others are left entirely to the reader to guess at. We think that the report would have been improved with a more explicit acknowledgment of the criteria behind the choice. Also, note how acknowledging the criteria would have provided a possible way of organizing the third and fourth paragraphs of the report.

How about the solution? Do you know what the solution is? We think you do: Attach key to handle of jack with spring clip fastener.

How about the steps toward implementation? Again, we think you know what needs to be done. The report is nicely focused on actions necessary, and the format as well as the use of attachments makes the actions easy to see and to follow.

How about limitations? We think you do not know what they are. Are there any you can spot? For example, consider this: the writer proposes drilling a hole in the jack handle. What effect will that have, if any, upon the structural integrity of the jack handle, which presumably was designed to lift the car in a safe and effective manner? Will the hole reduce the margin of safety and effectiveness? It is a reasonable question to ask, but there is no hint that it has been considered.

Also consider the issue of the spring clip fastener's possible interference with the functioning of the jack. Will the spring clip stay in place when the jack is used? Or is it to be removed and reinserted when the key is removed? Either situation might interfere with the effective use of the jack. Was that considered, and is it an issue?

Finally, consider alternatives: Were alternative locations considered and rejected? What are other possible locations? And if others were considered,

5 October 19XX

To: P.H. Hopkins, Manager, Corsair Body Design Group

From: S.T. Carruthers, Summer Intern *S.T.C.*

Subject: Proposal for Retainer to Store a Lug Nut Key in the 19XX Corsair

The 19XX Corsair has one special lug nut on each tire for theft prevention purposes. Currently, the dealer installs these lug nuts with a special key, and then stores this key loose in the front floor console. However, a new lug nut key design has been released, and because this key is significantly larger than the old key, it cannot be conveniently stored in the console. Jeff Watling, the Corsair Program Controller, asked me to propose a new storage location for the new key and to submit my proposal for your evaluation.

The purpose of this memo is to propose a spring clip fastener to attach the lug not key to the jack handle. The total cost involved in producing this clip is $12,500 in tooling cost, plus a $0.15 piece cost.

Attached is a sketch of the proposed spring clip. (See Figure A.) The clip would secure the key directly to the jack handle. This storage location was chosen because the key would always be used in conjunction with the jack handle, which is stored on the right wall of the rear storage compartment. This location is an easily accessible location, with ample storage space for the new lug nut key. Note that an estimate of the itemized costs involved in producing the clip is included on Figure A. These estimates were provided by Tom Ault.

The removable clip which I propose would be positively located by a four millimeter hole in the jack handle and a corresponding indentation in the clip. A 0.5 mm plastisol coating would cover the clip to eliminate exposed, sharp metal corners and to prevent rusting.

In order to release this part into production, three steps would have to be taken:

1. The exact dimensions and material specifications of the clip would have to be given to the drafting department. (Figure A provides a concept drawing, but this drawing would have to be turned into a specification drawing. Also the material specifications should be verified in testing.)

2. The Y-Components Department would have to authorize retooling of the jack handle (Part Number 149704769) to include a 4.0 mm hole to locate the clip. (Figure B shows the positioning of the hole.)

3. The Technical Publications Department would have to revise two documents:

    - The instruction sheet which the dealers use to install the lug nuts.
    - The Owner's Manual, which explains how wheels can be removed.

    (Figure C explains the necessary language changes.)

I believe that the three attachments give all of the necessary details for you to evaluate my proposed design. If you have any questions, I would be happy to talk with you. I can be reached at 5-9844.

FIGURE 7.11 Problem/Solution Report by Summer Intern.

Figure A: Design of the Proposed Spring Clip Fastener

Material: 1070 Carbon Steel, Spring Strip of 25 Thousandths, AG 181-T2 Standard

Hardness: 44 Rockwell C

Finish: Phosphate Organic, AG 6174 or 6047, 0.5 mm Plastisol coating

Costs: Retooling Handle: $2500; Tooling Clip: $10,000; Piece Cost: $0.15

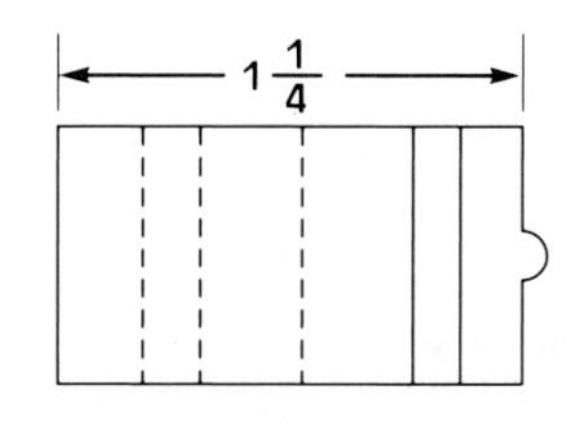

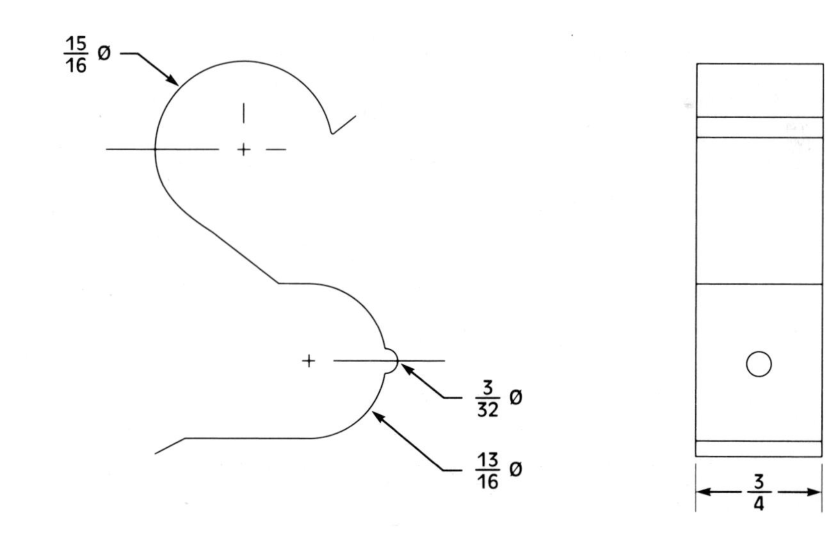

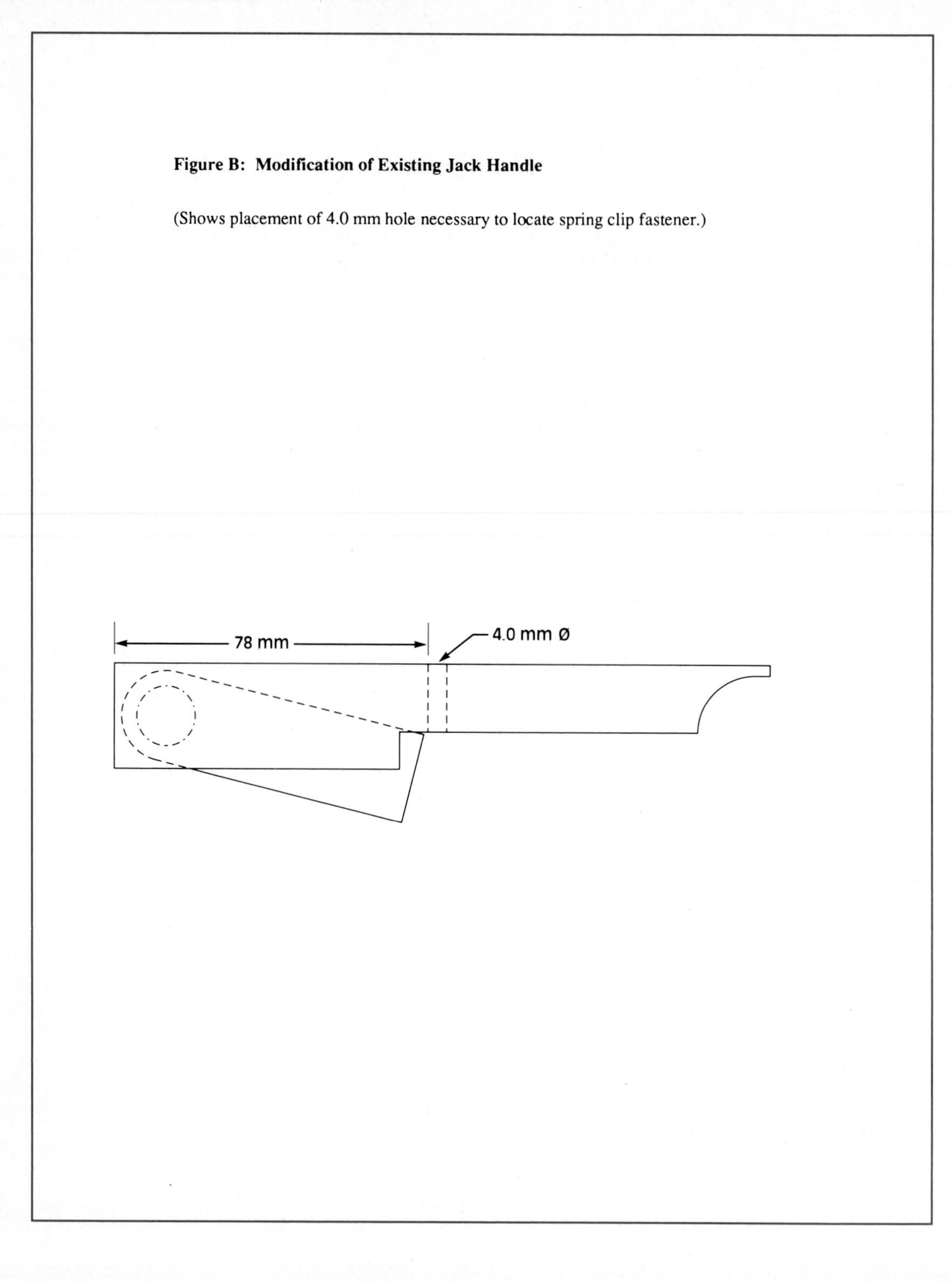

Figure B: Modification of Existing Jack Handle

(Shows placement of 4.0 mm hole necessary to locate spring clip fastener.)

Figure C: Necessary Modifications of Documents by Technical Publications Department

Instruction Sheet to the Dealer (Part Number 17543714)

Change Step 10 of existing instructions:

From: "After installing the lug nuts, place the lug nut key in the front floor console."

To: "After installing the lug nuts, place the lug nut key in the spring clip which is attached to the jack handle on the right wall of the rear storage compartment."

Owner's Manual (Part Number 14365971)

On page 35, under the heading "Changing a Flat Tire" revise the text as follows:

From: "The special lug nut on each tire must be removed with the lug nut key. This key can be found in the front floor console."

To: "The special lug nut on each tire must be removed with a lug nut key. this key can be found in the rear storage compartment, attached to the jack handle on the right wall of the storage compartment."

why were they rejected? What is it about the chosen location that makes it better than other possible locations? We think you do not know about alternatives because no alternatives or reasons for their rejection are discussed in the report. That leaves us wondering—is it safe to trust this recommendation? Are there better recommendations?

The point is that the example report looks pretty decent to most readers when they first read it. The writer thought it was OK. We thought it looked OK when it was handed in during our class. Yet, looking more closely, by using the outline as a sort of checklist, you can see how potentially important information is either not well presented in the report or not presented at all. Had the writer used the checklist as a preliminary guide, perhaps the report would not have missed these points. In fact, notice that had the writer consulted the checklist, perhaps she even could have used several of the headings of the outline directly in the process of writing her report. That would have been a time saver for her, and a good cue of structure for her readers. At least it is a possibility to consider.

Conclusion

In our discussion of Audience in Chapter 2, we pointed out that a report's success cannot be assured unless the lines of communication are sufficient for its transmission to the right readers and unless it is presented in such a form as to be persuasive. Those points apply to all reports in organization.

However, for many reports and sections of reports that are intended to produce specific decisions and actions on the part of readers, the persuasive aspect of the reports is especially important to remember. That is, the persuasive nature of the writer's purpose will work best if it is explicitly acknowledged and implemented in a way that leaves no doubt as to the thoroughness and logic of the support. This acknowledgment can be accomplished by use of any of the four persuasive patterns discussed in this chapter. Each of the four explicitly states the judgments of the writer, and each then explicitly presents the supporting logic to defend those judgments. The outlines of these four patterns should provide useful points of departure as you think about your purpose and consider structures that will help you to accomplish it. (See Guides, Figures 7.12, 7.13, 7.14, and 7.15 for the four outlines.) Remember, however, that the outlines are points of departure for planning, not fixed templates. Use them, but use them in ways that adapt to your specific purposes, audiences, and subject matter.

Guide for Persuasion Pattern

Use this pattern to justify a conclusion or recommendation—to present the case that can be made in support of a judgment.

Conclusion: Statement of conclusion.

Support: Justification for conclusion arranged in descending order of importance.

Alternatives: Anticipation and refutation of objection or alternatives (if needed).

Summary: Restatement and explanation of conclusion (if the segment is long enough to need a restatement).

FIGURE 7.12 **Guide for the Persuasion Pattern.**

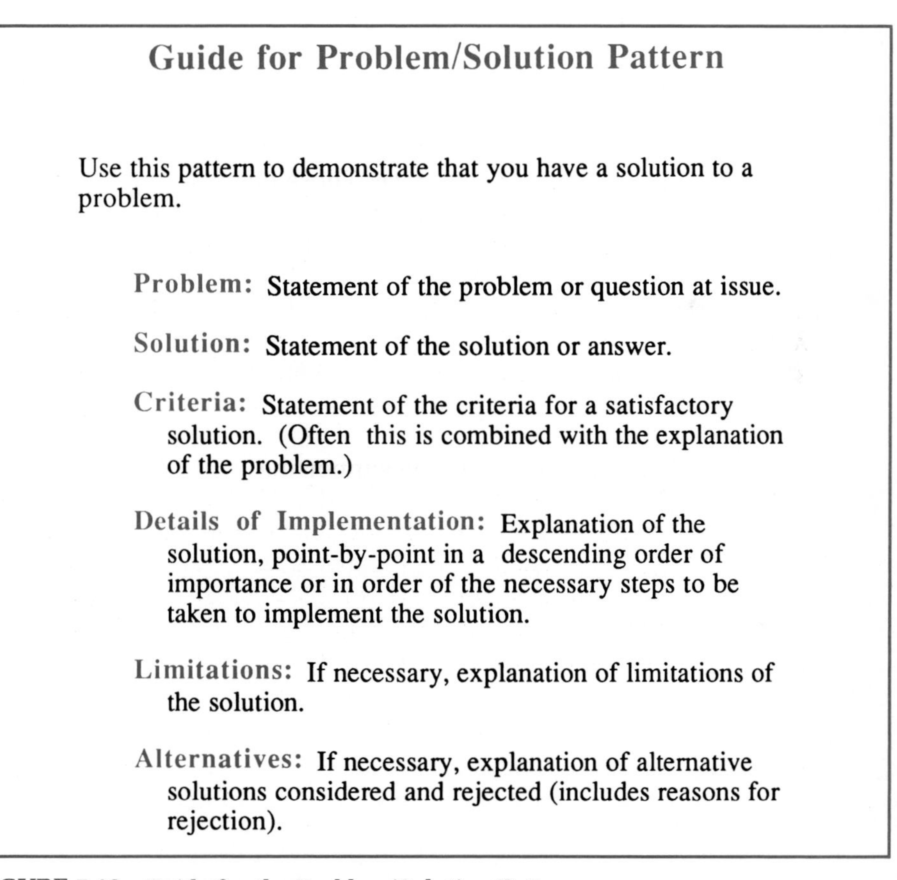

Guide for Problem/Solution Pattern

Use this pattern to demonstrate that you have a solution to a problem.

Problem: Statement of the problem or question at issue.

Solution: Statement of the solution or answer.

Criteria: Statement of the criteria for a satisfactory solution. (Often this is combined with the explanation of the problem.)

Details of Implementation: Explanation of the solution, point-by-point in a descending order of importance or in order of the necessary steps to be taken to implement the solution.

Limitations: If necessary, explanation of limitations of the solution.

Alternatives: If necessary, explanation of alternative solutions considered and rejected (includes reasons for rejection).

FIGURE 7.13 **Guide for the Problem/Solution Pattern.**

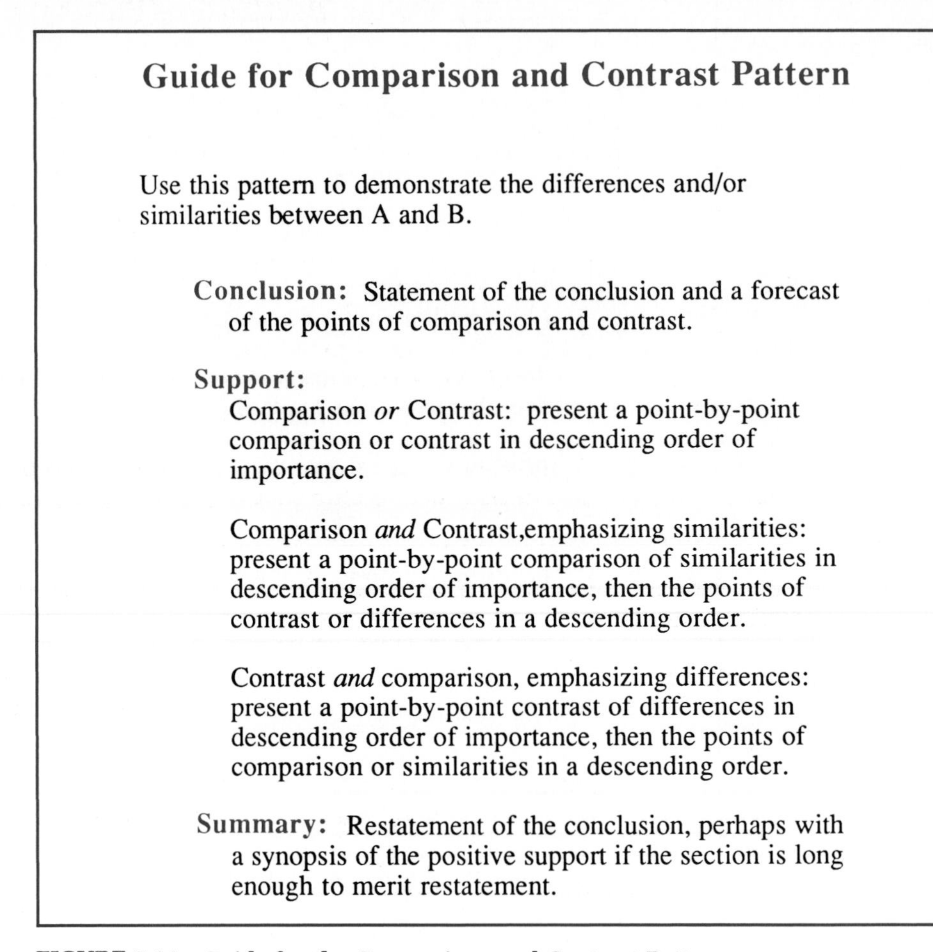

Guide for Comparison and Contrast Pattern

Use this pattern to demonstrate the differences and/or similarities between A and B.

Conclusion: Statement of the conclusion and a forecast of the points of comparison and contrast.

Support:

Comparison *or* Contrast: present a point-by-point comparison or contrast in descending order of importance.

Comparison *and* Contrast,emphasizing similarities: present a point-by-point comparison of similarities in descending order of importance, then the points of contrast or differences in a descending order.

Contrast *and* comparison, emphasizing differences: present a point-by-point contrast of differences in descending order of importance, then the points of comparison or similarities in a descending order.

Summary: Restatement of the conclusion, perhaps with a synopsis of the positive support if the section is long enough to merit restatement.

FIGURE 7.14 **Guide for the Comparison and Contrast Pattern.**

Guide for Cause/Effect or Effect/Cause Pattern

Use this pattern to answer the question "What caused X" or "What will happen because of X?"

Issue: Statement of the issue—cause? or effect?—and statement of the conclusion.

Causes (Effects): Support for the conclusion; that is, analysis of causes or effects, arranged in descending order of probability or importance.

Alternative Causes (Effects): Anticipation and refutation of alternative possibilities.

Summary: Restatement of the conclusion, if the section is long enough to warrant restatement or if alternatives have been rejected.

FIGURE 7.15 **Guide for the Cause/Effect or Effect/Cause Pattern.**

Notes

1. In traditional rhetoric, these patterns are exploratory. They are used by the writer to discover meaning. To write an essay on the impact of product packaging technology on urban waste disposal systems, for example, you would use the patterns to explore the implications of your thesis, not just to organize information already clear in your mind. You might discover that advertising drove the development of product packaging as much as the technology of plastics and other synthetic materials did. Your simple thesis becomes complicated as you explore cause/effect relationships.

   When you use these patterns in a technical report, there also will be some element of discovery—some clarification of your purpose and development of your subject. However, much of the time you as a professional have the conclusions of your analysis clearly in mind when you write the report. Your choice of patterns of arrangement depends on your purpose and audience analysis rather than helps you clarify your purpose. Thus, we discuss these patterns as purpose-driven and audience-oriented: addressing their expectations, their values, and their cognitive styles to accomplish your communication purpose.

CHAPTER 8

Outlining and Writing Informative Sections

Although many organizational reports must be structured so that they persuasively present the conclusions and recommendations of their writers, many other organizational reports must be structured primarily to present information rather than judgment. After all, it is information which sustains ongoing processes and assists in the implementation of changes within organizations.

For example, once managers have been convinced to alter a production procedure, line personnel must be instructed in how to perform that altered procedure. And there are others to inform about it as well. Vendors must be told of new specifications for materials. Maintenance personnel must be told how to sustain the new procedure. Supervisory personnel must be told how to monitor the effectiveness of the new procedure. And marketing personnel must be told about the consequences of the new procedure. Convincing the managers to change the procedure in the first place required a certain approach to the topic; that purpose suggested appropriate principles of selectivity and arrangement. Yet presenting information about the new procedure requires different approaches and suggests other appropriate principles of selectivity and arrangement.

But don't think of informative documents merely as alternatives to persuasive documents. Most organizational reports in fact combine persuasive and informative purposes. Even within dominantly persuasive documents, usually there are sections which are primarily explanatory or informative, as we suggested in Chapter 6 when we discussed the hierarchical structure of reports.

For example, you may recall the report on the Vendor Quality Index. Its overall purpose was to convince readers that a proposed procedure was better than an existing procedure and to get them to act to implement the proposed procedure. Yet within the dominantly persuasive structure of that report, there are sections which merely give information. One section explains how the current procedure is done. Another explains how the proposed procedure would be done. And yet another explains what specific actions would have to be taken to implement the proposed procedure. In other words, the report is primarily persuasive, but it also contains sections which are primarily informational. We find that to be the usual situation. Even if most of your writing as a professional is primarily persuasive, you must still be able to handle informative patterns as well as persuasive patterns.

Looking at a wide variety of technical and professional documents, we see four types of informative purposes as particularly frequent:

the writer wants to explain something by helping readers to see its constituent elements;
the writer wants to explain how something is done or should be done;
the writer wants to explain how a device functions;
or the writer wants to explain the procedure and outcomes of an investigation.

Certainly these are not all of the informative functions within technical reports, but they are frequent ones.

We identify these four frequently used informative patterns as follows:

Analysis
Process or Instructions
Description
Investigation

Unlike persuasive patterns, which have an underlying structural similarity, informative patterns are structurally dissimilar. Except for description, which combines elements of both analysis and process patterns, these four informative patterns can be treated as unique rather than as variations on a theme. For each pattern, as we did for persuasive patterns, we can suggest a simple outline that can serve as a helpful point of departure in planning what to include, what to leave out, and how the information might be arranged. Let us look at each pattern in turn.

❑ 8.1 ❑ The Analysis Pattern

The analysis pattern is used to explain a concept or thing by breaking it down into its constituent parts. This is a common pattern for professionals because much of their professional work involves the use of analytical methodologies. In fact, analysis perhaps is the most commonly used pattern within sections of reports, although it is not the most commonly used as a pattern for arranging entire reports.

When it is used to arrange an entire report—as it sometimes is—we have labeled this pattern "organization by subject matter." (Recall our discussion in Chapter 6, where we introduced three alternative high-level architectures for reports: organization by persuasive purpose, organization by subject matter, and organization by the procedure of the investigation.) When the pattern is used as an organizer for sections of a report, we use the term, "analysis," because the term represents the cognitive activity of breaking something down into its constituent parts, which you often do in sections of a report, whatever the report's overall organization and purpose.

As appropriate, the pattern consists of the following:

- Overview: Summary statement of the concept or object to be analyzed, with identification of its components.
- Analysis: Point-by-point presentation of the components of the whole, arranged in descending order of importance or in another purposeful sequence.
- Summary: Restatement of the concept or object, with an interpretation, if needed.

Begin with an overview of the concept or object to be analyzed. For a concept, this overview usually includes a statement of the purpose of the analysis, but the basic questions to answer are "What is this concept all about?" and "Why would you want to know about this?" For an object, this overview usually includes a statement of the object's purpose or function, perhaps as an introductory phrase or sentence. In technical reports, you usually analyze an object in order to explain its function. Therefore, the overview explains the function of the object, and then introduces the components as means by which the function is accomplished. The introduction of components also serves as a forecast of the pattern that follows.

Having defined the overall concept and having identified the constituent parts, then present and explain the constituents, item by item or point by point. This requires you to determine a principle of arrangement as well as a principle of selectivity. Descending order of importance might work, and if so that is often a good organizer in technical documents. Quite frequently, however, you will have to discover or invent some other purposeful order. Sometimes this order will be inherent in the subject matter. (For example,

"The Graphics Menu gives you four alternatives for showing your data: a Raw Data Plot [data against time]; a Fast Fourier Transform Plot; an ARMA Plot [autoregressive moving average plot]; and a Cross Index Plot. The following discussion explains each of these options." The menu itself dictates the order.)

Other times an appropriate order can be based upon assumptions about your readers, such as when you arrange things in an order of decreasing familiarity. (For example: "There are five laws applicable to the proposed dam construction: the first is probably familiar; the second and third might be familiar; but the fourth and fifth are almost certain to be new." Here the writer attempts to guess about what the readers know and organizes accordingly.) And still other times an appropriate order is a function of your overriding purposes in presenting an analysis at some particular stage of a report. (For example, "Plans for Network development include four types of initiatives: developing additional downlink sites; developing additional uplink sites; expanding programming; and expanding Board of Directors membership." Here [from a business plan for a television network] the sequence of initiatives includes a time element moving from near-term initiatives to more distant initiatives. However, the sequence also includes a strategic element—a matter of the writer's judgment about what network developments should occur first.)

You often conclude an analysis pattern with a summary and, depending on your purpose, with further interpretation or explanation. This is a matter of synthesizing a concept after you have explained its constituent points. Or, this is a matter of clarifying the function of an object, now that you have explained its parts.

As an example of an effectively arranged analysis pattern, examine the section on "Disk-based software" (Figure 8.1). The first paragraph introduces the software: there are five types of software. Then each type of software is discussed in the order forecast in the overview.

The explanation of the first four types of software starts with a paragraph stating the function of the software: what it does. MS-DOS provides, the SETUP utility program enables, the other utility programs perform, and TELCOM enables. The writer broke this pattern with the last item, "Disk-based documentation," which is treated as an analytic paragraph. It could have been introduced functionally with a statement such as, "The disk-based documentation provides information about ProSpeed 286 operation, MS-DOS, MS-DOS Shell, and TELCOM at your fingertips. This software includes four disk-based documents."

Two of the items are analytic units within this analytic section of the manual. The "SETUP utility program" and the "Other utility programs" items are analytically broken down into particulars. The overview paragraph provides a functional generalization that is followed by a list of tasks or programs. These two items thus are general-to-particular analytic units.

The report on "First Publication of the Slide Catalog" (Figure 8.2) also has an analytic pattern. The two items of the discussion, "The Slide Catalog" and "Tasks to Be Completed," are summarized. The first item, The Slide Catalog, has an overview paragraph; then it is broken down into two parts. The second item, Tasks to Be Completed, has an overview paragraph; then it explains the particulars of the tasks. Although the report has an instrumental purpose, it is arranged analytically.

The brief section on the water distribution system (Figure 8.3) again has a clear analytic pattern. With the referenced map, it provides the reader with an explanation of the three water transmission systems in the city. The introductory paragraph provides an overview and identifies the three systems. These then are explained in brief paragraphs with identical types of detail arranged in parallel patterns. Notice that the explanation is restricted to the relevant detail introduced by the pattern. The writer does not introduce tangential information, such as identifying proposed changes. These are discussed in other sections of the report.

The monthly report on "August 1990 Investment Portfolio Activity" (Figure 8.4) further illustrates analytic patterns in reports and sections of a report, regardless of subject matter. The discussion of the report has two items: current activity of the month and the portfolio review. The current activity is for the prior month; the portfolio review is for the prior year. The two current activity items are arranged from more to less amount of dollars. The portfolio review has a general-to-particular arrangement; it has comments on the portfolio as a whole and then on several items in the portfolio.

As you can see from these examples, the analytic pattern is a simple and frequent organizer. Yet it may even be deceptively simple in the sense that dividing a topic and dividing it logically are not always the same thing. For example, we remember a report sent to us by a former student several years after his graduation. On first glance, the report—a status report on a product—appeared to be a good example of use of a pattern, and we were feeling pretty smug about our former student's apparent ability—until we read the report carefully. There were six problems, each of the problems had a proposed solution, and for each of the six solutions there was a predicted time of completion. The writer therefore had used a simple problem–solution pattern as a basis for organizing the report. The outline of the report looked like this:

1. Problems Identified (sections 1.1–1.6)
2. Solutions Planned (sections 2.1–2.6)
3. Time to Completion (sections 3.1–3.6)

As you can see, the structure did have a kind of logic to it, but a logic based on subject matter (problem–solution) rather than on his purpose and his readers. Notice that to read this report one would have to remember from section to section what had been said before. That is, a reader would find out about a problem—for example, 1.3—in one part of the report. Later, in section

The memory-management software can configure the top 384 kilobytes of the standard 1-megabyte RAM as extended memory or expanded memory. Alternatively, if you add an optional memory expansion board to the ProSpeed 286, this software can configure this additional memory as extended memory, expanded memory, or both.

When the ProSpeed 286 is running the OS/2 operating system, extended memory enables you to use applicaiton programs that require more than 640 kilobytes. Some programs can use extended or expanded memory to process larger files or process them more quickly. You use the SETUP utility program to configure extended and expanded memory. See Chapter 4, "Setting the System Configuration Parameters", for information on configuring memory. See Appendix B, "Memory Maps" for information on the memory addresses.

Password protection

To prevent unauthorized use of your ProSpeed 286, the ROM-based software enables you to specify an optional password. After you set up a password for your system, you must enter the password before you can use the system. Also, to protect others from changing your password after you start your ProSpeed 286, you must reenter the password before you can change it. See Chapter 4, "Setting System Configuration Parameters" for instructions on how to set passwords.

Disk-based software

The disk-based software includes MS-DOS, the SETUP and general utility programs, the TELCOM telecommunications program and the disk-based documentation.

The ProSpeed 286 Hardware and Software

2-18

MS-DOS operating system

The MS-DOS operating system provides a software environment in which your application programs can operate. MS-DOS also manages the operation of the computer system hardware, including the display, memory, disk drives, and peripherals. See Chapter 5, "Using MS-DOS", for more information.

SETUP utility program

The SETUP utility program enables you to set the system parameters that configure various features of the ProSpeed 286. For example, some of the system parameters you can control through the SETUP utility program enable you to:

- Control how colors generated by application programs translate to shading levels on the black-and-white built-in screen.
- Change the CPU clock speed.
- Activate automatic battery-power-saving features that turn off the hard disk drive and the backlight for the display.
- Set up the RAM disk, disk cache, and expanded/extended memory features.
- Specify a password to protect your system from unauthorized access.

CAUTION: Although you can change the parameters at any time, the system automatically resets itself (clearing memory) when you change certain parameters. You should close all open files before changing these parameters. See Chapter 4, "Setting the System Configuration Parameters", for details on setting up system parameters.

2-19

FIGURE 8.1 **Analysis Section from a User's Manual.**

Other utility programs

The other utility programs included with the ProSpeed 286 perform a variety of functions. Some of these programs enable you to quickly make temporary changes to the system parameters. The remaining utility programs perform special functions. In addition to SETUP.COM, the ProSpeed 286 System disk includes the following utility programs:

- BL.COM enables you to conserve the battery power used by the backlight for the built-in screen. The ProSpeed 286 turns off the backlight if the screen doesn't change (in response to a program or a keystroke) within the time limit you set.
- CLICK.COM controls the volume of the key click.
- CRT.COM switches the display from the built-in screen to an external CRT monitor connected to the CRT port of the ProSpeed 286.
- DAA.COM enables a modem installed in the LTX slot to use the digital-access arrangement (DAA) circuitry of the optional ProSpeed 2400 bps Modem to perform communications through the telephone system.
- DCHARGE.COM enables you to discharge a Power-Block battery completely so that recharging the battery provides maximum operating time.
- EMULATE.COM enables you to configure the ProSpeed 286 for applications programs that specifically require an EGA, CGA, MDA, or Hercules display adapter.
- HDD.COM enables you to conserve the battery power used by the built-in hard disk drive motor. The ProSpeed 286 turns off the motor if the disk is not accessed within the time limit you set.

- LCD.COM switches the display back from an external CRT monitor to the built-in screen.
- PALETTE.COM enables you to prevent an application program from changing the shading levels you select to represent colors.
- SPEED.COM sets the clock speed of the ProSpeed 286 CPU to 16 megahertz (the default) or 8 megahertz.

TELCOM telecommunications program

The TELCOM telecommunications program and a modem (such as the optional ProSpeed 2400 bps Modem) enable you to use your computer to communicate with remote computers through the telephone system. For more information about TELCOM, see Chapter 6, "Using TELCOM".

Disk-based documentation

The software provided with the ProSpeed 286 includes two disk-based documents: the *MS-DOS Guide and Reference* manual and *TELCOM User's Guide*. After you install the documentation files on your hard disk, information about MS-DOS and TELCOM is always at your fingertips.

Optional Components

Many optional components are available for customizing the ProSpeed 286 to fit your needs.

IMS

Information Management Systems, Inc.
100 Computer Lane
Spring Rapids, Michigan 48042

DATE: May 3, 1990

To: Strategic Planning Consultants

From: Sandra Conrad
Graphics Coordinator

Subject: First Publication of the Slide Catalog

Attach: Worksheets and Examples Format

Introduction

As consultants, you spend a lot of your time developing innovative ways to convey information to your customers graphically. The exchange of these ideas within our group is encouraged to add to the quality of a study — a good idea developed in one study can be shared to enhance other studies. However, we currently do not have an efficient way of sharing this information. We need to gather and catalog existing ideas in a way that will make them available and useful to the group and we need to establish a process for making new ideas readily available to others. To this end, we are developing a Slide Catalog and would like to ask your help in putting the initial stages together. This memo describes our vision for the Slide Catalog and outlines the tasks to be completed for the initial stages.

Summary

The Slide Catalog will be a book of graphic concepts and ideas pertaining to our planning processes. It will help both new and experienced consultants to work more efficiently. It will be divided into two parts: stamp charts for each planning process cataloged by individual icons; and a miniature version (16 slides per page) of presentations representative of the planning process.

The attached worksheets should be completed as soon as possible. They will become part one of the Slide Catalog and serve as the basis for compilation of part two of the Catalog.

The Slide Catalog

The Slide Catalog will be a book of graphic concepts and ideas pertaining to the Information Technology Planning Process, the Strategic Systems Planning Process, and the Information Planning Process. This book will be a useful tool that will help the novice consultant to produce impressive presentations and the seasoned veteran to become much more efficient at his/her work.

The Slide Catalog will be divided into two parts. In part one, the stamp charts for each planning process will be cataloged by the individual icons so that you will be able to quickly identify the slides that are related to each step of the planning process. This will allow you immediate access to the icons that best suit your particular study, thus saving you time in searching through old files.

FIGURE 8.2 Project Memorandum Arranged Analytically.

Part two of the Slide Catalog will showcase a miniature version (i.e., 16 slides per page) of the presentations that are representative of the planning process so that you will be able to see entire presentations at a glance. This will allow you the convenience of comparing several presentations at once to see which presentations contain the information most appropriate for your current study. An example of this format is attached.

We will periodically update and republish the Slide Catalog when new concepts and ideas are developed. This will ensure that everyone will have access to the most current information.

Tasks To Be Completed

Attached is a series of worksheets that pertain to each icon of the planning process stamp charts. These worksheets, when completed, will become part one of the Slide Catalog. Part two of the Catalog will then be compiled based on the results of the worksheets.

To complete the worksheets for your studies, please go through your presentations and write down the file numbers of slides that should be cataloged under each process step icon. To eliminate unnecessary work, the studies conducted thus far have been assigned to the consultants who were the project managers for each particular study.

The following chart shows which studies you should look at:

Consultant	**I/T Planning Process**
James Whittaker	AEF Corp., Decker Computer Studies
Steve German	Murray Electronic Study
Consultant	**Strategic Systems Planning Process**
Dawn Michaels	Packer Automotive Study
Chris Swanson	Generic SSPP Study
Consultant	**Information Planning Process**
Jose Burnet	Westmine Computers Study

Please return the completed worksheets to me. Although we are not working against a deadline, a quick turnaround would ensure an early publication date.

The first publication of the Slide Catalog will be an intensive group effort; however, the results will be very beneficial. I will keep you updated on the project as we get closer to publication. Thank you for your help in putting this together. If you have any questions, please let me know.

The Present Distribution System

The present distribution system consists of an underground network of 4- to 12-inch water mains. The system consists of three major transmission subsystems: the Sherman Street System, the Park Street System, and the Division Street System (Map 1).

The Sherman Street System. The Sherman Street System, consisting of 10-inch cast iron pipe, begins at the pumping station and follows Sherman Street south to Carpenter and east on Carpenter to a point east of Bridge Street. This system provides an adequate water supply to the southwestern portion of the city.

The Park Street System. The Park Street System, also constructed of 10-inch cast iron pipe, begins at the pumping station and follows Park Street to the intersection of Park and Bridge Streets. This system serves as the feeder main to the existing elevated storage tank.

The Division Street System. This system, constructed in 1964, consists of 12-inch cast iron pipe. It serves as the major transmission main in the northern part of the city. When this system was constructed, numerous connections were made to existing 4- and 6-inch mains to provide better pressure distribution north of the River.

Two river crossings supply water to the north side of the city. One crossing is a 10-inch line at Bridge Street, and the other is a 6-inch line near the U.S. Coast Guard station.

FIGURE 8.3 Section with Analysis Pattern from Consultant's Report.

2.3, the reader would learn the solution. And still later, in Section 3.3, the reader would find out about the time for the solution's implementation. That structure forced readers to keep six problems, six solutions, and six calendars in mind while reading the report.

We think that the report would have been more effective if it had been organized analytically in six sections, each with subsections defining the problem, the planned solution, and the schedule for implementation. That is, the subsections would be organized according to simple problem–solution patterns, but the report's basic structure would have been analytic. The report after all was a status report on problems with the product, and so its basic structure should have consisted of a breakdown or list of the six problems.

Perhaps, then, a key question with the analysis pattern is not "what are the parts?" but rather "why are the parts logically presented in this order?" There should be a good answer to that question, and the reader should be able to figure out what the answer is.

❑ 8.2 ❑ The Process and Instructions Pattern

The process pattern is used to explain a sequence of events or a sequence of actions over time. In a sense, this pattern is the converse of an analysis pattern. A process pattern is dynamic; an analysis pattern is static. In a process pattern you focus on the interrelationships and changes among parts, elements, and states; in an analysis pattern you focus on the parts, elements, and states themselves. The purpose of explaining a process is to explain how the output of the process is achieved—the output is the end result of the process.

As appropriate, the pattern consists of the following:

- Objective: Objective of the process, introduction of the method or principle involved, and, especially, identification of the basic stages of the process.
- Background: Description of any parts, objects, or concepts necessary to explain the process and statement of any warnings or precautions that apply to any stage of the process.
- Procedure: Stage-by-stage or step-by-step explanation of the process, procedure, or instructions, presented in terms of the function of each stage.

Process patterns are especially difficult to present clearly wherever they are found—in an explanation of a chemical process or in a user's manual for a computer program. The difficulty usually is a result of immersing the readers in details without providing them with a clear idea of the overall pattern

NBIn

MEMORANDUM

NATIONAL BANK OF INDIANAPOLIS

October 11, 1990

TO: Frank E. Stewart, Vice-President, NBIn Wisconsin
John S. Turner, Vice-President, NBIn Indiana
Lee A. Carlson, Vice-President, NBIn Ohio

FROM: Frank B. Angelo, Municipal Trader, NBIn Corporation

SUBJECT: August 1990 Investment Portfolio Activity

SUMMARY OF ACTIVITY

Current Activity:

- Completion of the 110 million leverage program approved at the May Corporate ALCO.
- Sale of 60 million in Asset-Backed Securities for reinvestment in Agency-issued CMO product.

Portfolio Review:

- The on-balance sheet portfolio surpassed $4.0 billion for the first time. MBS represents about 63% of portfolio securities. 76% of this is in CMO/REMIC product.

CURRENT ACTIVITY

August saw the completion of the 110 million leverage program initially approved a the May Corporate ALCO. The 110 million in purchases done in the three months following the May ALCO were primarily in Agency-issued CMO product, except for approximately $28 million in seasoned 15-year FNMA and FHLMC paper described in last month's activity memo. In total, the purchases averaged 9.65% yield, with an average life of about 3.6 years. The maturity sensitivity of the purchases is not expected to shorten to less than 2.0 years or extend much beyond 7.0 years.

The other major activity during the month centered around the sale of just more than $60 million in Asset-Backed Securities (ABS), with reinvestment in Agency-issued CMO product. The swap was done for three reasons. First, CMO spreads to AMB spreads were relatively attractive. Second, we were able to pick up yield, 5-10 basis points,

FIGURE 8.4 Monthly Report Arranged Analytically.

without adding much maturity sensitivity risk, if any at all. And third, we were able to improve liquidity by selling a corporate-issued security and buying an Agency-issued security.

Other activity centered around some smaller relative value trades, primarily within the Government/Agency and MBS sectors of the portfolio. There was very little off-balance sheet activity in August.

PORTFOLIO REVIEW

I thought it important to highlight a few of the numbers in the attached Portfolio Review.

First, the size of the on-balance sheet portfolio surpassed $4.0 billion for the first time, primarily as a result of the $850 million leverage program and the shifting of other balance sheet assets (credit card loans and automobile receivables) into portfolio securities. MBS, at 2.5 billion, represent nearly 63% of portfolio securities, up dramatically from the $1.2 billion or 36% of the portfolio on 8-31-89.

Some characteristics of the MBS portfolio (excluding MBS used as money market alternatives): $2.2 billion, or 80% of total MBS, is in CMO/REMIC product. Of this $2.2 billion nearly 1.7 billion or 76% is in Agency-issued CMO/REMIC product. Of the $2.5 total MBS holdings, 2.2 billion, or nearly 90%, is Agency-issued. A recent study (September 9) by the First Detroit Corporation on the yield and maturity sensitivity characteristics on our MBS holdings indicated that in an unchanged interest rate scenario the average yield would be 9.,37% and the average life 3.14 years. In a +300 basis point yield environment, the yield would improve slightly to 9.44% and the average life would extend to 4.89 years. In a -300 basis point yield environment, the yield would drop to 9.23% and the average life would shorten to 1.50 years. All in all, we feel this is about as stable an MBS portfolio in terms of yield and average life as we could have, given we do not concentrate on buying PAC tranches.

Treasury and non-MBS Agency securities total $199 million or 5% of the total portfolio, down from their peak of $730 million (24% of the portfolio) on 4-30-89.

ABS total just in excess of $120 million, or less than 3% of the total portfolio, down from their peak of $207 million, or 6% of the total portfolio.

Should you have any questions about what has been presented or the numbers attached, please don't hesitate to give me a call at (313) 764-1426.

Attachments

FBA:ef
BO/SC035

behind the details. The result is that the details overwhelm the reader. We have three suggestions to help you avoid this problem:

- First, write in terms of function or output of the process. Your readers will understand a process or a set of instructions much more quickly if they know what the result is supposed to be than if they are merely told to do a series of acts in a particular sequence. They need to know the purpose of the process or why they are doing the acts.

- Second, provide the readers with a clear statement of the basic stages or steps of the process. That is, instead of presenting a process with eleven steps, one through eleven, for example, present the reader—if you can—with a 3-stage process or a 4-stage process. Impose groupings and higher-level structure on the discrete steps of a process. (Obviously this suggestion applies to processes with numerous steps. As a rule of thumb, if the process has more than six or seven or eight discrete steps, look for possible ways to group the steps into larger stages. This keeps the reader from becoming lost in the details.)

- Third, within each stage (and at times within each step) repeat the process pattern or present a truncated version of the pattern. That is, each stage or step should have its own objective, background, and procedure subpattern. This enables the reader to sort out the details within an immediate framework rather than to interpret them in terms of the process as a whole.

An effectively written manual or set of instructions often presents a hierarchy of process patterns. Such a hierarchy serves two functions. It helps the readers to sort out the details of the process, procedures, or operations and to establish relationships among them. Further, it enables readers to refer to a discrete stage or step of the process and repeat the procedure without having to read more than necessary. In other words, it serves a reference function in explanations which will be used repetitively.

Begin with a statement of the objective or output of the process. This often includes a statement of the method or principle behind the process. Also, it should include an identification of the basic stages of the process or steps of the instructions. This helps clarify the objective and the method and provide a forecast of the details that follow. As mentioned above, a forecast enables the reader to keep the overall process in mind while still being immersed in the details of any stage or step.

Next, provide any background necessary for the reader to understand the process. This includes a description of any parts, objects, or concepts necessary to understand a stage or event of the process. With instructions, it includes any tools or equipment that will be required. Also, it includes any warnings and precautions which must be taken.

The purpose of providing a background unit is to avoid having to interrupt your explanation of the process once you have started. Because you are focusing on sequence, any interruption can cause readers to lose track of where they are. With instructions, any interruption can cause delay or even confusion. (For example, suppose you are replacing a valve in the humidifier on your furnace. You know which valve needs replacement, and you have already gone to the hardware store and bought it. You disassemble the humidifier, keeping the parts more or less carefully arranged. Everything is going along fine until you get to the actual installation of the new valve; then the instructions tell you that you need a special kind of mini-wrench, which of course you don't have. Three hours later, after you finally have found a hardware store which carries the wrench, you no longer see humidifier parts arranged logically in front of you in the order in which you disassembled them. Instead, you see a random collection of parts, a jigsaw puzzle. Time to call a repair person.)

A background unit is especially important when it comes to warnings. These need to be up-front in a process section, not just put in the middle of the instructions where they actually might apply (although they must be there as well). If the main circuit breaker needs to be turned off for a crucial operation within a procedure, your instructions need to inform the reader of that fact well ahead of time—and again at the specific time of that operation. Otherwise, the reader—already well into the operation—might find that he or she must turn off the main circuit breaker at an inconvenient or impossible time—after dark or during the hour or so before dinner when it would be inconvenient to have the electricity off. After starting the process of following a set of instructions, he or she might not be able conveniently to just wait until tomorrow or to retrace the steps back to the starting point.

Having presented the objective, the background, and the forecast, then present the stage-by-stage or step-by-step explanation of the process, procedure, or instructions. This usually requires at least a 2-level structure. The process is divided into stages, which in turn are divided into substages. Then, each stage is explained separately.

Each stage and substage must be clearly delineated, set off by itself so that the reader can grasp it as a discrete event or action. Often this means treating each stage or substage as an independent process pattern. That is, state the objective of the stage or substage, set up the background if necessary, and then divide the stage or substage into its component steps.

As an example of an extensive process pattern, examine the section on "The Experimental Test Process" (Figure 8.5). The first subsection (3.5) states the objectives of the test and provides an overview of the entire test process. It also includes some background information.

The first subsection (3.5.1) explains the test setup and the test procedure. Then the next two subsections (3.5.2 and 3.5.3) explain the two stages of the actual tests themselves. Notice that each subsection has its own functional overview. Thus, the section has a hierarchy of process patterns.

Particularly important to the effectiveness of the section is the segmentation of material. Background and explanatory details are separated from the

3.5. The Experimental Test Process.

To test how reliable use of instrument by an operator is, two questions must be answered:

(1) Given a well-defined peak in the output curve, will different operators of the instrument perceive the same mode?

(2) Will the perceived differences in the exposed and nonexposed tremor modes be significant enough to allow an operator to differentiate between these two groups of employees?

To answer these questions a two-stage experimental test process was devised. The process involved four test subjects as potential operators of the instrument. These subjects read and interpreted data supplied to the instrument by a magnetic tape recording. Before providing the data input, the tape recording supplied the subject-operator with instructions for operating the instrument. The entire experimental test process took about one hour with each subject-operator.

3.5.1. Testing the Operation of the Instrument.

The magnetic tape recording first provided operating instructions and then supplied seventeen sets of data to the instrument. These seventeen sets of data were of two types: oscillator signals and employee tremor signals. These signal types formed the two-stage experimental process needed to answer the two questions.

The tape recording first explained the purpose of the experiment and told the subject how to operate the instrument. After this introduction, the tape was stopped to allow the subject to ask questions to make certain he knew exactly what to do. Then the tape presented a sample set of data for the subject to interpret on the instrument meter. If the subject used the instrument to interpret the data correctly, the tape then proceeded to supply the seventeen sets of data for interpretation. (If the subject had difficulty interpreting the data correctly, the tape was rerun as often as necessary for the subject to learn how to use the instrument correctly.)

The tape then supplied seventeen sets of data for the subject to interpret from the instrument meter. Each data set contained ninety seconds of signal input followed by forty-five seconds of rest period to reduce the chance of mental fatigue in the subject-operator. The subject-operator had to analyze each data set as follows:

(1) State the signal frequency he judged to cause the maximum deflection on the meter;

(2) State the degree of certainty (1% to 100%) he felt about his judgment;

(3) Comment briefly on any distinctive characteristics of the data set.

The time required for the subjects to state judgments with a high degree of certainty was monitored to determine the amount of data required for an operator to test an employee for exposure to mercury poisoning.

FIGURE 8.5 Section on Test Procedure with Process Pattern.

3.5.2. Stage One of the Experiment.

This stage addressed the question: Given a well-defined peak in the output curve, will different operators of the instrument perceive the same mode? This stage involved the first six and the seventeenth sets of the subjects tested.

The first six data sets were put on tape from an oscillator. These oscillator signals were used to determine how reliable and reproducible the output of this instrument was when interpreted by different operators. The signal-to-noise ratio instrument was then interpreted by different operators. The signal-to-noise ratio (S/N ratio) was varied from 100% signal to 50% signal-50% background noise. These S/N ratios were used because they span the range of S/N ratios on actual tremor recordings. The performance of the operators was monitored as the S/N ratio increased.

The seventeenth data set also came from the oscillator. This set was used to determine the increase in skill of the subject-operator after he had analyzed sixteen sets of data.

3.5.3. Stage Two of the Experiment

This stage addressed the question: Will the perceived differences in the exposed and nonexposed tremor modes be significant enough to allow an operator to differentiate between these two groups of employees? This stage involved the seventh through sixteenth sets of data. These sets were employee tremor signals.

These ten data sets formed a random sequence of tremor responses from five nonexposed employees and from five mercury-exposed employees (the subject-operators, of course, did not know ahead of time which data sets were which). These sets of signals had been previously obtained in the original Institute study of mercury poisoning. They were from employees matched with respect to such characteristics as age and weight. The ability of the operators to distinguish between signals from exposed employees and from nonexposed employees was evaluated.

To help further evaluate the four subject-operators' analyses of these data sets in stage two of the experiment, their judgments on differences in signal characteristics were compared to differences in signal characteristics as determined by computer analysis of the original signals.

actions, which usually are presented in functional terms. The basic arrangement is based on the segmentation, not on the sequence of activities. Sequence is used to order the information in the subsections.

The instructions for Crossword Bingo (Figure 8.6) follow a simple process pattern, which serves to outline this entire "report." The introductory paragraphs state the objective of the game and the method and type of game it is. These paragraphs essentially summarize the game. The next two items, Number of Players and Equipment, specify the players and the equipment, which essentially form the background section of the pattern. Then the Object, Set-Up, and Play items present the actual procedure for playing the game: first the Set-Up, then the Play. (Note that in the actual instructions the objective is reintroduced.) The play consists of "rounds," each of which consists of the 3 steps listed in order.

After the 3-step procedure, three items (The Arrow Piles, When No Word Is Made, and New Face-Down Pile) are presented before the final item, the result of the game: The Winner. The three items following the procedure are very specific details that describe subordinated events within the Play procedure itself. The writers, however, wisely separated these three items from that procedure so as not to confuse the reader, who is trying to follow the basic procedure. If these three items had been integrated with the procedure, they almost certainly would have interrupted the reader's train of thought. (These three items, of course, could have been included in the "Background" section. However, the reader needs to read the Play section in order to understand them.)

Two sections from a computer user's manual illustrate adaptions of the process pattern to segment and present operations clearly. "The [Exit] command" (Figure 8.7) follows the process pattern, but the section contains considerable material before it gets to the actual instructions at the end. The first paragraph presents a functional overview of the program; it can either incorporate or discard the SETUP parameters the user has specified. Then the section has two "CAUTION" units. These are the background step of the process pattern, but this step is emphasized here so that the user does not lose files and data. In terms of your reader, warnings and precautions can be more important than actual instructions, especially when the instructions are simple. You adapt the process pattern to this context. The actual instructions consist of four simple steps at the end.

"The [File] command" (Figure 8.8) from the same manual omits the background step of the process pattern. The overview simply introduces the four operations the user can perform from the file menu: display, erase, rename, and display. The steps to do each of these then are presented in that order. The four discrete operations are clearly separated by the formatting. "The [Guide] command" which follows has the same clear and simple presentation. The sections of this manual are clearly segmented (with an analytic pattern). Each section is presented in a general-to-particular framework with a functional overview. Then the parts of each section are clearly segmented and formatted.

It might be worth mentioning here that many procedural explanations are written according to detailed specifications of content and arrangement which go considerably beyond the preceding simple architecture which we have described. They are written this way because individual procedures are parts of large procedural collections; if there were no specifications for content, they would inevitably be dissimilar, a significant problem for an organization in which there are numerous procedures and in which a written description of those procedures is required for quality control and legal purposes. For example, the National Committee for Clinical Laboratory Standards (NCCLS)[1] stipulates that procedural explanations used to document procedures in hospital laboratories follow this outline:

Title
Method as Subtitle

I. Principle

II. Specimen
   A. Patient Preparation
   B. Type Specimen
   C. Unacceptable Specimens
   D. Storage of Specimens
   E. Rejection Criteria

III. Reagents, Controls, Special Supplies, Equipment
   A. Reagents and Controls
   B. Special Supplies and Equipment

IV. Calibration

V. Quality Control

VI. Procedure

VII. Calculations

VIII. Reporting Results
   A. Reference Ranges
   B. Sensitivities
   C. Results

IX. Procedure Notes

X. Limitations of Procedure

XI. References

XII. Supplemental Materials

(Figure 8.9 shows one page from a procedure written according to this outline.)

RANDOM GAMES & TOYS

416 W. HURON / ANN ARBOR, MI 48103 / PHONE (313)-761-9140

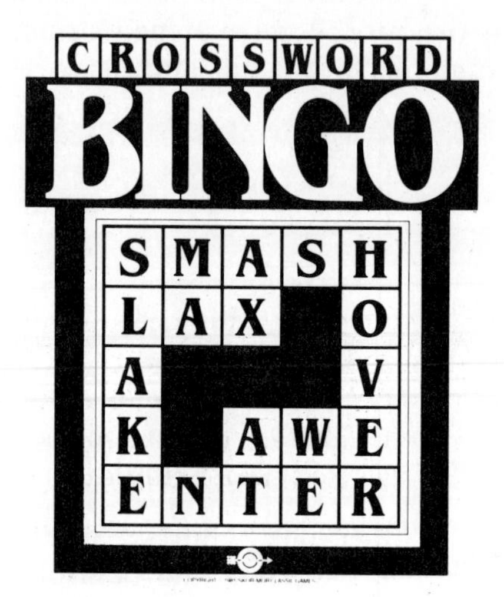

"CROSSWORD BINGO" GAME

Two of America's favorite pastimes were the inspiration for this challenging new game of wordplay. That's what makes "CROSSWORD BINGO" twice as exciting and double the fun!

When the timer starts, everyone plays. The first player to cover all the letters on his or her card is the winner. But, there's a twist! Words must be formed before tiles can be placed on the card.

"CROSSWORD BINGO" truly features family fun. The child has an equal chance against the adult, because the game has 12 cards for adults and 12 easier cards for children. Two to four players, ages eight and up, can get in on the action, and the typical game lasts 30 minutes.

FIGURE 8.6 Instructions for a Game Follow a Process Pattern.

CROSSWORD BINGO

NUMBER OF PLAYERS: Two to four.

EQUIPMENT: 24 CROSSWORD BINGO cards (12 green cards for adults and 12 yellow cards for children), one 60-second timer, and 144 letter tiles, as follows: 12 each of A, E; 11 each of I, O; 6 each of D, N, R, S, T, U; 4 each of B, G, H, L, M, P, W, Y; 3 each of C, F, J, K, Q, V, X, Z; and six Arrow tiles which may be used as any letter.

OBJECT: The first player to cover all of the letters on his or her CROSSWORD BINGO card is the winner.

SET-UP: Each player chooses a CROSSWORD BINGO card, green for adults or yellow for children. The letter tiles are placed face down in the center play area and mixed. Each player draws one tile and places it face up, to form a face-up pile near the face-down pile. The player with the highest draw ("Arrow" is highest, then A, B, etc.) is the keeper of the timer for the first round of play. In case of a tie for highest draw, the players in the tie draw again, adding more letters to the face-up pile.

PLAY: Play proceeds in rounds. Each round is made of three steps as follows:

STEP 1: Beginning with the keeper of the timer and proceeding to the left, each player in turn may, if desired, take one tile from the face-up pile, and then must take enough tiles from the face-down pile to total seven tiles. It is important that players do not turn up their face-down tiles at this time.

STEP 2: When all players have their seven tiles, the keeper of the timer says, "GO", and starts the timer. Each player immediately turns up his or her face-down tiles and has 60 seconds to make one word out of two or more of the seven tiles. The word can be any standard dictionary word and is not restricted to just the words on the player's cards.

STEP 3: When the timer runs out, each player, starting with the keeper of the timer and proceeding to the left, announces his or her word to see if it is accepted by all as a legitimate word. If so, the player places any letters in the word on the matching but not yet covered letter spaces of his or her card. For example, the player makes the word PIG and has not yet covered "I" and "G" spaces on the card. The player places the "I" and "G" from PIG on the "I" and "G" spaces on the card.

The player next looks at all of his or her remaining tiles (from the word or otherwise) and discards to the face-up pile and tiles that the player does not want to keep. The player should keep for the next round any remaining letters that match letter spaces not yet covered on the card. For example, the player may have a letter "T" left over and needed for his or her card. In the next round, if the player can make a word using the "T", the player will be able to add the "T" to the card.

Upon completion of the Three Steps, the keeper of the timer passes it to the player on his or her left. This player becomes the new keeper of the timer and starts a new round of play, with the drawing of more tiles, and so on.

THE ARROW TILES: An Arrow tile may be used as any letter. However, if an Arrow tile is used as a certain letter in a player's word, it can only be used to cover the same certain letter on the player's card. SPECIAL RULE: If an Arrow tile is used to spell a word, but cannot be placed on the card, it must be discarded to the face-up pile.

WHEN NO WORD IS MADE: When the timer runs out, if a player could not make a valid word or chose not to make one, then the player may discard any number of tiles he or she wishes and keep the rest for the next round.

NEW FACE-DOWN PILE: If the face-down pile becomes used up, the face-up pile is turned face down and mixed to become the new face-down pile. However, before the face-up pile is turned over, each player is allowed to choose one face-up tile for the next round of play.

THE WINNER: Rounds of play continue until a player wins the game by covering all of the letters on his or her card. If two or more players cover their cards in the same round, then the player who made the longest word in that final round is declared the winner. If two or more players are still tied, then they are declared co-winners.

3. Select the type of your external floppy disk drive, according to the size and capacity of the disks it accepts, and press <Enter>.

Reversing the floppy disk drive names

1. Select [Drive]. A menu appears.
2. Select [Drive A:]. A menu of disk drives appears.
3. Select the floppy disk drive you want to designate as drive A, either [Built-in] or [External], and press <Enter>. The other drive becomes drive B.

The [Exit] command

The [Exit] command exits from the SETUP utility program and removes the program from memory. The [Exit] command lets you make permanent the SETUP parameters shown on the SETUP main screen, or lets you discard the changes.

CAUTION: If you used any of the following commands to change parameters, selecting the [Exit] command and pressing <Y> to answer the "Update the system parameters? (Y or N):" prompt causes the system to reset, clearing the memory of its contents (including all memory-resident programs and RAM disk files).

- [Drive]
- [Memory]
- [Port]
- [File] (only if you changed any parameters of the [Drive], [Memory], or [Port] commands)

Setting the System Configuration Parameters

4-14

- SETUP /R or SETUP /L followed by a file name at the MS-DOS prompt (only if the named file changes any parameters of the [Drive], [Memory], or [Port] commands)

For each of these commands, the system displays a warning about the potential loss of data.

CAUTION: Data loss may occur in the following situations:

- If you are using a RAM disk, a system reset erases memory-resident programs and the data they contain, and data stored on the RAM disk.
- If you have loaded SETUP as a memory-resident program and then loaded other memory-resident programs, using the [Exit] command to terminate memory-resident SETUP erases these programs and any data they contain. (It also aborts any application program you are using.) However, if you load SETUP as the last memory-resident program, using the [Exit] command doesn't erase the previously loaded memory-resident programs.

To use the [Exit] command:

1. Select [Exit]. The "Update the system parameters? (Y or N):" prompt appears.
2. Press <Y> to update the SETUP parameters (and possibly reset the system); otherwise press <N> to leave the parameters unchanged.
3. Press <Enter>.
4. If you choose [Exit] to terminate memory-resident SETUP, SETUP displays a message warning you that this command will remove memory-resident SETUP from memory, and abort the application program you were using when you activated SETUP. SETUP then asks you to confirm that you want to continue.

Setting the System Configuration Parameters

4-15

FIGURE 8.7 Section from User's Manual Emphasizes Precautions.

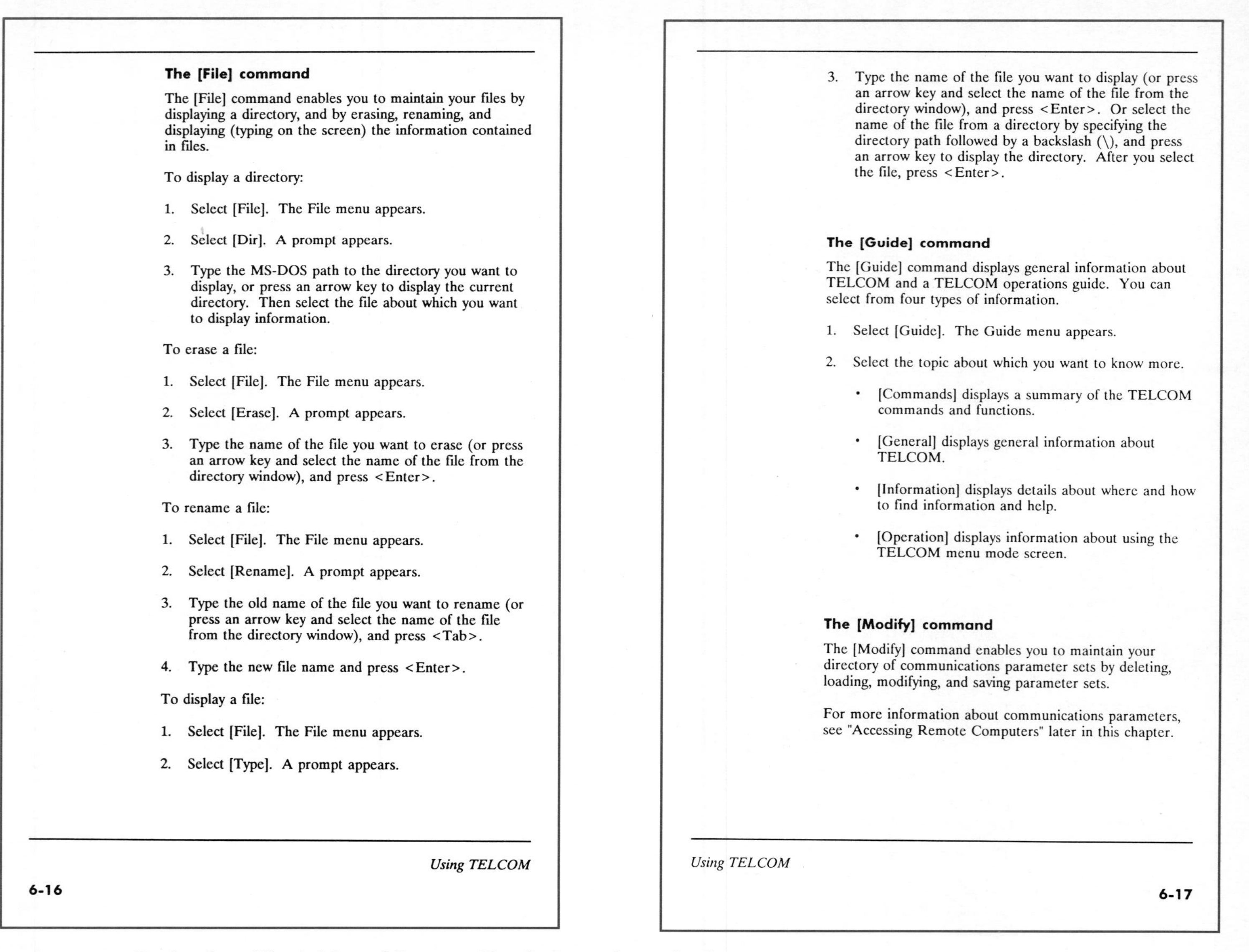

The [File] command

The [File] command enables you to maintain your files by displaying a directory, and by erasing, renaming, and displaying (typing on the screen) the information contained in files.

To display a directory:

1. Select [File]. The File menu appears.
2. Select [Dir]. A prompt appears.
3. Type the MS-DOS path to the directory you want to display, or press an arrow key to display the current directory. Then select the file about which you want to display information.

To erase a file:

1. Select [File]. The File menu appears.
2. Select [Erase]. A prompt appears.
3. Type the name of the file you want to erase (or press an arrow key and select the name of the file from the directory window), and press <Enter>.

To rename a file:

1. Select [File]. The File menu appears.
2. Select [Rename]. A prompt appears.
3. Type the old name of the file you want to rename (or press an arrow key and select the name of the file from the directory window), and press <Tab>.
4. Type the new file name and press <Enter>.

To display a file:

1. Select [File]. The File menu appears.
2. Select [Type]. A prompt appears.

Using TELCOM

6-16

3. Type the name of the file you want to display (or press an arrow key and select the name of the file from the directory window), and press <Enter>. Or select the name of the file from a directory by specifying the directory path followed by a backslash (\), and press an arrow key to display the directory. After you select the file, press <Enter>.

The [Guide] command

The [Guide] command displays general information about TELCOM and a TELCOM operations guide. You can select from four types of information.

1. Select [Guide]. The Guide menu appears.
2. Select the topic about which you want to know more.
   - [Commands] displays a summary of the TELCOM commands and functions.
   - [General] displays general information about TELCOM.
   - [Information] displays details about where and how to find information and help.
   - [Operation] displays information about using the TELCOM menu mode screen.

The [Modify] command

The [Modify] command enables you to maintain your directory of communications parameter sets by deleting, loading, modifying, and saving parameter sets.

For more information about communications parameters, see "Accessing Remote Computers" later in this chapter.

Using TELCOM

6-17

FIGURE 8.8 Section from User's Manual Presents Simple Operations Clearly.

University of Michigan Hospitals
Clinical Immunopathology Laboratory

Immunoglobulin Quantitation (IgG, IgA, IgM, Kappa, Lambda) Beckman Array

I. Principle

The Beckman Array uses a rate nephelometric system. The method measures the rate of increase in light scattered from particles suspended in solution as a result of complexes formed during an antigen-antibody reaction.

In the performance of the following immunoglobulin testing, IgG, IgA, IgM, Kappa, and Lambda, each specific antibody is added to the patient sample. The increase in light scatter resulting from the antigen-antibody reaction is converted to a peak rate signal which is a function of the sample concentration. Following calibration, the peak rate signal for a particular assay is automatically converted to concentration units by the analyzer.

II. Specimen

A. Patient Preparation
No special preparation of patient. Patients suspected of having macroglobulins or cryoglobulins should have specimens drawn and held at 37 degrees C. Samples suspected of containing cold agglutinins should not be refrigerated prior to serum separation from the clot.

B. Type Specimen
Blood, with no anticoagulant added. Minimum volume is 0.5 ml of serum.

C. Unacceptable Specimens
Fluids other than serum and specimens collected in Corvac tube.

D. Storage
The specimen is allowed to clot for 4 hours, then is spun at 2000 rpm for 10 minutes, and the serum is separated from the red cells. Serum may be maintained at 4 degrees C for 72 hours. Sera not assayed within this time should be frozen at -20 degrees C.

E. Rejection Criteria
Specimens with incorrect or insufficient identification.

III. Reagents, Controls, Special Supplies, Equipment

A. Reagents and Controls
IgG Antibody, Beckman (#449400): Human IgG antibody with 0.1% (w/v) sodium azide. Store at 2-8 deg. C.

FIGURE 8.9 First Page from Hospital Laboratory Procedure Written According to the Specifications of the National Committee for Clinical Laboratory Standards (NCCLS).

On the surface, this structure might look significantly different from the preceding, simpler principles we have suggested. (Note that only one of the twelve elements in the outline is identified as "Procedure.") However, if you look more closely, you can see that most of what we have suggested fits the NCCLS format very well. You can see that the NCCLS format stipulates that a procedure always begin with an overview of the principle lying behind the procedure. (The example in Figure 8.9 nicely explains the "why" and "how" lying behind the actual procedure.) It further stipulates that preliminary matters of test specimen, supplies, and equipment should always be explained before the procedural steps are explained. (Again, the sample in Figure 8.9 illustrates this principle.) And finally the NCCLS format stipulates that a process is actually a series of subprocesses in which calibrating something, although part of a process, is also a separate process with stages of its own. Similarly, reporting results is part of a process, yet the reporting process itself also consists of necessary steps and topics. About the only element missing in the NCCLS format seems to be a forecast of the structure to be followed. However, because the overall format is a matter of a well-established specification within hospital laboratories, perhaps no such forecast is necessary on the overall level. Within the details of Section VI, "Procedure," however, it would be quite a simple matter to write-in such a forecast—as indeed writers of these procedures often do.

As the NCCLS format suggests, the process pattern is similar to the analysis pattern in fundamental respects. It has a clearly segmented general-to-particular structure, with steps of a process clearly delineated just as the parts of an object are. It includes functional overviews and, when appropriate, provides a forecast of the steps to be followed.

❑ 8.3 ❑ The Description Pattern

The description pattern is used to explain how a physical device functions. An effective descriptive pattern essentially combines an analysis pattern with a process pattern, adding a graphic aid such as a circuit diagram, a cutaway sketch, a flow chart, a schematic drawing, or a photo. A descriptive pattern should not be difficult to write, especially if the graphic is well used to illustrate the parts and operations of the object. In fact, to a very large degree, good technical description depends upon good graphic representation.

As appropriate, the pattern consists of the following:

- Overview: Overview of the object's function.
- Description: Physical description of the object, with a simplified visual aid.
- Explanation: Functional description of the object: a part-by-part explanation of how the object works.

Begin with an overview of the object's function. As with a process pattern, in a descriptive pattern the functional overview is extremely important. Readers can often intuitively understand an object which they have never seen before if they are merely told what the object does, what its function is. In technical reports, you almost invariably use functional description. That is, you aren't so much interested in explaining what an object looks like as how it functions. You explain the object in order to explain how it accomplishes some task. Thus, the function, use, or purpose of the object is the essence of the overview, the first sentence or paragraph of the description. At times, as with a process pattern, the overview includes a statement of the theory or principle of operation or design of the object.

Next, provide a verbal description of the object and its parts, and illustrate these with a simplified graphic. The graphic identifies all of the parts of the object that the reader must be familiar with in order to follow the explanation of the operation of the object. (When we say that you should use a simplified graphic, we mean that you should show only the detail necessary to explain how the object works. Parts or details not necessary to understand how a device operates should not be included in the graphic because they can distract the reader's attention from the significant detail.)

The verbal description of the object follows an analysis pattern. You itemize the parts one by one in a functional sequence, that is, the sequence of their operation.

Finally, explain how the object functions by presenting a part-by-part explanation of its operation. Unfortunately, that isn't always as easy as it sounds. Some devices function in serial fashion, one action after another, while others function in parallel fashion, with several actions going on concurrently. Serial action is easier to describe than parallel action, particularly integrated parallel action in which there is interaction among otherwise separate functions. Again, however, as with the analysis pattern, the issue of what sequence to follow is a critical factor for you to consider. If the device functions in serial fashion, you probably have no real difficulty. If it functions in parallel, you may want to impose groupings upon functional descriptions—as we discussed when we were talking about creating stages within processes consisting of numerous steps.

The description pattern, then, is similar to a process pattern. Notice that the second element of the pattern, the physical description of the object, is analogous to the background element of the process pattern. Then the part-by-part explanation that follows explains the sequence of interactions of the parts, so the focus is on the operation rather than on the parts themselves.

As an example of an effectively arranged description pattern, examine the section on "The Separation System" (Figure 8.10). It opens with a statement of the function of the system and then gives an overview of how it accomplishes its function.

Then a physical description of the system is presented, with a visual aid that identifies the important parts of the system. These parts are then explained one by one in functional terms. This subsection has an analytic pattern.

The Separation System

The separation system provides for the separation of the spent stage of the launch vehicle from the satellite. This is done by firing explosive bolts to sever the steel bands which hold the satellite to the fourth stage. The explosion severs the bands and allows springs to push the satellite and the fourth stage apart with sufficient force so that the two cannot collide.

Here is a cross-sectional view of the system:

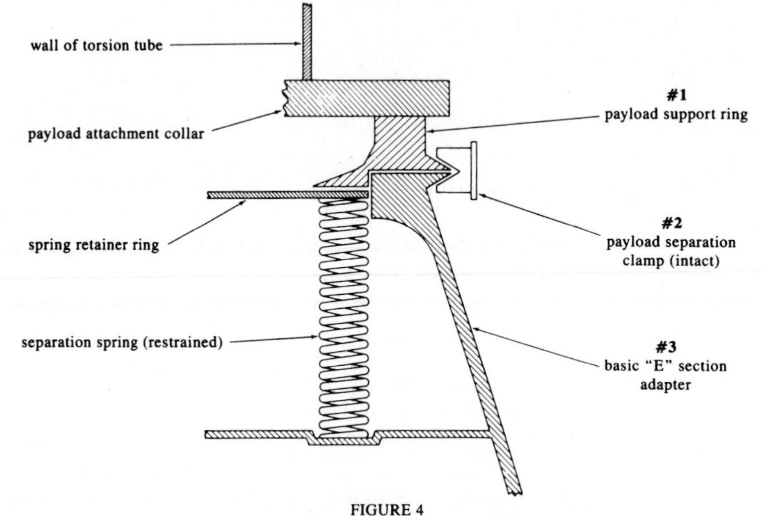

FIGURE 4
Three Basic Parts of Separation System

The three basic parts of the "*E*" payload separation system chosen for Project SCOPE are:

1. The *payload support ring*, which mates the top of the fourth stage to the bottom of the satellite.

2. The *payload separation clamp*, steel bands with V-block which hold the section adapter (3) and the payload support ring (1) together prior to separation.

3. The "*E*" *section adapter*, a conical magnesium structure in which are mounted the springs which supply the force for the separation.

Once the payload has reached the desired orbit, the explosive bolts attached to the steel bands of the payload separation clamp are fired. This severs the bands and allows them to drift harmlessly off into space. As soon as the bands are severed, the previously restrained springs in the section adapter force the satellite and the fourth stage apart. Once the two are sufficiently separated, the springs are also completely free and drift off into space. The separation is completed.

The severed parts of the bands, the springs, and the fourth stage should be quickly clear of the satellite. When they are, the solar paddles and antennas can be opened . . . .

FIGURE 8.10 Section with Description Pattern.

Finally, an explanation of how the parts interact to perform the object's function is presented. This is an abbreviated process subsection, with the overview having been provided in the first paragraph of the section itself.

The explanation of how the Front-end Cross-correlator functions (Figure 8.11) illustrates the importance of the use of a figure. Without the figure, we doubt that even an optics specialist would be able to understand how the object functions. The first paragraph provides an overview of the object's function. The second paragraph then introduces the three subsections of the object and summarizes its operation. The following three paragraphs explain both the components of the subsections and their operation in detail. The three paragraphs have parallel patterns: each starts with the components of the subsection, then explains its operation.

The descriptive section in Figure 8.11 contrasts with that in the report on the Maritime Venture (Figure 8.12), which lacks a figure. This report was sent to a congressman, who had requested it so that he could publicize his success at bringing federal funds into his district (the construction of the ship was subsidized by the government). Despite the nontechnical audience, the writer immediately buries the reader in excessive detail without providing a functional overview of the ship. A graphic is needed to enable the reader to identify the parts of the ship and sort out the details. The important points the writer wants to emphasize are obscured by the detail.

If the writer had followed the descriptive pattern and had arranged the information in decreasing order of importance, he would have communicated effectively with his audience. The tabular information at the beginning of the report would have come at the end. The final paragraph (with an additional sentence stating the function of the ship) might have come at the beginning to identify the important characteristics of the ship, which would have been illustrated and then explained in detail in a revised order of decreasing importance.

❑ 8.4 ❑
The Investigation Pattern

The investigation pattern is used to explain how an investigation was conducted and judgments formulated. We reintroduce the investigation pattern here because it is used to organize information in short reports and sections of reports as well as the discussion of long reports as a whole. In a section of a report, however, the pattern is not as complete as it usually is when used to arrange an entire report. That is, in a section of a report the pattern often is abbreviated, with some steps, such as "Background," omitted.

As appropriate, the pattern consists of the following:

- Objective: Statement of the purpose of the investigation.
- Background, if necessary: Explanation of the specifications, criteria, or technical background.

Front-end Cross-correlator Overview

The Front-end Cross-correlator (FECC) of the Real Time Image Analysis System functions as a preprocessor for image analysis. It accomplishes this by optically cross-correlating any incoming electrical signal with any second electrical signal. This results in an array of correlation spots that can be analyzed by the Back-end.

The FECC is composed of three subsections, and operates in two alternating modes (see Diagram #1, *Front-end Cross-correlator Design*). During both modes the first subsection converts light from a pulsed laser source into the form necessary for carrying the signals. During the first mode the second subsection converts the two electrical signals into optical signals, takes their cross-correlation's Fourier transform, and stores the transform. During the second mode it emits the Fourier transform. During the first mode the third subsection is not used. In the second mode it takes the inverse transform of the cross-correlation's Fourier transform, producing cross-correlation spots that are analyzed by the Back-end.

The first subsection is composed of a pulsed laser source, a collimating lens, and two gratings. The source and lens together produce a collimated beam of coherent light. The first grating splits this beam in two. The second grating redirects the two beams into the second subsection, where the lower beam carries the incoming signal and the upper beam carries the second signal.

The second subsection is composed of a double-read/write eidaphor, a transform lens, and a pneumatic liquid crystal. During the first mode the eidaphor converts both the electrical incoming signal and the electrical second signal into two optical signals, which are carried on the lower and upper beams, respectively. The lens then takes the Fourier transform of each signal. The two beams meet again at the light sensitive pneumatic crystal, which records the interference pattern of the two transforms. This pattern contains the Fourier transform of the cross-correlation of the two signals. During the second mode the eidaphor shuts off the upper beam, and causes the lower beam to diverge. The lens collimates the lower beam, illuminating the crystal, which then emits the cross-correlation's transform.

The third subsection is composed of single lens. During the second mode, when the cross-correlation's transform is produced, this lens takes the inverse transform, producing the correlation spots of the two cross-correlated signals. These spots are formed at the detector array plane of the Back-end.

Once the Front-end is connected to the Back-end, and the source is turned on, the system will be operational. Care must be taken, however, to stay within the bandwidth tolerance limits; otherwise, the output of the Front-end, and thus of the entire RTIAS, will not be accurate.

FIGURE 8.11 Description Section from a Design Report.

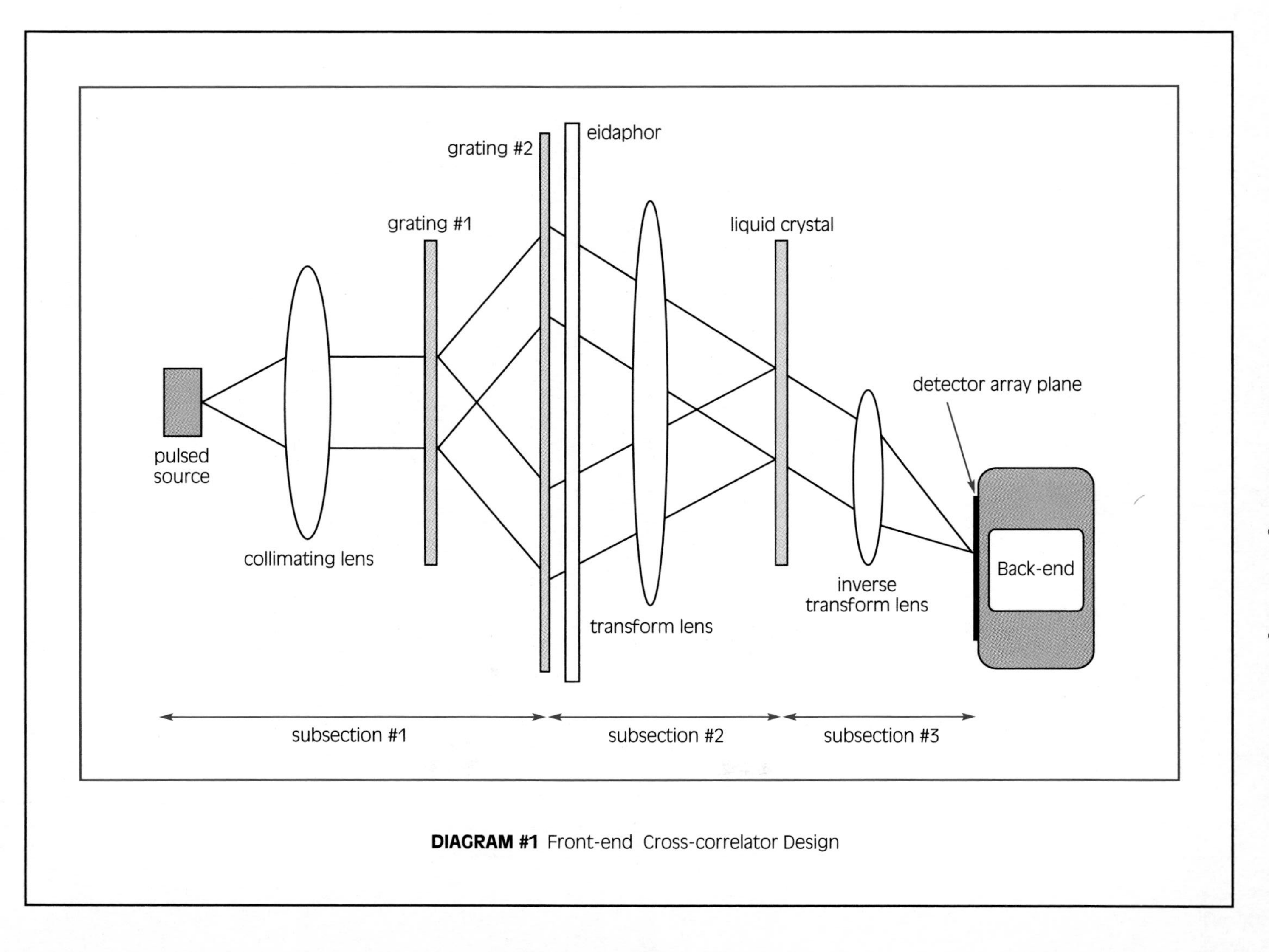

DIAGRAM #1 Front-end Cross-correlator Design

The following are physical characteristics of the Northern Seahorse class, of which the Maritime Venture is the first to be built:

Length	250'-0"	
Beam	50'-0"	
Depth to Main Deck	22'-6"	
Designed Draft	18'-0"	
Displacement at Design Draft		3565 Tons
Cargo Deadweights:		
Deck Cargo		1070 Tons
Liquid Cargo		0450 Tons
Dry Cement		0250 Tons
	TOTAL	1770 Tons

Crew: 6 Officers, 9 Seamen

Passengers: 6

Propulsion: Four medium-speed diesels, each pair driving a 10-foot diameter ducted, controllable-pitch propeller. Total continuous shaft horse power, 14,000. Maximum bollard pull, 120 tons.

Service Speed: 16.5 Knots

Maximum Range at Service Speed: 5,500 Nautical Miles

These new vessels are to be built by North Sea Shipyards for Caladonian Exploration to meet the added demands of near-Arctic offshore drilling service. They will have the ability to remain with the rig for extended periods of time and will be able to singlehandedly tow the rig from one drill site to the next as well as handle anchoring duties at the new location.

The Northern Seahorses are notable in several respects. They are to date the largest offshore tug/supply vessels in the world. Their increased dimensions will enable them to carry over 700 tons more cargo than the typical 190 footer in service today. They are also among the first tug/supply vessels to be built to American Bureau of Shipping Ice Class 1-A standards, which will give them virtually an unlimited latitude of operation.

These new vessels have many automated features. The machinery has been designed for a one-man engine room watch. Bridge control of engine speed and computer correlated propeller pitch settings have been provided. Critical pressures and temperatures are automatically recorded and can be monitored from the control station console. Bilge alarms and fire retardation systems are all automatic.

Ship's power is provided by four shaft-driven generators, each clutched off an engine, giving a variety of power capabilities to meet the varying needs of anchor and tow winch operations at the drill site. An independent generator will be used for ship's services during standby at the rig or at dockside.

The cargo deck is equipped with tandem hydraulic anchor handling and towing winches mounted on the centerline. Each winch has a static load of 250 tons, and can be operated from either the aft command station in the pilot house or from controls on the supply deck. The winches are located as far forward as possible to allow maximum utilization of the supply deck. The forward half of the supply deck is equipped with a rolling deck platform which enables the forward cargo to be moved aft previous to being unloaded from the ship, which is anchored in a stern-to position during this operation. This feature enables the vessel to maintain a greater distance, thus reducing the chance of a collision.

Dry cement pressure tanks are located low and forward in the ship. Ballast tanks are interspersed through the length of the ship to facilitate proper trim at different load conditions. Fuel oil, which is transferred by hose for use at the rig, is located in centerline deep tanks to afford maximum protection in the event of a collision.

Spacious air-conditioned accommodations are furnished for all crew members and passengers. Officers are provided with private staterooms, and the remaining crew and passengers will be berthed in double cabins. All staterooms will be attractively wood paneled and fully carpeted to afford the most in luxury and comfort.

The Northern Seahorses are the finest example of offshore service vessel design and construction. This versatile class will combine unmatched cargo capacity, towing ability, and range with the maximum in safety and comfort.

FIGURE 8.12 Report that Should Have Followed the Description Pattern. *Negative Example.*

- Method and Equipment: Explanation of the procedure; the equipment, facilities, or materials used; the parts or the objects tested.
- Results: Presentation of significant results.
- Analysis: Analysis of results and derivation of conclusions.
- Implications, if necessary: Discussion of conclusions to demonstrate their significance or to formulate recommendations.

Although most of your reports will be working reports to help an organization solve problems and operate efficiently, in these reports you often need to explain an investigation or a test in order to present results and support a conclusion. Thus, in some section of a report you will use an investigation pattern. For example, you might have to present the results of a test by explaining the test procedure used to produce them, as different procedures could yield different results.

In a section of a working report, however, the investigation pattern often is used inefficiently because the writer isn't selective. The outline for the investigation pattern includes elements that lead writers to include a lot of unnecessary detail. This detail is unnecessary because often you are not presenting the results of original experimental research; instead, you are presenting the results of fairly routine experiments or tests. Thus, the purpose and focus of the investigation pattern should be on the results and analysis, not on the methodology.

Begin with a simple statement of the purpose of the investigation or the test. If any background is necessary, that information often can be included in this paragraph.

Then explain the method and equipment used. In short investigation sections, this often consists of simple statements of procedure, equipment, facilities, materials, and parts or the objects tested. Not all of these items are needed. When mentioned, they need to be mentioned as succinctly as possible. For example, a phrase, such as "by regression analysis we. . . ," can suffice for a statement of the procedure used. Equipment, facilities, and parts often only need to identified.

Then, present your results. Do so in a list or a table if at all possible. These results are final results, not test or experimental data. Keep the presentation of results separate from the analysis that follows.

Finally, analyze your results by explaining your interpretation, not just by reporting the numbers. The interpretation of the results is phrased in terms of your conclusions. Because this is just a section of the report, these usually are intermediate conclusions. That is, they are not the final conclusions on which you base your recommendations. Thus, an investigation section ends with the analysis rather than discussion of implications. The implications provide the content of other sections of the report.

Reports arranged in terms of a complete method of investigation pattern are outlined in Chapter 6. However, short reports often follow an investigation pattern. The test report on front end sheet metal mounts (Figure 8.13) follows a method of investigation pattern, although in abbreviated form. The

CHRYSLER
CORPORATION

Inter Company Correspondence

To—Name & Department		Division	Plant/Office	CIMS Number
File Code			Date 8/1/89	
M. A. Bowen, Manager	Truck Engineering	E&RO	Chrysler Center.	415-04-31

From—Name & Department		Division	Plant/Office	CIMS Number
B. L. Ross, Test Engineer	Truck Development	E&RO	Proving Grounds	422-01-09

Subject:

TB-3056: APPROVE RELOCATION OF LIGHT DUTY CONVENTIONAL FRONT END SHEET METAL MOUNTING

OBJECTIVE

To approve use of common Front End Sheet Metal (F.E.S.M.) mounting, with Dayton A/C condenser, for all Conventional and Sport-Utility vehicles.

CONCLUSIONS

1. Standardized F.E.S.M. mounting should be satisfactory on Conventional and Sport-Utility vehicles with structural changes made as a result of endurance tests.

2. Dayton A/C condenser P/N 4039595 mounting is satisfactory for all Conventional and Sport-Utility vehicles.

RECOMMENDATION

Approve use of standardized F.E.S.M. mounting along with Dayton A/C condenser P/N 4039595 for all Conventional and Sport-Utility vehicles.

BACKGROUND

Sport-Utility and Conventional vehicles are now being built in the same plant. Standardization in location of F.E.S.M. mounts has been proposed to simplify plant handling. This standardization should also result in a cost reduction on some models. For these reasons, F.E.S.M. mount standardization was tested as outlined in the following "Discussion."

DISCUSSION

Two Sport-Utility vehicles, E501 (N8AW1) and E512 (NBAW1), plus Conventional vehicle E622 (N8W2) have been endurance tested at the Proving Grounds with standardized F.E.S.M. mounts installed. Both Sport-Utility vehicles have now completed testing, E501 with 23,338 miles of Schedule F-4 (Accelerated Off Road Endurance) and E512 with 33,739 miles of Schedule F-1 + G (Off Road Endurance plus Brake Steps). Conventional vehicle E622 has completed only 6,982 miles of Schedule F-4 and is awaiting design information on chassis brackets before continuing.

FIGURE 8.13 **Test Report Illustrates Investigation Pattern.**

Bowen
Re: "...F.E.S.M. Mounting Relocation"

Standardized F.E.S.M. mountings should be satisfactory on Sport-Utility vehicles. Problems noted during tests on these vehicles have been corrected (as stated in replies to Reference 4, "Deficiencies." The most significant of these problems (Deficiency No. 1680002-3) resulted in the redesign of the radiator grille support extensions and of the lower outer reinforcements for the grille support.

Standardized F.E.S.M. mountings should be satisfactory on Conventional vehicles. This conclusion is based on previous feasibility tests on Proving Ground vehicles E124 (L803), E229 (M3D2), E230 (M8W2), and E981 (K6D1).

The standardized mounting location will be approved prior to final testing on Conventional endurance vehicle E622 (N8W2). Revised components (Reference 5, PCN 60823-400A) have now been installed on E622. Testing has not started, however. Any problems uncovered during subsequent testing will be reported by a Product Evaluation Report.

Use of the Dayton A/C condenser P/N 4039595 is acceptable for both types of vehicles. Change B of TB-3056 requested evaluation of this condenser mounting. This mounting was tested for 29,994 miles of Schedule F-1 + G on endurance vehicle E512. The Stress Lab also investigated the Dayton mounting and determined the mounting to be satisfactory for all light duty Conventional trucks with all configurations of F.E.S.M. mounting (see Reference 3).

With issuance of this letter, TB-3056 is considered closed.

References

1. Program Item 1680002.
2. Letter, B.L. Ross to M. A. Bowen, "Light Duty Conventional F.E.S.M. Mounting Relocation Test Program," dated 11/15/88.
3. Report, D. McKenzie, "L. D. Conventional Air Conditioning Condenser Fix," dated 11/17/88.
4. Deficiencies 1680002-1, -2, and -3.
5. PCN 60823-400A, "Revised Rod Grille Supt. Lower Reinf. and Mtg. Extensions."

cc: R. Chapman, Manager, Truck Development
W. M. Doerr, Supervisor, Truck Development
D. D. Freese, Test Engineer
R. W. Geisler, Engineering
R. E. Lutoesky, Engineering
D. McKenzie, Manufacturing

report is distinctive for its selectivity. The "Background" section states the objective of the investigation. Then the first paragraph of the "Discussion" states the vehicles tested, the type of test, and the test procedures.

The next two paragraphs present intermediate conclusions of the tests of the two types of vehicles. No problems exist; the tests resulted in modifications of the sport-utility vehicles so that front-end sheet metal mounts could be standardized. The final two paragraphs present the basic conclusions and recommendations.

This short test report omits most of the test results and many details of the mountings, as these are in background documents. Other short test, design, and experimental reports, however, often include test results and analysis sections. In short reports and sections of reports, therefore, you use the method of investigation pattern selectively, adapting it according to your purpose.

Conclusion

As in the preceding chapter, we outline the four informative patterns separately for reference purposes (Figures 8.14, 8.15, 8.16, and 8.17). These, along with those outlined in Chapters 6 and 7, present various strategies for organizing information. We present these outlines primarily as guidelines that you

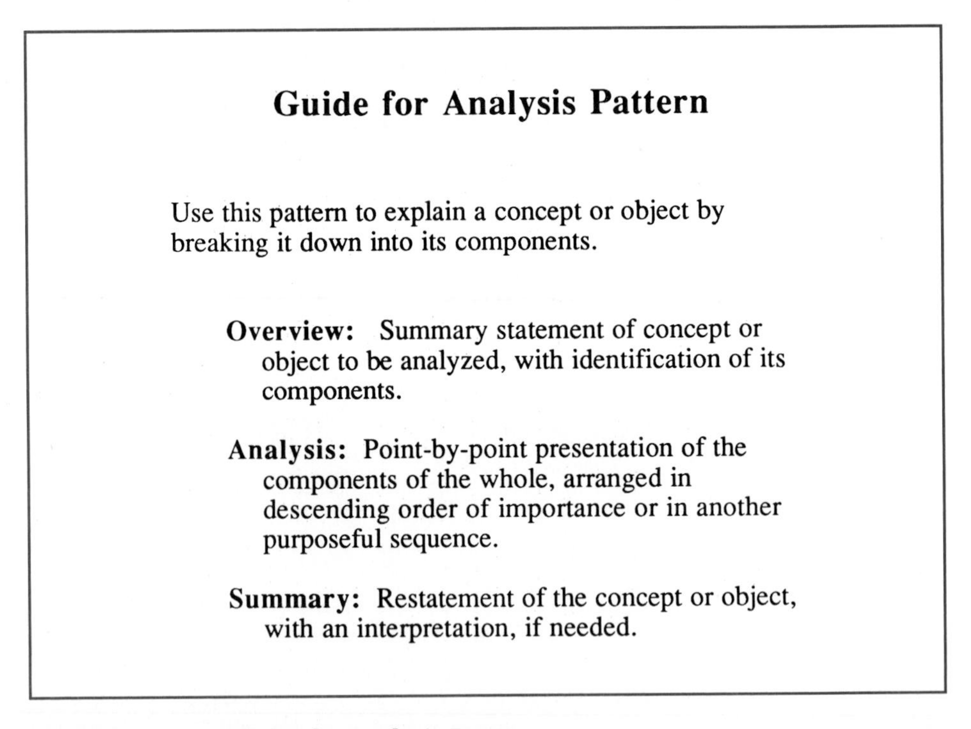

Guide for Analysis Pattern

Use this pattern to explain a concept or object by breaking it down into its components.

Overview: Summary statement of concept or object to be analyzed, with identification of its components.

Analysis: Point-by-point presentation of the components of the whole, arranged in descending order of importance or in another purposeful sequence.

Summary: Restatement of the concept or object, with an interpretation, if needed.

FIGURE 8.14 **Guide for the Analysis Pattern.**

can adapt for your purposes in any particular report, as "generic outlines" that can be of help as you grapple with ways to organize a report or one of its parts.

The premise behind our approach is that, although the content of an individual document will always be unique, the structure (and substructures) of the document often can be recognized as generalized, familiar patterns that you and your readers encounter again and again. These patterns—which cognitive psychologists and linguists call "schemata"—can be powerful aids to comprehension.[2] They help people to understand new things by connecting them to old, familiar things. They also help readers to infer missing information. And finally, they help readers to remember information. If you use generic patterns as ways to help structure your document, they can aid the reader (and yourself as well when you compose your document).

Our only cautionary note is that you must treat these patterns as guidelines rather than as prescriptions. Used improperly, these patterns can cause you not to think at all. Used properly, however, they can help you to think well about what to include, what to leave out, and how to arrange either the individual sections (which combine to form long reports) or the 1-page and 2-page reports (which constitute so much of the "bread and butter" writing of professionals in a variety of fields). They can cause you to play with alternatives and to look for variations.

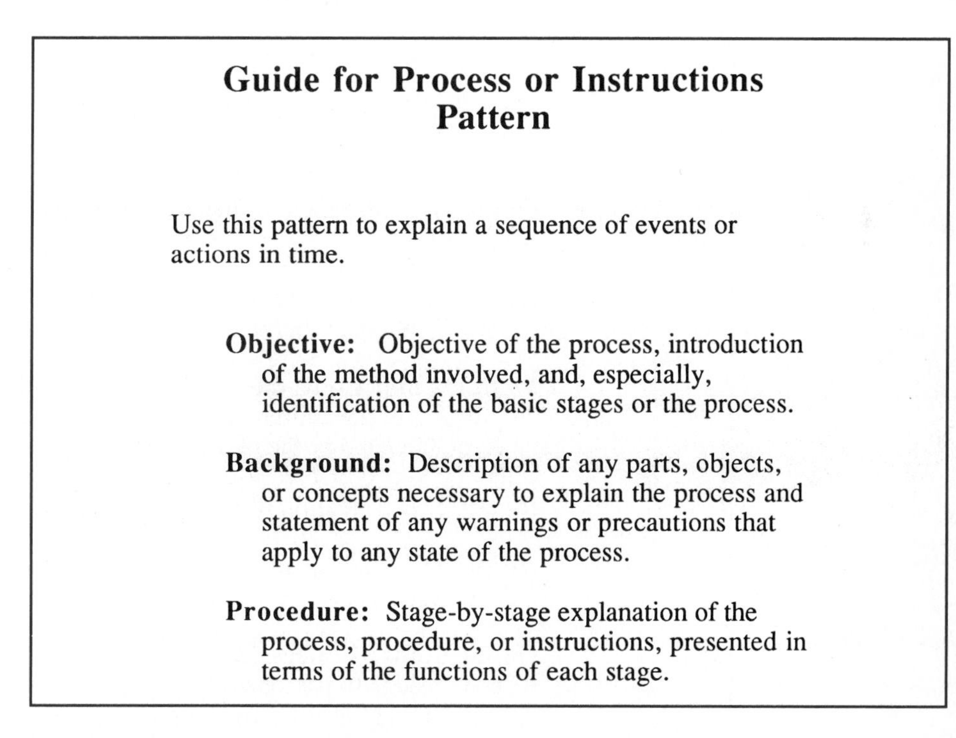

Guide for Process or Instructions Pattern

Use this pattern to explain a sequence of events or actions in time.

Objective: Objective of the process, introduction of the method involved, and, especially, identification of the basic stages or the process.

Background: Description of any parts, objects, or concepts necessary to explain the process and statement of any warnings or precautions that apply to any state of the process.

Procedure: Stage-by-stage explanation of the process, procedure, or instructions, presented in terms of the functions of each stage.

FIGURE 8.15 **Guide for the Process and Instructions Pattern.**

Guide for Description Pattern

Use this pattern to explain how a physical device functions.

Overview: Overview of the object's function.

Description: Physical description of the object, with a simplified visual aid.

Explanation: Functional description of the object: a part-by-part explanation of how the object works.

FIGURE 8.16 **Guide for the Description Pattern.**

Guide for Investigation Pattern

Use this pattern to explain how an investigation was conducted and judgments formulated.

Objective: Statement of the purpose of the investigation.

Background, if necessary: Explanation of the specifications, criteria, or technical background.

Methods and Equipment: Explanation of the procedure; the equipment, facilities, or materials used; the parts or the objects tested.

Results: Presentation of significant results.

Analysis: Analysis of results and derivation of conclusions.

Implications, if necessary: Discussion of conclusions to demonstrate their significance or to formulate recommendations.

FIGURE 8.17 **Guide for the Investigation Pattern.**

And one final comment: the patterns in Chapter 6 and those in Chapters 7 and 8 overlap. The distinction between the entire discussion and sections of the discussion is not clearcut, as reports range from one page to five pages to thirty pages and more. Therefore, we need to point out that just as the guidelines in Chapter 6 can be used for sections of a report as well as for an entire report, the guidelines in Chapters 7 and 8 also can be used for entire reports as well as for sections of a report. For short reports, especially, you will find these guidelines for persuasive and informative sections useful and, often, sufficient.

Notes

1. National Committee for Clinical Laboratory Standards, Villanova, Pennsylvania. As an example, NCCLS publication H3-A defines the procedures for the collection of diagnostic blood specimens by venipuncture. It is an Approved Standard, published in 1983. There are numerous other NCCLS-published standard procedures. When new procedures, such as that shown in Figure 8.8, are written in hospitals, they are written according to the NCCLS standard in terms of content and arrangement.

   Other fields have similar specifications. For example, in the field of architecture and construction, an excellent and useful book is by Thomas C. Jellinger, *Construction Contract Documents and Specifications* (Reading, MA: Addison-Wesley, 1981). Jellinger's book presents information regarding the content, preparation, application, and interpretation of construction contract documents.

   Similarly, the federal government publishes an extensive list of standards governing technical publications. Just a bibliography of these requires 62 pages in a publication of the Society for Technical Communication, Joseph M. Kleinman's *List of Specifications and Standards Pertaining to Technical Publications* (Washington, DC, 1983).

2. Thomas N. Huckin, "A Cognitive Approach to Readability," *New Essays in Technical and Scientific Communication: Research, Theory, Practice*, ed. P. V. Anderson, R. J. Brockmann, C. R. Miller (Farmingdale, NY: Baywood, 1983) 90–108. In his article, Huckin makes the following point: "The power of schemata in the communication process resides largely in their ability to induce inferences from the reader (listener, viewer, etc.). This process of schema-based inferring works as follows. When the writer and reader share a schema, that is, when their respective schemata for a particular concept are essentially similar, the writer does not have to refer explicitly to all of the details of that schema for those details to be conveyed to the reader; the reader will simply supply any missing details by inference. In other words, a schema is conceived of as a mental construct with fixed place-holders for its constituent features; when a schema is evoked but not all of its features are explicitly mentioned, the reader can simply 'fill in the slots.' In this way, a single word or phrase can actually call up an entire constellation of images in the reader's mind. . . . For one thing, it [use of schemata] enriches the imagery of concepts, making them easier to perceive and easier to remember. Secondly, it can increase the coherence of a message by giving the reader a means of filling in various missing links. Writers do not normally spell out everything they mean; they usually count on the reader's being able to infer unstated details, including the details that tie the different parts of a message together. And much of this inferring is based on shared schemata. This is why subject-matter familiarity is so advantageous to a reader trying to comprehend a message: it provides a set of schemata that help him or her integrate the various parts of the message and thus create a meaningful whole." Huckin, 92–93.

Writing, Editing, and Formatting the Communication

C H A P T E R 9

Writing and Editing Paragraphs

To this point, we have concentrated on the larger issues of document design: identifying the needs of your readers, defining your specific purposes in a particular organizational context, and designing the basic structure and substructure of your document.

We have done this because writers who plan well on both the whole document level and the section level tend to produce generally good first drafts at the paragraph and sentence level as well. They work from a plan which tells them what the key points are, what the main support must be, and what the order of the presentation should be to accomplish their purposes for their intended readers. Thus, when they actually write paragraphs and sentences in their first draft, ideas flow purposefully from one to the next and major discontinuities or excessive false starts are minimized. They produce a first draft which is ready to be edited.

Now we turn from designing the document to suggestions for actually writing the document or, more accurately, for revising and editing the first draft of the document at the paragraph level. Here too your awareness of principles of effective design can make major differences in the readability of your document. We have six suggestions for effective writing and editing of paragraphs.

❑ 9.1 ❑ Recognizable Patterns for Paragraphs

Readers can make connections between ideas, and even fill in gaps between ideas, because they understand generalized patterns behind detailed explanations. That is, many paragraphs are only specific instances of some familiar pattern which readers have seen many times before: the lyrics may be different, but the tune behind them is generally familiar.

These patterns were established in our culture by Greek and Roman rhetoricians and now have become embedded in the unconscious understanding of readers. For example, as explained in Chapter 6, we recognize and anticipate the basic "plot" of persuasion: conclusion, support in descending order, anticipation and refutation of counter-arguments, and restatement. Similarly, we recognize the patterns of problem/solution; cause/effect; and comparison and contrast. In Chapter 7 we explained familiar informative patterns: analysis, process, description, and investigation.

These familiar patterns operate not only at the whole document and section levels, but on the paragraph level as well. For example, here is a paragraph which follows a simple and familiar analytic pattern:

> Three sources of nutrient load in West Bay must be identified before we can consider how these nutrient sources may be reduced. The first source of nutrients, the natural source, is the decay of organic materials in marshes, lakes, and topsoil which releases the constituent elements to the ground and surface waters of the watershed. The second source of nutrients is the inorganic fertilizers applied to crops and lawns, nutrient which is either washed away in surface runoff or leached into the subsoil and which eventually reaches the bay. This source is very difficult to control because the fertilizer is applied in many places by many well-meaning individuals. The third source of nutrients is the effluent from the wastewater treatment plant. Because it is concentrated at one point, this source is most amenable to control by treatment for nutrient removal before release to the receiving stream.

In skeletal form, the paragraph has this structure:

> Topic idea: there are three sources of nutrient load in West Bay, sources we must identify before attempting to reduce the pollution. They are
>
> 1. First source, the natural source: decay of organic materials.
> 2. Second source, a human-made source: inorganic fertilizer applied to crops and lawns.
> 3. Third source, another human-made source: effluent from the wastewater treatment plant.

The paragraph is really a simple list headed by a generalization. There are three sources. They are as follows: 1, 2, 3.

In addition, notice that the list is ordered in a logical sequence. The basic, most widespread source is a natural source, decay of vegetation. The next source, inorganic fertilizers, is a human-made and widely distributed source. The third source, wastewater, is again human-made, although a localized source. In other words, the paragraph moves from natural to human-made sources, and—more significantly—from untreatable to treatable sources.

As the first sentence suggests, the analytic pattern of the paragraph is actually part of a larger problem/solution pattern. The example paragraph establishes the three problems. The paragraphs which follow it (not shown) discuss the possible solutions. Thus you can anticipate that the next paragraphs will say that

1. there is not much we can do about the first source;

2. the second source, widespread as it is, resists action beyond making the public aware of the negative effects of applying fertilizer and encouraging them to avoid using fertilizer whenever possible;

3. the pollution in West Bay can be significantly reduced by treating the effluent from the wastewater plant.

The analytic pattern of the example paragraph is nested within the problem/solution section of several paragraphs. By itself, however, it has its own evident structure.

The pattern lying behind the sentences of the example paragraph is one which you can both recognize and predict. You can even recognize that the basic pattern is varied twice by subordinate sentences, each of which elaborates on only the sentence immediately preceding it without altering the basic pattern:

- (Sentence 4) This source is very difficult to control because the fertilizer is applied in many places by a great many well-meaning individuals.
- (Sentence 6) Because it is concentrated at one point, this source is most amenable to control by treatment for nutrient removal before release to the receiving stream.

An outline of the paragraph, including these two sentences, shows it to have this structure:

Main idea
1. First subpoint
2. Second subpoint
   - Subordinate idea
3. Third subpoint
   - Subordinate idea

As with these patterns in sections of reports, these patterns in paragraphs are independent of subject matter. Almost precisely the same structure can be seen in a second example:

> In determining the total cost per lost casting, we must consider six factors. First, there is the cost of melting the casting, risering, and gating material. This represents $2 of the total lost cost of $102 per casting. Second, there is the cost of mold room pouring ($20 of the total). Third, there is the labor cost of cleaning the castings ($16). Fourth is the cost of radiographic inspection ($24). (Two radiographic exposures are needed for each casting, at a cost of $12 each.) Fifth is the machining cost, which includes not only labor time but allowances for shipment of the parts to and from the machining facility ($17). And finally there is the cost of pressure testing each casting ($23).

An outline of this paragraph looks like this:

Main idea
1. First subpoint
   - Subordinate point
2. Second subpoint
   - Subordinate point in parentheses
3. Third subpoint
   - Subordinate point in parentheses
4. Fourth subpoint
   - Subordinate point in parentheses
   - Subordinate idea in parentheses
5. Fifth subpoint
   - Subordinate point in parentheses
6. Sixth subpoint
   - Subordinate point in parentheses

As you can see, the second example has the same pattern as the first example. Of course, it differs in subject matter and has six points of analysis rather than three. Also, its order is based upon the chronological sequence of melting, pouring, cleaning, machining, and inspecting castings—not on significance. And finally, it is slightly more complex on the subordinate level than the first example. But in its basic pattern, the second example is recognizably more like the first example than different from it: both are simple analyses.

To illustrate other patterns which operate on the paragraph level, here are some examples of the other seven basic patterns which we discussed in the previous two chapters.

Persuasion (from an OSHA Review Commission opinion in an industrial accident case):

> After reviewing the record, I am now convinced that the complainant has failed in his proofs. He has failed to produce evidence that any part of the crane structure, its hoist line, or its load came to within ten vertical feet of the trans-

mission line. That is the standard of proof required by the relevant OSHA regulations cited above. Although the complainant has presented three witnesses—all of them present at the time of the accident, including the victim—none of these witnesses was able to supply direct testimony that the crane boom, hoist line, or load came within ten feet of the vertical plane of the power line. Furthermore, the testimony presented by these witnesses, as well as that evidence presented by the police reports taken at the accident site, leads me to conclude that no part of the crane or its attachments came any closer than 20 feet of the line. In conclusion, I find that the complainant has failed to prove his allegation that the accident resulted from violation of the OSHA standard for cranes working near high-power transmission lines. Accordingly, I must find for the respondent.

An analysis of the persuasive example shows it to have this structure:

- Conclusion: Complainant's proofs have failed (sentences 1 and 2).
- Criterion for conclusion: Relevant OSHA standard (sentence 3).
- Support 1: Failure of eyewitness testimony to support complainant's allegation (sentence 4).
- Support 2: Lack of contradictory evidence (sentence 5).
- Restatement of conclusion (sentences 6 and 7).

Problem/Solution (from a laboratory memorandum):

Because mercury spills pose the most serious health hazard, provisions must be made for cleaning up mercury spills in the laboratory. Mercury often splashes and spills while being poured because of its great density, surface tension, and low viscosity. Droplets of mercury form and scatter in all directions and roll into cracks and corners. Therefore, each lab bench where mercury is used should be enclosed to contain spills and to simplify cleanup. A sheet-linoleum surface covering with sealed joints and coved edges should be used on the bench to eliminate cracks, which trap mercury droplets. A suction apparatus then should be used to recover all spilled mercury.

Analysis of the problem/solution example shows it to have this structure:

- Two-part topic sentence: (problem) mercury spills pose hazard; (solution) provisions must be made.
- The problem (sentences 2 and 3): mercury splashes; mercury scatters and is trapped.
- The solution (sentences 4, 5, 6): enclose benches; use sealed surface covering; use suction apparatus.

Cause/Effect: (from an environmental impact study):

Regulation of streamflow will cause widespread and permanent environmental changes along the banks of the Gambia River. Periods of high erosion during the wet season and low erosion during the dry season will be replaced by a

> constant low level of erosion throughout the year. The annual flood no longer will inundate much of the upstream river bank, while tidal flooding no longer will keep the bank downstream moist. In turn, these changes will change the environmental habitats of the river banks and significantly alter their function as a mechanism for the exchange of nutrients and material between the land and water environments. Forest along the river will be jeopardized by erosion. Wildlife and waterfowl will lose preferred habitat. Floodplain vegetation will be lost, and considerable biological nutrients will be lost to the river and lakes. In sum, regulation of streamflow will have long-term, potentially negative environmental consequences.

Analysis of this cause/effect paragraph shows a chain of causes and effects over time. (In this regard, the paragraph mixes elements of cause/effect and process patterns.) In the chain, causes produce effects which in turn become new causes to produce new effects. The topic sentence at the beginning of the paragraph establishes the overall cause and its overall effect. The subsequent sentences trace the chain reaction of causes and effects. The paragraph ends with a restatement of the topic sentence:

- Topic sentence: (cause and general effect) Regulation of streamflow will cause significant environmental changes.
- Regulation (cause) will replace cyclical erosion patterns with a constant low-level erosion pattern (effect).
- Regulation (cause assumed) will eliminate the annual flooding (one effect), which (effect becomes new cause) will eliminate the floodplain upstream (new effect).
- Regulation (cause assumed) will eliminate tidal flooding (effect), which (effect becomes new cause) will cause the banks downstream to dry out (new effect).
- These changes (the effects become new causes) will alter the environmental habitats of the river bank and affect its function as a mechanism for exchange between land and water (new effects).
- Forest will be jeopardized (effect).
- Wildlife will lose habitats (effect).
- Floodplain vegetation will be lost (effect).
- Nutrients will not enter the river and lakes (effect).
- Restatement of topic sentence: Regulation of the flow of the river will have long-term environmental effects.

Comparison and Contrast (from a report of the effects of format on differences in reading speeds):

> The four documents used for this study of reading speed were selected primarily because of their differences in content and format, although the four also shared important features frequently found in conventional business memoranda. The documents differed in the following respects:

- They contained differing degrees of numerical data, ranging from none (document one) to a modest amount (documents two and three) to a great deal (document four).
- Document one used no numbering and had neither headings nor subheadings.
- Document two used main-level headings, one level of numbering, and one level of block indentation.
- Document three used no headings or subheadings but had both numbering and block indentation.
- Document four used headings and subheadings, numbering, block indentation on two levels, differences in vertical spacing between units and sub-units, and two tables of numerical data (one in two columns, one in four columns.)

Despite these content and format differences, the four documents were alike in three significant respects:

- All were business memoranda actually used on the job.
- All were short (between 338 and 1154 words).
- None of the four contained any graphic material.

Analysis of this contrast and comparison paragraph cluster shows that it emphasizes the differences between the four documents, but also identifies the areas of similarity. (Hence it places contrasts first and then follows with similarities):

- Topic sentence: the four documents were selected because of differences in their content and format, although they shared some key features (sentences 1 and 2).
- Differences itemized (sentences 3 through 7).
- Subtopic sentence repeated: the four were alike in key respects (sentence 8).
- Similarities itemized (sentences 9 through 11).

Description (from a fisheries study):
In the technique called "chemofishing sampling," a blocking net encloses a measured area so that all the fish can be trapped and chemically killed. The net, placed where the depth, vegetation, and current will not impair the efficiency of the sampling, is 150 meters long and 8 meters deep. It is held in place by bamboo stakes at the corners, with the ends staked at the shoreline from 25 to 50 meters apart. The net is then staked in the reservoir to form a rectangular enclosure of 1250 square meters. On the surface, the net's float line is supported by 12.5 centimeter cork floats placed 50 centimeters apart. On the bottom, the net is held down by .05 kilogram lead sinkers placed 75 centimeters apart. The net itself is made with a 2.54 centimeter square mesh which traps all juvenile and adult fish so that they can be killed with Rotenone.

Analysis of this descriptive paragraph shows a general-to-particular movement:

- Topic sentence (an overview of the function of the object being described): blocking net encloses measured area so that fish can be trapped and chemically killed.
- General description of net's dimensions: net is 150 meters by 8 meters.
- Particulars of net's configuration (sentences 3–6): net is staked at corners from 25 to 50 meters apart to form a rectangular enclosure of 1250 square meters, supported by floats, and held down by sinkers.
- Specific explanation of how the net functions to trap fish (restatement): mesh of net traps all juvenile and adult fish, which can then be killed with Rotenone.

Process (from a manual for a home water softener):

As your water softener works, the beads contained in the resin tank become coated with minerals from the incoming hard water. The beads are cleaned for reuse by an automatic cleaning process called "regeneration."

Regeneration is a timer-controlled, automatic process consisting of four stages or cycles: brine soaking, brine rinse, backflush, and fast rinse. First, brine from the brine tank is allowed to flow into the resin tank and to remain there for just under one hour. This brine causes the resin beads to release the minerals which have been attracted to them during operation of the water softener; the accumulated minerals are then washed down the drain when the brine is pumped out at the end of the cycle. Second, in a 5-minute cycle, fresh water is washed downward over the resin beads to flush any remaining brine from the beads. Third, for about 10 minutes, fresh water is pumped upward through the resin tank with sufficient force to lift any accumulated sediment from the bottom of the tank and to loosen the packed resin beads. Fourth, a fast flow of clean water, again a 5-minute downward flow cycle, repacks the resin bed. Thus, in a little over an hour and a quarter, the resin beads in your water softener are automatically cleaned of accumulated minerals and are prepared to resume their normal water softening function.

Analysis of this process paragraph cluster shows the following pattern:

- The process is named and its objective is identified (sentences 1 and 2).
- The four stages of the process are identified (sentence 3).
- Each of the four stages of the process is explained, one at a time (sentences 4 through 8).
- The process and its results are summarized (sentence 9).

Investigation (from the abstract of a water resources report):
Hydrophobic organic pollutants (HOP) such as polychlorinated biphenyl (PCB) are serious environmental hazards whose fate in aquatic systems is controlled by their sorptive behavior. To determine the effect of temperature on sorption, we conducted equilibrium isotherm studies of 2, 4, 5, 2', 4', 5' hexachlorobiphenyl (HCB) over a temperature range from 5° C to 25° C. Our results, which

contradict accepted theory, show that sorption of HCB increases with increasing temperature. We attribute this effect to temperature-induced changes in the molecular arrangement of water in ways which affect its solvent capacity for HCB. We recommend further study to confirm and quantify the trend observed in this investigation. We also recommend investigation of the effects of other factors such as the presence of organic and inorganic species in solution, as these could affect sorption by altering water solvent properties.

The pattern of this paragraph is essentially the pattern of an investigation report in miniature, with one sentence or so devoted to each of the key elements in the investigation:

- Problem and purpose of investigation: Determine effect of temperature on sorption of hydrophobic organic pollutant, hexachloro-biphenyl.
- Method: Conduct equilibrium isotherm studies over a range of 5° C to 25° C.
- Results: Sorption of HCB increases with increasing temperature.
- Conclusion: Sorption of HCB increases with increasing temperature because of temperature-induced molecular changes in water.
- Recommendations: Confirm and quantify the trend; investigate contributory role of other factors such as presence of other organic and inorganic species.

The patterns in the preceding illustrations are not the only patterns you will use, nor are they absolute.[1] For example, there is a pattern that might be called "topic and illustration." In fact, this section itself has been an example of that simple pattern. You state a topic, then you support it with an example or series of examples. Then perhaps you restate the topic sentence.

There are also patterns based upon definition. For example, you might identify a term, then classify it into a genus and differentiate it into species. This is the pattern of classical Aristotelian definition, a pattern which you can extend by explicating the terms and concepts used in your definition or by adding examples to it. Or you might extend the pattern by adding a contrast of the thing being defined with other things similar to but different from it. Or you might add a negative definition, an etymology, an operational definition, or a stipulative definition—the restricted meaning you intend to use in the context of a specific discussion.

We all have a series of strategies that we commonly use for defining terms and concepts. Any or all of these might be used to form generalizable patterns on the paragraph level to define terms, no matter what the subject matter of the specific instance. Readers will recognize your patterns because they have seen them before.

As you look for patterns based upon purpose, look also for patterns based upon formal relationships among sentences, patterns which are analogous to the grammatical relationships which operate on the sentence level. That is, paragraph patterns are form-based as well as purpose-based. For example,

Francis Christiansen argues that paragraphs use fundamental patterns based upon coordination, subordination, or a mix of the two.[2] Here, for example, is a coordinate pattern:

- Topic sentence
  - support
  - support
  - support

Here is a subordinate pattern:

- Topic sentence
  - support for previous sentence
    - support for previous sentence
      - support for previous sentence
        - support for previous sentence

Here is a dominantly coordinate pattern, but with an embedded subordinate element. (You might recognize that the paragraph about the three sources of nutrient load in West Bay is an example of this very pattern):

- Topic sentence
  - support
    - support for previous sentence
  - support
    - support for previous sentence
  - support

Here is a dominantly subordinate pattern, but with an embedded coordinate element:

- Topic sentence
  - support for previous sentence
    - support
    - support
      - support for previous sentence
        - support for previous sentence

You can look at the pattern of paragraphs essentially in terms of their purpose. And you can look at them in terms of their formal relationships among sentences (coordination, subordination). Yet, in neither case is a paragraph an arbitrary sequence of sentences. Paragraphs are real structures which readers recognize and expect.

The reality of paragraph structures has been convincingly demonstrated by research which used passages of text from which all the paragraph breaks had been removed.[3] The researchers asked readers to indicate where the paragraph breaks should occur. Not only did the researchers find agreement among readers, they found that readers could still identify appropriate paragraph breaks even when nonsense words replaced content words in the original text. And surprisingly, they even found some agreement among readers and writers when all of the words in the text were replaced with "squiggly" lines representing the sentences. They conclude that paragraphs are indeed psychologically real units of structure for readers, that repetition of content words and pronouns are significant cues of coherence within paragraphs, and that spatial and/or sentence count information seems to influence readers' choices of when paragraph breaks are "appropriate."

In developing paragraphs and in editing and revising them, then, your first strategy is to look at paragraphs and clusters of related paragraphs to see if you can recognize a structure behind them. If you can, you need to decide if it is the pattern best suited to your purpose. If not, you need to develop and revise the paragraph to establish a pattern that will accomplish your purpose. Ask yourself what the main message is. Ask yourself what the purpose is. Ask yourself how the sentences of the paragraph relate to one another. See if you can summarize the paragraph or paragraph cluster.[4] Only then, only when you are satisfied that there is in fact a clear pattern behind the paragraph, should you move on to other revising and editing strategies on the paragraph level.

❑ 9.2 ❑ Functions of Topic Sentences and Core Paragraphs

With perhaps the exception of narrative paragraphs and transitional paragraphs, most paragraphs in technical and professional texts appropriately include a single sentence statement of the main idea. Usually this is called the topic sentence, or—in the case of a topic sentence which is separated from the body of its paragraph—we might call it the core paragraph.

Unfortunately, these topic sentences and core paragraphs sometimes get left unstated in early drafts and need to be inserted during the revision and editing process. Probably the reason is that in the mind of the writer, the generalization is so obvious that it does not seem to need statement. However, writers cannot be certain that ideas which readers infer to be main ideas are in fact the ones the writers intend. Readers might not grasp the main point, or—worse still—they might interpret it incorrectly. As cognitive psychologist David Kieras points out after extensive experimental studies of paragraph reading behavior,

> Providing the generalization explicitly almost guarantees that the pattern will be recognized. But, in the construction-integration passages [i.e., those with no explicitly stated main idea], the reader must be able to deduce or recognize the chain of antecedents and consequences in the argument being presented, or the

> final outcome of a sequence of events. This reasoning is more complex compared to the generalization passages, and so is slower, and it depends more on the idiosyncratic knowledge and reasoning process of the individual [experimental] subjects, and so is less consistent.[5]

In other words, writers who do not explicitly state their main ideas in paragraphs will slow their readers down, work them harder, and risk their inaccurate interpretation of the message.

For example, consider the following "paragraph" from an automobile company test report:

> The fuel system did not leak at impact or during the thirty-minute observation period following the test. There was no loss of fuel during a post-test pressure check of the fuel system. Furthermore, a visual inspection of the system after the test revealed that although the fuel return line was in close contact with the oil consumption valve, the line itself was undamaged.

Here the writer has presented four discrete facts observed after the impact test of an automobile's fuel system. In the writer's mind, the conclusion that can be drawn from these four facts no doubt was implicitly obvious: Fact A + Fact B + Fact C + Fact D leads to Conclusion E. But Conclusion E is unstated. Thus, there is no clear signal that the appropriate pattern of the paragraph is essentially persuasive rather than merely informative. The paragraph invites misinterpretation. What conclusion can be drawn from these four facts? Or more precisely, what conclusion does the writer expect you to draw? You have to read "between the lines," guessing that Conclusion E is something like the following: "The test vehicle's fuel system performed satisfactorily during impact testing." The fuel system passed the test. Is that what the writer expected you to conclude? You can't be sure.

In most of your first drafts you will find topic sentences or core paragraphs for most of your paragraph-level units, but in some cases they will be missing. If so, it is a simple matter to decide what the point of this paragraph is and to state that point explicitly.

In many other cases, you might find either imprecise, inaccurate, or "pseudo" topic sentences. These are the signs of your groping for the idea as you wrote the first draft. For example, you might find a sentence that states, "Several significant facts to consider should be mentioned." This looks like a topic sentence, but how many is "several" and precisely what is it that makes these particular facts significant? Perhaps before you actually wrote the paragraph you couldn't answer those questions. Now, having written it, you can replace the pseudo topic sentence with a sentence that really summarizes the main point you want to make. After all, the original says nothing more than "I'll keep writing and you keep reading."

In a sense, topic sentences are to paragraphs what main clauses are to sentences. Everything else in the unit merely supports, explains, or elaborates the core idea. Thus, just as you would find it unacceptable to present a subordinate clause as if it were a complete sentence, you should find it unacceptable to present most paragraphs without explicit statements of their controlling ideas.

❑ 9.3 ❑ Placement of Topic Sentences and Core Paragraphs

If you begin your paragraph with a controlling generalization rather than ending with it, you will help your readers to spot your pattern and to read quickly and purposefully. Again, experimental research on reading behavior supports this bit of advice. Kieras says,

> If an obvious candidate for the main idea appears first, then the reader need only adopt it, and then test it for adequacy while reading the remainder of the passage. If not, the reader must attempt to formulate a main idea while reading, and be prepared to reformulate it whenever a poor fit is noticed. As a result, some sentences may appear important when first read, but then later turn out to be merely details or irrelevant. Thus, the initially presented explicit main idea 'protects' the reader against the irrelevant material or alternative possible main ideas, and so simplifies arriving at the main idea.[6]

In other words, without an explicitly stated main idea that appears first, readers are slower and less accurate in their understanding and recall of the message. A good editorial strategy, then, is to make sure that most of your paragraph units begin with an explicit topic sentence or core paragraph.

Unfortunately, you will often find that your first draft paragraphs have topic sentences at the end rather than at the beginning. This is because your act of writing the draft was also an act of discovering the idea. The idea eventually gets stated, but it is buried. Accordingly, unless you move the topic sentence during your editing, you will force your readers to read the details of the paragraph without knowing in advance what the organizing concept is. They will grope along, trying to remember details long enough to arrive at a statement that integrates them.

For example, here is the skeleton of a paragraph in which the writer begins without a topic sentence, groping toward the eventual conclusion:

> In analyzing the commercial potential of the proposed motel location, first we have noted that . . . We have also noted that . . . in addition, we have noted that . . . And finally, we have noted that . . . On the basis of these facts, we conclude that . . .

The sentences unfold in a chronological tracing of the writer's thought process leading up to the conclusion. You do not know in advance what that conclusion is, nor do you know how many points of analysis are to be listed.

Of course, you can quickly recognize that the "meat" of the paragraph is probably at the end, so you may jump to the end. But probably you will plow through the analysis until you get to the significant point, just as the writer did. Your reading task would be more purposeful and probably faster if you knew in advance that "The proposed motel location promises to be a good choice because it offers a unique potential to attract three types of guest patronage: business travelers, tourists, and small conventions. First, the proposed location lies within 25 miles of three . . ."

Putting the topic sentence first in a paragraph is usually a simple matter of moving an existing sentence from the end of the paragraph to the beginning. Often you will have to recast that sentence so that it can stand more clearly upon its own. However, you should begin most of your paragraphs with a single-sentence opening which explicitly presents the main idea of the paragraph and which clearly suggests to the reader what pattern of development logically follows.

The only notable exception is that sometimes the topic sentence will actually follow a transitional sentence which cannot really be seen as part of the individual paragraph structure. These sentences are parts of a larger structure of relationships among paragraphs, and you do need to be alert to the need for these transitions among paragraphs.

❑ 9.4 ❑ Maintaining and Shifting Subject Focus

So far, we have recommended taking a global view of your paragraphs, looking at them in terms of overall architecture as well as in terms of both the presence and placement of your topic sentences. These three editorial strategies will help to clarify your underlying logic as you edit your rough drafts. However, in editing your paragraphs, you can also make significant changes on the sentence level. There you can look at the relationships between the sentence forms which you choose and the content which you intend to convey. You can also look at the relationships among sentences.

Basically you should hope to find two things. First, the grammatical subject slots of your sentences should be filled with content words (or with their familiar synonyms and pronouns) that really do tell the reader what you are talking about. Second, when shifts of focus do occur between sentences, they should be purposeful shifts which lead the reader logically from sentence to sentence. If you don't find those two things, some editing or rewriting is needed.

Take a close look at how you have filled the grammatical subject slots of your sentences. These slots contain the information on which you are focusing your readers' attention just as surely as if you were shining a spotlight upon one object among many and saying, "there, that's the thing I want to talk about." After all, the grammatical subject is what a sentence is about, and you should therefore hope that the *grammatical* subject equals the *conceptual* subject. The form should model the content.

For example, consider the paragraph which we saw before, the paragraph about the three sources of nutrients in West Bay:

> Three sources of nutrient load in West Bay must be identified before we can consider how these nutrient sources may be reduced. The first source of nutrients, the natural source, is the decay of organic materials in marshes, lakes, and topsoil which releases the constituent elements to the ground and surface waters of the watershed. The second source of nutrients is the inorganic fertilizers applied to crops and lawns, nutrient which is either washed away in surface runoff or leached into the subsoil and which eventually reaches the bay. This source is very difficult to control because the fertilizer is applied in many places by many well-meaning individuals. The third source of nutrients is effluent from the wastewater treatment plant. Because it is concentrated at one point, this source is most amenable to control by treatment for nutrient removal before release to the receiving stream.

As you can see by our underlining, the subjects of the sentences do reflect the topic being discussed: sources of nutrients. Further, you can see that the writer has consistently kept that term in front of us throughout the paragraph, thereby moving us along from sentence to sentence without losing the conceptual thread that ties the sentences together.

Contrast the effect of the original paragraph with the effect of a reworded version in which grammatical subjects are shifted arbitrarily. In this reworded version, each sentence has a new grammatical subject which neither models the content nor connects to the subjects of the preceding sentences:

> Solutions to the problem of nutrient load in West Bay require prior identification of the sources of nutrients. Constituent elements in the ground and surface waters are released by the decay of organic materials in marshes, lakes, and topsoil into the Bay. Surface runoff and leaching of inorganic fertilizers into the subsoil causes nutrients to eventually reach the bay. Many well meaning individuals apply fertilizer in a great many places, so this source of nutrients is difficult to control. The wastewater treatment plant releases effluent into the bay. Treatment for nutrient removal before its release into the bay is made possible because this source is concentrated at one point.

The reworded version is incoherent because each sentence introduces an apparently new topic, as the underlining indicates. We say "apparently" because you know better. That is, you know that behind the grammatical form of the sentences—in the mind of the writer—there really was a unifying concept. But that concept is not recognizable in the choices of grammatical subjects. Instead, each new grammatical subject focuses your attention momentarily on a noun or noun phrase that cannot immediately be seen to reflect the intended conceptual focus. The result is a "paragraph" in which you could almost move sentences around arbitrarily without doing harm to the relationship between the concept and the form of the paragraph or to the sequencing of the sentences.

Of course that was a manufactured example. Unfortunately, many first drafts contain "paragraphs" in which the subject focus is almost as inconsistent as it is in this manufactured example. For example, consider this "paragraph" from an actual shipyard document:

> To assure proper protection, the face piece shall be checked by the employee each time he puts on his respirator. To check the respirator, he should place his hand over the discharge valve and exhale. The face piece should not leak air. Appropriate supervision shall periodically check the respirator fit of all employees. If an improper fit is observed, the employee shall be removed from the work area by his supervisor until a proper fit is attained and demonstrated to the supervisor. Proper maintenance, cleaning, and storage are as essential as selecting the respirator for proper fit. Failure to follow through on these parts of the program makes the respirator protective devices worthless and, in fact, potentially dangerous. A poorly maintained respirator gives a false sense of security to the user.

As the underlining indicates, the subjects of these sentences are all over the map. Although there are some discernible relationships among most of the sentences, the paragraph is chaotic. You may even have trouble deciding what conceptual focus the writer intended. What is the purpose of the "paragraph"? Did the writer intend to focus on face pieces? On the employee? On the supervisor? Or on maintenance? It is really hard to tell, so if you attempt to edit this paragraph, you probably would have to revise it completely to establish a controlling purpose and an appropriate grammatical focus. Almost certainly you would need more than one paragraph if you wanted to keep all of the ideas in the original version.

The same is true in the following paragraph from an agricultural bulletin:

> Economic growth of a farming or ranching business may be enhanced by combining assets of two or more individuals into a partnership. However, the stability of the firm may be in question because of the possibility of willful termination by a partner or involuntary termination caused by the death or retirement of a partner. Involuntary termination may be guarded against through specific provisions in the partnership agreement which allow the partnership to continue with the estate as a partner or permit the surviving partner to buy out the interest of the deceased or retiring partner. A partnership may have great access to equity capital from retained earnings or from contributions from capital by new or existing partners from nonpartnership sources.

As the underlining shows, the real topic of the paragraph—partnerships in farming and ranching businesses—emerges as the grammatical subject only in the last sentence of the paragraph, and there in abbreviated form. In the other sentences of the paragraph, the topic is there, but instead of being featured in the subject slots of those sentences, it is tucked into prepositional phrases and other modifying elements. The net effect is that the thing actually being talked about is not boldly evident, and the sentences seem only loosely connected to one another.

For most writers, composing first draft paragraphs in a document is a slow process with many starts, stops, and trips to the coffeepot. It is a slow groping from sentence to sentence, a halting process in which the conceptual focus may easily be lost.

For readers, however, the act of reading the paragraph is a rapid matter. They pick up the concept of the subject in the first sentence. Then they follow that concept from sentence to sentence. Or at least they try to. Yet, if readers suddenly find new information, a new and unaccounted-for subject showing up in a subject slot, they are momentarily puzzled, if only unconsciously. Presuming they can pick up other cues of relationship, they can keep going. However, little increments of momentary uncertainty add up. As a result, readers might decide that a paragraph is incoherent even though, between the lines in the mind of the writer, there are logical connections from sentence to sentence.

In your editing process, then, take a close look at the grammatical subjects of the sentences in your paragraphs, especially those at the beginning. Ensure that what you have put into grammatical focus is really the conceptual focus you intend. If it is not, revise the sentences and the paragraph accordingly.

Now you might well be thinking, "Fine, I can make the grammatical subjects of the sentences model the content. But doesn't that imply that a paragraph can have only one grammatical subject for all of its sentences, as in the case of the nutrient load example, where the word *source* showed up in the subject slot of every sentence?"

Of course, the answer is no. Subject focus will shift from sentence to sentence much of the time. However, purposeful shifts in subject focus will connect the information of a new grammatical subject to established conceptual information from a previous sentence. It is only unpurposeful, abrupt shifts which you must avoid.

For example, consider the following paragraph from an industrial accident report:

> On July 9, 1989, Young Excavating company was digging a sewer-line trench for a mobile home park located in the southeastern portion of Webster Township. This trench was examined by an OSHA field inspector on that same date. He observed that the trench was open for a distance in excess of one hundred feet, that it was at least seven feet deep, and that it was nine feet wide at the top. A backhoe was continuing the excavation. Three employees were in the trench near where the backhoe was working.

At first glance, this paragraph looks very much like some of the preceding negative examples in that each new sentence introduces a new grammatical subject. But look more closely. You will find that only in the last two sentences is the new grammatical subject an unanticipated shift to new conceptual information. In the other sentences the new subjects are really not new information; they have been established as "old information" by what has gone before, as the following analysis indicates:

Company was digging trench
The trench was examined by field inspector
He observed. . .
A backhoe was continuing
Three employees were in the trench

The object slot of the first sentence (trench) yields the subject of the next sentence. The ending of the second sentence (inspector) yields the subject of the third sentence. Only with the fourth and fifth sentences do new subjects suddenly appear for which there has been no precedent in anything that has gone before. The backhoe and the three employees suddenly appear. In sentences 4 and 5 the reader needs to do a bit of mental gymnastics. How do these last two sentences connect to what has gone before?

The answer, of course, is that the final two sentences of the paragraph probably continue to report what the field inspector observed. The conceptual subject of each sentence probably is "the field inspector." He observed the trench; he also observed the backhoe and the three employees working in the trench. But notice that because the grammatical subjects shift abruptly to "backhoe" and "three employees," you have to guess that the final two sentences of the paragraph are reporting what the inspector saw. The sentences don't really say that, so you have to force a connection by accepting as fact something that is not said. That guess in any document would be risky, but here especially—in the context of a legal action—guessing about the intended meaning behind the actual sentences seems particularly dangerous.

To summarize, then, two additional editorial strategies at the paragraph level are to (1) establish and maintain a grammatical subject focus which directly reflects the conceptual focus which you plan to convey or (2) assure that shifts in grammatical subject focus from sentence to sentence are adequately justified by explicit use of familiar terms and concepts established in previous sentences. (Note: Be careful that you don't confuse readers by using synonyms which they might not recognize as synonyms.)

❑ 9.5 ❑ Use of Metalanguage Cues

Just as signs are used on highways to signal upcoming intersections, exit ramps, and merging traffic, metalanguage cues are words and phrases used to signal the upcoming "conditions" in a passage of text. Examples are *for example, therefore, however, consequently, first, second, third, in other words, on the other hand, further, finally, thus, in the same manner, also, next, to summarize, in conclusion.* These cues are what linguists call "metalanguage." They are "higher" (meta) language: language about language. They are words and phrases whose function is to guide the reader within a text rather than to inform the reader about the content of the text.[7]

Such terms can be powerful cues for readers as they move between sentences within a paragraph, so a useful editing strategy is to assure that you have used sufficient cues to keep readers "on track."

To illustrate, we can omit most of the content words from a passage of text and still see the basic design of that text merely from the metalanguage cues and a few content words:

> This memorandum summarizes the two main points which I wanted to stress in our conversation on Xxxxxxx. Xxxxxxxxxxxxxxxxxxxxxxxxxxxxxxx xxxxxxxxxxxxxxxxxxxxxxxxxxxxxxxxxxxxxxxxxxxxxxxxx.
>
> First, you are not xxxxxxxxxxx. Although I can understand that xxxxxxx xxxxxxxxxxxxxxxxxxxxxxxxxxxxxx, xxxxxxxxxxxxxxxx. For example, xxxxxxxxxxxxxxxxxxxxxx. Similarly, xxxxxxxxxxxxxxxxxxxxxxxx xxxxxxxxxxxxxxxxxxxxxxxxxxxxxxxxxxxxxxxxxxxxxxxxx. xxxxxxxxxxxxxxxxxxxx. In short, xxxxxxxxxxxxxxxxxxxxxxxxxxxxx xxxxxxxxxxxxxxxxxxxxxxxxxxxxxxxxxxxxx.
>
> Second, you have not xxxxxxxxxxxxxxxxxxxxxxxxxxxxxxxxxxxx xxxxxxxxxxxxxxxxxxxxxxxxx. Xxxxxxxxxxxxxxxxxxxxxxxxx xxxxxxxxxxxxxxxxxxxxxxx. For instance, xxxxxxxxxxxxxxxxxxxxxxxxx xxxxxxxxxxxxxxxxxxxxxxxxxxxxxxxxx. Xxxxxxxxxxxxxxxxx xxxxxxxxxxxxxxxxxxxxxx. Further, xxxxxxxxxxxxxxxxxxxxxxxxxxx xxxxxxxxxxxxxxxxxxxxxxxxxxxxxxxxxxxxxxxxxxxxx.
>
> These two criticisms are important for you note because to xxxxxxxxxxx xxxxxxxxxxxxxxxxxxxxxxxxxxxxxxxxxxxxxxxxxxxxxxxxxx xxxxxxxxxxxxxxxxxxxx. In closing, let me urge xxxxxxxxxxxxxxxxxxxxx xxxxxxxxxxxxxxxxxxxxxxxxxxxxx.

This example is from an actual memorandum written by a manager to a subordinate within a federal agency. Even without most of the content words, you can probably see that the manager has had a conversation with the subordinate, that he wants to put it on record, that he has two criticisms of the subordinate's work, and that for each he has given two specific instances. You can also see that the key paragraphs start with topic sentences, that the second paragraph ends with a restatement of its topic sentence, and that the memo ends by restating the fact that there are two criticisms and by emphasizing why they are significant.

As this example indicates, metalanguage cues can reveal structure and relationship, so you edit to assure that you have used such cues where they are needed. You also need to assure that you have used them consistently and accurately and that you have not overdepended upon them. Inaccurately or inconsistently used, these cues can send readers off in the wrong direction.

Overused, they can quickly become a wordy distraction rather than a useful guide.

For example, here is a paragraph in which there is so much metalanguage—and inaccurate metalanguage—that readers can almost lose sight of the content language:

> What, then, can be concluded from the results of the computer modeling? First of all, increasing the stinger width while leaving the router location fixed is essentially increasing the cross-sectional area of the cut at all router depths. As a consequence, the proportion of the power going directly to increasing the load will increase, while that being used by the routers will decrease. To the extent that the stinger width can be increased without reducing the router's effectiveness, this productivity gain can be realized. However, we would point out that as has been explained earlier, the present model is not capable of fully evaluating router effectiveness. Yet, our experience with the CB 301 would seem to suggest that a wider stinger may be possible without hindering the routers' ability to facilitate loading. If such is the case, the performance improvement indicated by this analysis may be realized at little cost. In light of the above, we can conclude that the analysis has shown the importance of experimentally determining the optimum router, stinger, and end-bit configuration.

Eliminating or reducing much of the underlined metalanguage here is both possible and desirable. As you can see, in its present form the metalanguage obscures the idea of what can be concluded from the investigation, and some of the cues of structure are actually misleading. For example, the phrase "first of all" suggests that there will be a second. Where is it? (It is an important point, but it is hard to spot in the original.) In most cases, the metalanguage is merely wordy. The phrase "to the extent that" is reducible to "if." The phrase, "However, we would point out that as has been explained earlier" is reducible to "However" or "But." The phrase, "In light of the above" can be reduced to "Therefore." "We can conclude that" can be eliminated.

When the excess is chopped out, the essential content emerges and can be stated much more directly. For example, here is a more direct version of the paragraph which uses only 100 words rather than the original 177 words, and yet it seems to focus more effectively upon the important conclusions:

> Computer modeling has shown that increasing the stinger width but leaving the router location fixed can increase the cross-sectional area of the cut at all router depths. This could allow more power to be used to increase the load rather than to drive the routers, and it could result in higher productivity at little cost. (Our experience with the CB 301 supports this conclusion.) However, the investigation has also shown that the present computer model is not fully capable of evaluating router effectiveness. Therefore, before proceeding, we must experimentally determine the optimum configuration of the router, stinger, and end-bit.

Well used, metalanguage can be a useful guide for readers. It can signal that an example *is* an example. It can signal that point three *is* point three. And it can do much to show subordination, coordination, causal relationship, chronological relationship, spatial relationship, and a host of other logical connections that exist in your mind as a writer but that might not be created in the mind of the reader. The trick is merely to signal these relationships accurately, economically, consistently, and only as needed to make the text flow smoothly from sentence to sentence.

To return to the highway sign analogy, you certainly wouldn't want inaccurate signs, nor would you want too many signs or ones with so many words on them that you couldn't read them in passing by at normal highway speed. Nor would you want signs which were inconsistent in shape despite consistent function. Good metalanguage cues, like good highway signs, can be helpful, but in your editing of paragraphs you need to assure that the cues you have given are really the ones intended and needed.

❑ 9.6 ❑ Formatting the Paragraph for Easy Reading by Hurried Readers

Writing paragraphs with clear structural patterns, providing clear topic sentences at the beginning, maintaining subject focus or shifting it logically, and using metalanguage cues skillfully will help you to ensure that your readers' task is an easy one. But even in the format of your paragraphs, you need to remember the hurried reader you inevitably will have for most of your documents. Here too you can perform simple editorial tasks that will make readers' jobs easier. In fact, there are three things you can do in your editing to improve the paragraph format.

Use Block Indentation

First, consider block indentation within paragraph clusters to make physically evident the hierarchical relationships that are there. For example, Figure 9.1 shows two different representations of a 4-paragraph cluster. On the left, there are no physical uses of subordinate relationship, so the paragraphs appear to be of equal importance. On the right, the use of horizontal white space—block indentation—signals that three paragraphs are parallel (coordinate) in their relationship and that they are subordinate to the first paragraph.

Physically signaling the relationships among paragraphs in this way provides readers with an effective cue of how things fit together within clusters of related paragraphs. It also provides a kind of "fast path" through the text in the sense that the degree of indentation signals relative significance. Readers can pay more attention to material that is signaled to be most important. Without block indentation, a skim-reader would have no way of knowing which paragraph to read most closely; however, with block indentation, the reader would receive a clear cue that the first paragraph merits more attention

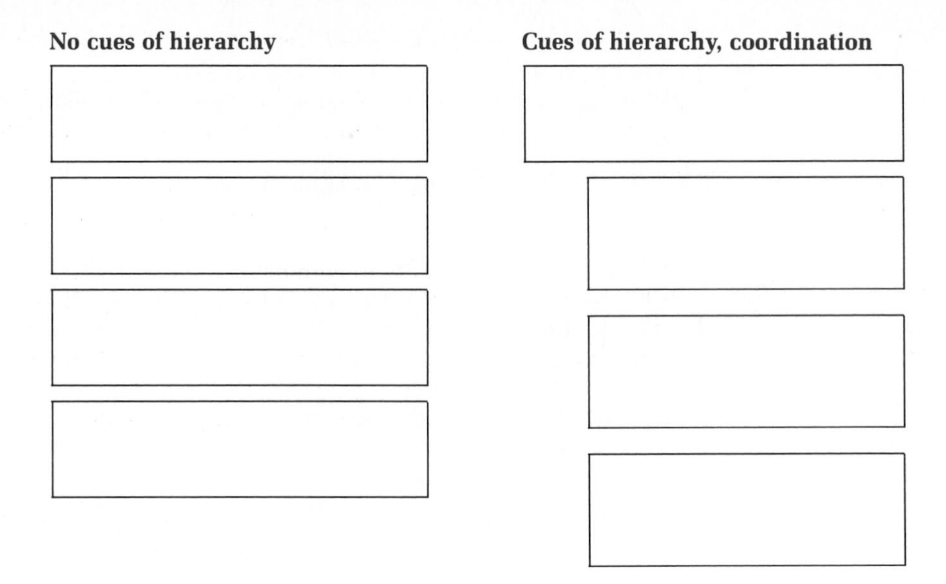

FIGURE 9.1 Use of Block Indentation to Signal Hierarchy.

than those that follow it. Admittedly, as you can see, there is a slight space penalty for using the block indentation: it takes slightly more pages to format text in this way. However, from your readers' point of view, that slight penalty is well worth it.

Use Formatted Lists

Second, consider using formatted lists—numbered or bulleted as appropriate—to signal coordinate relationships within paragraphs. (Numbers are used when the sequence is ordered and inclusive; bullets are used when the sequence is not ordered and when the list is merely illustrative.)

For example, here is a simple coordinate paragraph in which a 6-part list is presented without format cues:

> The next step will be to assess the impacts of the river basin development upon different sectors of the national economies. Such impacts include technological changes in agriculture with a shift to a more intensive agriculture with a much greater use of mechanization. They also include decline in percentage of food imports; much greater diversity in employment opportunities as a result of the creation of new enterprises in light industry, transportation and commerce; decline in fossil fuel imports due to the availability of hydroelectricity; increase in traffic and commerce both across the river and along it as a consequence of bridge and reservoir construction; and greater income from fisheries. This evaluation of the development of the river basin will examine the probable impacts of development in terms of the economic goals of the Member States.

Reformatted, the paragraph is easier to read more quickly:

> The next step will be to assess the impacts of the river basin development upon different sectors of the national economies. Such impacts include the following:
>
> - technological changes in agriculture with a shift to a more intensive agriculture with a much greater use of mechanization;
> - decline in percentage of food imports;
> - much greater diversity in employment opportunities as a result of the creation of new enterprises in light industry, transportation, and commerce;
> - decline in fossil fuel imports due to the availability of hydroelectricity;
> - increase in traffic and commerce both across the river and along it as a consequence of bridge and reservoir construction;
> - greater income from fisheries.
>
> This evaluation of the development of the river basin will examine the probable impacts of development in terms of the economic goals of the Member States.

In this second example, the use of a formatted list rather than an unformatted list makes for quick reading. Note that the text is the same, but the second format is almost certainly one you would find easier to read. There is a slight change in meaning, however. The unformatted paragraph subordinates five items to "technological changes in agriculture" by a transitional cue ("also"), whereas the list presents all six items in decreasing order of importance.

By the way, notice the use of the bullets. Here it would have been inappropriate to use numbers because the list is mostly illustrative ("include the following"). The items in the list are not inclusive and vary in degree of generality.

Of course, the urge to put everything into a formatted list can easily be overdone. If it is, the text will become less readable, not more readable. But look for opportunities to use formatted lists rather than unformatted lists. In instances like the preceding example, where there are more than three or four items in the list, a numbered or bulleted format might be a good choice.

Neither Overload nor Underload Paragraphs

A third suggestion is that you keep the number of sentences within any paragraph to a reasonable limit. Of course, interpreting this guideline is a highly subjective matter. You can't just break paragraphs in terms of some absolute count of sentences, even though in industry or business you probably will encounter someone who cites a "rule" that paragraphs should have x number of sentences. Yet there is truth in the fact that we all know an overloaded paragraph when we see one ("I don't want to read that"), and we can spot a series of underdeveloped paragraphs when we see them ("Looks like notes for a document rather than the document itself"). Again, research indicates that readers apparently do use paragraph length as one of their cues for when to break a paragraph into subparagraphs:[8] In a study mentioned earlier, researchers found the following:

1. Readers asked to insert paragraph breaks in unparagraphed text would not break paragraphs even at primary cues (such as a significant shift in subject focus) if doing so would result in a paragraph that was "too short." Sometimes they did break single sentences out into paragraphs, but the sentences in these instances tended to be relatively long (about 34 words in one experimental text and 42.5 words in another). In other cases, they did break pairs of sentences out into paragraphs, but again the sentences tended to be a little longer, with an average of 26.4 words in a passage averaging 22.8 words per sentence in one text, and 31.9 words in another text averaging 22.6 words per sentence.

   Generally, readers tended to break paragraphs in one experimental text at between one and five sentences, with a mean of 3.9 sentences per paragraph. In another experimental text, although the normal paragraph length was between two and five sentences, readers often broke at longer paragraphs, with from six, seven, or eight sentences and even a sprinkling of longer ones. (What people see as acceptable paragraph length is apparently influenced by context, content, and a variety of other factors that make it impossible to quantify an "appropriate" length. Still, it is clear from these experiments that paragraph length was one of the three significant cues that readers relied on in paragraphing unparagraphed text.)

2. When a paragraph began to seem too long to readers, they reread the text above it to look for secondary cues at which the paragraph might be broken. Topic shift was the primary cue, but content such as the parts of a list provided acceptable secondary cues.

3. If a paragraph was not too long, readers tended not to act upon the secondary cues, although as they read they did make mental note of where those cues were.

The basic point is that readers apparently do respond to paragraph length. They see shifts in subject focus as a primary cue, and they do react to secondary cues. But the one thing that triggers their decision on whether or not to break a paragraph is their apparent subjective decision on the appropriate length of the paragraph. From this we can conclude that as you edit, you should reassess paragraphs with too many sentences. You should also look again at skimpy paragraphs with only a sentence or two, especially if the sentences in those paragraphs are significantly shorter than the average for the text.

Conclusion

Although many writers tend to ignore paragraph-level editing and concentrate only on sentence-level editing, the readability of text can be greatly improved by applying the simple editorial strategies that we have discussed in this chapter.

Notes

1. As pointed out by Richard E. Young, Alton L. Becker, and Kenneth L. Pike, *Rhetoric: Discovery and Change* (New York: Harcourt, 1970) 317–26.
2. Francis Christiansen, "A Generative Rhetoric of the Paragraph," *College Composition and Communication* 16: 3 (October 1965): 144–56. Christiansen's analysis has been further extended by Willis Pitkin, who argues that there are really four patterns of relationship: superordination, subordination, coordination, and complementation. Although Pitkin's analysis is persuasive, for our purposes here the Christiansen analysis adequately suggests that paragraph patterns are form-based as well as purpose-based. See Willis L. Pitkin, Jr., "Discourse Blocs," *College Composition and Communication* 20: 2 (May 1969): 138–48.
3. Sandra Bond and John Hayes, "Cues People Use to Paragraph Text," *Research in the Teaching of English* 18: 2 (May 1984): 147–67. See also Frank Koen, Alton L. Becker, and Richard E. Young, "The Psychological Reality of the Paragraph," *Journal of Verbal Learning and Verbal Behavior* 8 (1969): 49–53.
4. In technical prose a conceptual paragraph is sometimes divided into several orthographic "paragraphs." There is no name for these groupings of miniparagraphs, so we call them "paragraph clusters." You should edit conceptually at the cluster level, not at the miniparagraph level.
5. David E. Kieras, "Thematic Processes in the Comprehension of Technical Prose," *Understanding Expository Text: A Theoretical and Practical Handbook for Analyzing Explanatory Text*, ed. Bruce K. Britton and John B. Black (Hillsdale, NJ and London: Lawrence Erlbaum, 1985) 98.

   Kieras has published several additional experimental studies relating to reading behavior and paragraph design. Among them are the following: "Good and Bad Structure in Simple Paragraphs: Effects on Apparent Theme, Reading Time, and Recall," *Journal of Verbal Learning and Verbal Behavior* 17 (1978): 13–28; "Initial Mention as a Signal to Thematic Content in Technical Passages," *Memory and Cognition* 8: 4 (1980): 345–53; "Component Processes in the Comprehension of Simple Prose," *Journal of Verbal Learning and Verbal Behavior* 20 (1981): 1–23; "Topicalization Effects in Cued Recall of Technical Prose," *Memory and Cognition* 9: 6 (1981): 541–49; "The Role of Major Referents and Sentence Topics in the Construction of Passage Macrostructure," *Discourse Processes* 4 (1981): 1–15; "A Model of Reader Strategy for Abstracting Main Ideas from Simple Technical Prose," *Text: An Interdisciplinary Journal for the Study of Discourse*, Special Issue 2 (1982): 47–81; and "Representation-saving Effects of Prior Knowledge in Memory for Simple Technical Prose," *Memory and Cognition* 11: 5 (1983): 456–66.
6. Kieras, 101–02.
7. Joseph M. Williams, *Style: Ten Lessons in Clarity and Grace* (Glenview, IL: Scott, 1985) 81.
8. Bond and Hayes, 159.

Chapter 10

Editing the Sentences

Developing an effective style requires more than merely meeting the demands of a handbook. Indeed, if you are serious about developing an effective style, you need to work at it over time, pay attention to what skillful writers do, and study the lessons in handbooks devoted exclusively to the subject.[1] Nonetheless, you can learn to recognize and deal with a few particularly troublesome stylistic habits. The purpose of this chapter is to identify frequent stylistic problem areas in technical and professional writing and to help you to know how to correct these problems. Specifically, we summarize and illustrate basic advice for dealing with problems associated with achieving correctness, clarity, and economy and directness.

10.1 Correctness

Many technical and professional people adopt questionable positions relating to correctness in language. They say, "Well, I never was very good in English," as if that somehow made them not accountable for mistakes in punctuation, diction, and spelling. Yet clearly they are accountable, and indeed careless style merely suggests careless professional work. After all, if readers cannot trust the writer to be exact with words, how can they trust the writer to be exact with anything else?

Moreover, to say that "I never was good in English" suggests that somehow language skill is not really part of one's professional responsibilities, that the technical and professional work can be separated from the communication aspects of that work. Yet, as we have seen, communication is not a minor complement to a professional person's role; it is an important, integral part of that role.

Correctness does matter; professionals are responsible for their use of language. When you send out a document over your signature, you assume responsibility for its language as well as its content. Even if you rely on a secretary to "correct mistakes"—a dangerous practice, as the evidence of any organization's files indicates—your readers take your words to be your own.

Therefore, you should take the time to learn to recognize some common problems of correctness and to develop an effective style.

Diction and Spelling

Basic errors in diction and spelling undercut your professional credibility. Accordingly, if your computer does not have a good spell-check program, get one and use it when you can. But don't assume that a spell-checker can recognize many of the errors that you may make. After all, spell-checkers recognize only words that don't exist in the program's dictionary; they will treat any correctly spelled word as "correct" no matter what its context. This means that you must carefully proofread your text for diction and spelling as well as use a spell-checker to catch typos.

Probably you have your own small number of diction and spelling problems. We all do. But, as any editor will tell you, many writers keep having trouble with a few basic words. Watch for the frequent confusions listed in Table 10.1.

In addition to avoiding confusions in diction, you do need to know how to spell even though there are such things as spell-checkers. After all, you won't always have a spell-checker available, and you shouldn't be a slave to the machine anyway. We frequently see words misspelled in technical and professional documents and have compiled a list of the most common (Table 10.2). If you don't already know how to spell these words, learn this list; you will be well on your way to control of spelling. And as you notice words you misspell, add them to this list or create your own list of troublesome spellings.

Punctuation: Commas

Commas are a mystery to many writers who never bothered to learn the rules. Instead, these writers rely upon punctuation by sound. Unfortunately, this practice causes them to make numerous comma errors, a few of which will

TABLE 10.1: Frequent Diction and Spelling Problems

affect/effect	*Affect* means "to influence, to have an effect on." As a verb, *effect* means "to cause." Obviously, these are totally different meanings, but we often see the words used interchangeably. The confusion is increased because *effect* can also be used as a noun meaning "result." Look closely at every use of either word.
analyze/analysis	Often misspelled as "analize" or "analisis" or "analyzsis."
choose/chose	*Choose* is a verb meaning to select. *Chose* is the past tense of that verb.
cite/site/sight	*Cite* is a verb meaning "to draw attention or to reference a source." *Site* is a noun meaning "location." *Sight* is either a noun or a verb meaning "something viewed" or "the act of viewing."
criterion/criteria	Many writers use the word *criteria* as both a singular and a plural. Actually, the singular is *criterion*. *Criteria* is plural.
datum/data	Again, many writers use *data* as both singular and plural, but in this instance they are probably safe except in formal writing. *Datum* is actually the correct singular, but it is so seldom used in informal writing that you may actually confuse readers by using it.
e.g./i.e.	The abbreviation e.g. stands for the Latin *Exempli gratia*, meaning "for example." It is often confused with *i.e.*, the Latin *Id est*, meaning "that is." But why write in Latin anyway? And why abbreviate? But do note the correct punctuation: (e.g., or i.e.,), if you choose to use either term.
foreword/forward	*Foreword* refers to the part of a publication which comes before the main text, literally a "word before." It is often confused with *Forward*, which means "onward" or "Charge!"
its/it's	*Its* is a possessive pronoun, like *his* or *hers*. *It's* is a contraction meaning "it is."
lead/led	*Lead* is either a base metal or a verb meaning "to conduct." *Led* is the past tense of that verb.
loose/lose	*Loose* is an adjective meaning "not rigidly fastened." *Lose* is a verb meaning "to misplace."
personal/ personnel	*Personal* is an adjective meaning "private" or "relating to an individual." *Personnel* is a noun, usually meaning "employees."
precede/proceed	*Precede* is a verb meaning "to go before, to take precedence." *Proceed* is a verb meaning "to go ahead" or "to begin."
principal/ principle	*Principal* is an adjective meaning "main" or noun meaning "chief" or "head of a school." *Principle* is a noun meaning "general truth or rule."

quiet/quite/quit	*Quiet* is an adjective meaning the opposite of noisy. *Quite* is an adverb meaning "very, totally, fully." *Quit* is a verb meaning "to stop or abandon."
then/than	*Then* is an adverb meaning "at that time." *Than* is a conjunction used after an adjective or adverb to express comparison.
who's/whose	*Who's* is a contraction, meaning "who is." *Whose* is the possessive case of *who* or *which*.
who/whom	Many writers are unsure about these words. Just remember that if the word is an object, any object, *whom* is correct in formal English (informal English allows *who* as an object except after a preposition). Otherwise, use *who*.
would/had	In speech, many people incorrectly express the subjunctive by saying something like "If I would have . . ." (often pronounced "woodov"). (A subjunctive expresses a condition contrary to fact.) Their confusion leads them to write the subjunctive as "would have" or, worse still, as "would of." (Negative example: "If I *would have been able* to get there, I would have come." The first "would have" is incorrect, the second correct. The sentence should read, "If I had been able to get there, I would have come." Or, in more formal English, "Had I been able to get there, I would have come.")

cause genuine confusions in meaning and most of which will cause readers who know proper punctuation to lose confidence in the writer.

Fortunately, the list of problem areas is quite short. Indeed, there are only four comma rules you really must learn to be relatively safe. The first three concern the use of commas with main clauses (also called "independent clauses"), which are clauses that can stand alone as sentences when all subordinate clauses and modifying phrases are deleted. A clause is a construction with a noun or noun phrase as a subject and a verb that describes an action or condition; a main clause is a clause that can stand alone as a complete sentence.

Comma Rule 1 Use commas to separate main clauses joined by coordinating conjunctions (*and, but, for, or, nor, so,* and *yet*).

When you join two main clauses, you have a "compound sentence." The comma alone is not strong enough to bear the load of connecting the clauses in a compound sentence. The difficulty is that many writers confuse a *compound element* in a sentence with a *compound sentence*. "A letter grading system and a numeric grading system are both in use in the bank's commercial loan department" is a simple sentence with a compound subject. In contrast,

TABLE 10.2: Commonly Misspelled Words

acceptable	environment	omission	recommendation
accidentally	equipment	omit	referring
alternative	equipped	omitted	remember
among	excellent	omitting	repetition
analysis	existence	opposite	reservoir
appearance	experience	originally	resistance
approximately	explanation	parallel	safety
argument	finally	participation	satellite
article	generally	particle	schedule
attached	height	penetrate	secretary
attachment	immediately	perceive	separate
availability	interest	perform	several
average	interference	permanent	significance
basically	irrelevant	phase	significant
benefit	judgment	pollute	sincerely
business	knowledge	pollution	source
chief	length	possible	specimen
commitment	license	possibly	strength
committee	maintenance	practical	succeed
consensus	manageable	preparation	summary
consistent	management	prevalent	susceptible
controlled	manual	procedure	technique
convenient	material	professional	temperature
dealt	mathematics	programmer	tendency
definitely	nuclear	programming	transferred
description	occasion	publicly	typical
desirable	occasionally	quantity	until
divide	occur	receipt	visible
emission	occurred	receive	wholly
entirely	occurrence	recommend	yield

Others? Add your own list of troublesome spellings.

"We currently use a letter grading system, but we have been asked to convert to the numeric grading system" is a compound sentence—two sentences joined with a comma and coordinating conjunction.

Just remember, if each of the two halves of the sentence can stand alone, and you have connected them with *and, but, for, or, nor, so,* or *yet,* you must put a comma before the coordinating conjunction. This actually serves two functions: the coordinating conjunction provides the additional strength to join the two main clauses; the comma signals to the reader that the connective is being used as a coordinating conjunction. (We recommend you use the comma with the coordinating conjunction all of the time, although "newspaper style" omits the comma in short, uncomplicated compound sentences.)

Examples

The pump was tested at the Research Center, and the test results must be sent to the supplier.

The preliminary design is complete, but releases have not been obtained for all of the property in question.

Negative Examples

The bearing was found to be damaged, and was replaced. (Compound verb, not compound sentence.)

The engine has two oil filters, and a 50-quart oil refill capacity. (Compound element, not compound sentence.)

A loan account and a deposit account are used, but combined in an innovative way. (The second part of the sentence cannot stand alone because it lacks a subject, which is implied; the comma probably has been used because the subject is clearly implied. To be correct you should either add the words "these are" and keep the comma or just omit the comma.)

Programs have been written to handle this problem but neither the hardware nor the software has been documented. (Needs comma before "but.")

Comma Rule 2 Use commas to set off introductory sentence elements (such as transitional words and phrases, adverb clauses, introductory phrases, and subordinate clauses) which come before the main clause.

Sentences often begin with miscellaneous introductory elements which effectively "put the main clause on hold" until the preliminaries are out of the way. In most cases, these preliminary elements cause no real confusion whether they are punctuated or not. But sometimes, especially in difficult technical text, it is hard to tell when the preliminary element ends and the main part of the sentence begins. Accordingly, it is a good standard practice to separate all introductory elements with a comma. Do it all the time, and you won't confuse readers in those few instances when it really matters.

Examples

After component checks, the tests were repeated.

To this end, we are developing a slide catalog so that new ideas will be readily available to other groups.

After taking off and climbing to test altitude, we used a 6-step procedure to collect test data.

Negative Examples

Since we are primarily interested in lift-off accelerations operating the data-collection equipment only during lift-off makes sense. (Probably you misread the sentence because a comma after "accelerations" is missing. Hence, you couldn't tell where the introductory element ends and the main clause begins.)

In an attempt to rectify this problem school trustees requested that an alternative budget be prepared. (Needs comma after "problem.")

Once in the area inspectors are conducting extensive interviews in each building. (Needs comma after "area." Again, you may have misread the sentence because you couldn't tell where the main clause begins.)

Although I am confident that this report will answer your important questions about the field testing contact me at any time if I can be of help. (Needs comma after "testing," to set off introductory subordinate clause. Note that the line spacing may throw readers off because the subordinate clause appears to end with "questions.")

Comma Rule 3 Use commas to set off nonrestrictive modifiers as well as other parenthetical sentence elements. (Note: Pay special attention to this rule. Your meaning can get really confused here if you do not know when to use commas and when *not* to use them.)

Modifiers (words or phrases) may be either "restrictive" or "nonrestrictive," depending upon whether you set them off with commas. If they are restrictive, they qualify the meaning of the subject or object in the main clause: they are essential to its meaning. If they are nonrestrictive, they add parenthetical information about the subject but do not qualify its meaning.

For example, the sentence "Laborers who lack seniority will be laid off first" means something completely different from the sentence "Laborers, who lack seniority, will be laid off first." The first sentence contains a modifier (who lack seniority) which qualifies or "restricts" the term to which it is attached (laborers). The first sentence is not about all laborers; it is about only a subset of laborers, those who lack seniority. The modifying phrase is a "restrictive modifier." In the second sentence, the same modifier is set off by commas. The second sentence is about all laborers, not just some laborers. The phrase "who lack seniority" becomes merely incidental, parenthetical information about all of the laborers; it does not qualify the meaning of the term. That makes the modifying phrase in the second sentence "nonrestrictive"; it can in fact be dropped from the sentence without changing the basic meaning.

This difference between restrictive and nonrestrictive modifiers is a significant difference. Learning to recognize this difference is crucial.

Examples

Supervisors, who are required to file monthly activity reports for their units, must follow the newly approved reporting guidelines. (The modifier is nonrestrictive: all supervisors, the subject of the main clause, must file reports and must follow the guidelines. The presence of the two commas makes this a nonrestrictive modifier.)

Supervisors who have not yet taken the CPR class are required to sign up before the end of the month. (The modifier is restrictive: some supervisors have already taken the course; only those who have not yet done so must sign up. The absence of the commas makes this a restrictive modifier.)

Negative Examples

All alumni, who are to be interviewed, live within a 40-mile radius of the University. (This sentence says that all alumni of the University live within a 40-mile radius—probably an illogical statement that would not apply to any university in the United States—and all will be interviewed. Without the commas, the sentence would refer only to a subset of alumni, those who are to be interviewed.)

This will create drag which allows the MOAT to match the orbital speed of the space station. (Is it "drag which . . ." or "drag, which . . ."? Actually everything in the modifying phrase "which allows . . . station" is nonrestrictive. Hence the comma is needed.)

The necessary gas flow which was found to be 9,900 cubic feet per hour does not meet the flow rate specified by FEC. (The phrase "which was found . . . hour" is really nonrestrictive, additional information. It needs to be set off by a pair of commas.)

Comma Rule 4 Use commas to separate items in a series.

Most people know that they should use commas to separate items in a series. However, there is a difference between the "newspaper-style" rule that most people know and the technical-style rule appropriate for technical and professional documents. That is, newspaper convention minimizes the number of commas, using them to separate all items in the list except the last item. (Example: Buy carrots, beans, potatoes and corn. Note that there is no comma before "and.")

Technical style differs from newspaper style in that it usually includes the comma before the last item. (Example: Calculations will be based upon the following assumptions of weight for the various models: 1313 Kg for Model 102, 1563 Kg for Model 103, and 1624 Kg for Model 104.)

The rationale for using this extra comma before the last item in a series is that sometimes in technical text it is difficult to tell what constitutes a single item. For example, is "ball and socket" one item or two items?

Punctuation: Semicolons

Like commas, semicolons confuse many writers, but again there are only a few rules that cause much difficulty for most writers. In fact, there are only two that cause a lot of trouble.

Semicolon Rule 1 Use semicolons to link closely related main clauses which are not linked by a comma and coordinating conjunction (*and*, *but*, etc.).

If you know that a semicolon essentially is interchangeable with a period, you can see how this rule applies. A semicolon is used in place of a period when you want to join closely related main clauses rather than separate them with a period. (Note the similarity of this rule to Comma Rule 1. The essential conditions are the same, but there is one difference—no coordinating conjunction. The semicolon can be used instead of a comma and coordinating conjunction.)

Again, the key is to recognize a main clause. Remember, a main clause is a complete sentence that can stand alone. When one such clause is connected to another such clause, but without the usual comma and coordinating conjunction, a semicolon is the linking punctuation. In fact, perhaps semicolons should be called "semiperiods" because, like periods, they often mark a full stop; yet, unlike periods, they indicate close relationship.

Examples

The forces cannot be attributed to the springs alone; they are the result of the entire linkage and valve system. (Each clause is a main clause, so a period could be used to separate the two; however, because the clauses are closely related, a semicolon is used.)

The seal wear evident during the two inspections was almost identical; the seals were flat, and wear bands covered approximately 60% of the surface. (Again, the two clauses are main clauses which are closely related in concept. Hence a semicolon is a reasonable choice of punctuation to show the close relationship. An alternative would have been to use a colon, which would have introduced a 2-item list.)

Negative Examples

During the inspection, the left grapple tube was found to be loose, probably it was not tightened during the initial construction. (Here there are two main clauses, no coordinating conjunction, yet close relationship between ideas. Hence a semicolon would have been correct after "loose." The writer's intuition was good: he or she recognized that a period was too much separation; however, a comma won't suffice here. In fact the use of a comma in this situation is a common mistake called a "comma splice.")

This initially would seem very advantageous to the customer, however, our obligation is to advise the customer as to how he or she can obtain the best return on the investment. (The comma before "however" should be a semicolon because these are two main clauses separated by a conjunctive adverb. Like *and*, *however* can be used as a conjunction. Also, it can be used as an adverb, and as an adverb, *however* is set off by commas. Hence, the source of confusion.)

Normally, the framing is attached to the inside layer of fiberglass, in this case, the structure will be in direct contact with the foam and the inside layer of fiberglass. (Another comma splice. The sentence needs a semicolon before "in this case.")

Semicolon Rule 2 Use semicolons to separate list items which themselves contain commas. (Note the similarity of this rule to Comma Rule 4; here the semicolon has to be used to differentiate items in a series because the items themselves have commas.)

Example

A 3-stage process is required: making a preliminary plot, called a "compressor map"; determining the conditions on the compressor map at which conservation of mass is satisfied; determining the conditions under which the turbine power and compressor power are equal. (Here there is a simple list with three elements. Of these three, the first includes a comma. Hence, to separate the three elements clearly, commas do not suffice; instead semicolons are required.)

Negative Example

Three wing configurations have been tried: the "swing wing," which can be swung from almost straight forward to almost straight back for high-speed flight, a "delta wing," which uses a large vortex on the top surface to achieve its low-speed characteristics, and a "standard" swept-back wing, which employs classical aerodynamics. (There are three elements in the list, but each element requires an internal comma. Hence to separate the items clearly, semicolons are needed after "flight" and after "characteristics.")

Punctuation: Colons

Colons are appropriately used to introduce lists, although many writers make the mistake of trying to introduce lists with semicolons.

Example

A 3-stage process is required: making a preliminary plot, called a "compressor map"; determining the conditions on the compressor map at which conservation of mass is satisfied; determining the conditions under which the turbine power and compressor power are equal. (Note the colon introducing the 3-part list.)

Negative Example

The proposed disposal site would accept only the following items; broken concrete, broken asphalt, gravel, and clean fill sand. (The semicolon is not appropriate here because there is a 4-part list following an introductory sentence. A colon is required. Remember, the semicolon is not a cousin of the colon.)

Punctuation: Hyphens

Hyphens are handy punctuation to help readers to see the relationships among clusters of related words. Unfortunately, many technical and professional writers almost completely ignore their use. For example, consider the phrase, "five fin per inch radiator." It really consists of two ideas expressed in five words. "Five-fin-per-inch" is effectively one adjective idea, as the hyphens can indicate; "radiator" is the second idea, a noun idea. Hence, with hyphens, the phrase is easier to understand: "Five-fin-per-inch radiator." (Of course, one could make it even easier by saying, "radiator with five fins per inch.")

Learn to look for opportunities to use hyphens within clusters of words which effectively form one adjective, one noun, or one verb.

Here are some examples:

"A low-temperature, high-humidity environment" rather than (as written by the original author) "a low temperature high humidity environment."

"Hard-starting problems" rather than "hard starting problems."

"Out-of-plane rotation" rather than "out of plane rotation."

"Injection-molded seals" rather than "injection molded seals."

"Four-disk arrangement" rather than "four disk arrangement."

"Load-carrying capability" rather than "load carrying capability."

An additional comment about hyphens. Often we see justified right text (squared off on the right margin) in which, for the sake of a neat right margin, words have been broken into syllables, hyphenated, and put onto two lines. This practice may cause the lines to look neat, but when hyphens separate the parts of a word onto two lines of text, readers have to slow down. Avoid this use of hyphens unless there are no alternatives.

Parallelism

Readers expect similar ideas to be presented in parallel form. Unfortunately, when we write, we move so slowly (and have so many other things to worry about) that it is easy to drop out of a consistent grammatical form. In fact, it is so easy that we can almost state the axiom that you will find lack of parallelism in almost any first draft. Hence, editing is necessary to assure that coordinate elements are put into consistent grammatical form.

Example

The following design changes have been made: the solenoids redesigned to operate at 10 volts rather than 12 volts, the switch covers silver-plated, the electronics hard-potted, and the actuators sealed to reduce leakage. (Note that the four elements of the list are in exactly the same grammatical form.)

Negative Example

Recommendations:

- Request the supplier to improve his switch design by adding potting, adding a positive mechanical connecting pin, and identify the cause of decreasing dead-band.
- Engineering should identify another switch supplier to develop a switch for the high-pressure range.
- Engineering should list maximum rate of pressure rise on the functional specification.

(Note: there are really two lists here, as the first item itself is a list. Both are inconsistent in form.)

Subject/Verb Agreement

Few professionals actually speak ungrammatical English by using singular verbs with plural nouns and the reverse. However, they easily get tricked into mixing singular and plural elements when writing. Specifically, you will make mistakes if you take a wrong cue from collective nouns and compound subjects, usually due to excessive separation of subject and verb.

Negative Examples

Originally, the Alpha grading system, A–EL, for commercial loans were used in our Indiana branches. ("Were" agrees in number with "loans," but the word with which it must agree is "system.")

Inspection of the modal analysis results show that there are two critical modes. ("Show" does not agree in number with the subject, "Inspection." However, you can see that the word "results" probably tricked the writer into thinking that a plural was needed.)

Plots of the mode shapes for a representative case of hydraulic cylinder spring rates is presented on Sheets C and D. ("Plots are presented" would be correct, but the excessive separation of subject and verb tricked the writer into thinking that "case" was the controlling word.)

A different set of labels are needed. ("Set" is a collective noun that requires a singular verb.)

The significance of the follow-on testing is that the repeatability and predictability of the screening test was demonstrated. (Both "repeatability" and "predictability" were demonstrated. Here the compound element (two things) requires a plural verb, not a singular verb.)

None of the results are available at this time. (*None*, which literally means "no one," takes a singular verb. Hence, the correct sentence is "None of the results is available at this time." Many writers think the correct sentence looks incorrect, but just remember that "none" takes the singular.)

Unclear and Dangling Modifiers

Writers make so many mistakes in this area that some have trouble seeing that there is a problem.

For example, many readers see nothing wrong with this sentence: "By lining the air cooler tubes with rubber, the corrosion problem can be eliminated."

But look again. The modifier "by lining the air cooler tubes with rubber" has nothing to modify. Someone has to line the tubes, but there are no humans in the sentence. Hence the modifier will (must) attach itself to the first available noun, "corrosion problem." The sentence therefore seems to say that "a corrosion problem lined the air cooler tubes with rubber." Obviously, that can't be.

Another example: "By choosing a particular value for x, a straight line is defined on the plot." Again, there is nobody in the sentence to "choose a particular value for x." Hence, again the modifier must attach itself to the first available noun, "a straight line." Thus the sentence seems to say, "a straight line chooses a particular value for x." Again, an illogical statement.

Recognize the symptoms of this problem, or you are quite likely to fall into this common dangling modifier trap.

First, notice that each of the example sentences begins with a participial or verbal phrase, a phrase that contains an action. (A cue is the *-ing* or *-ed* ending on the verb.) That phrase must modify a noun (or pronoun) which is the agent of the action described in the phrase; this noun is the subject of the main clause. Second, notice that the main clause of each sentence uses a passive verb (the corrosion problem "can be eliminated" and a straight line "is defined.") In each case, the passive verb puts an object of an action, not the logical subject of the action, in the subject slot of the sentence. Thus, the introductory phrase modifies a grammatical subject but not a logical subject. The source of the action, the real subject which the introductory phrase should modify, is not in its appropriate position, if it is in the sentence at all.

Notice another feature of each of these two sentences. There is no human being in either. This impersonality, combined with the passive verb, almost assures that the phrase will have nothing to modify.

To summarize, then, your introductory participial phrases will dangle if three conditions are met: (1) you begin your sentences with participial phrases, (2) you keep human beings out of your sentences, and (3) you write in the passive voice.

Examples

By lining the air cooler tubes with rubber, we were able to eliminate the corrosion problem. (This is correct. "We lined the tubes . . . and we eliminated the problem.")

By choosing a particular value for x, we define a straight line on the plot. (Again, this is correct. "We choose the value . . . and we define the straight line.")

Of course, many writers say, "But I'm comfortable with the passive verb here because I don't want to focus on the 'doer' of the action but upon the action itself." If that is your position, then a verbal phrase is not a good choice for beginning the sentence.

A workable solution is to recast the phrase as a conditional clause: "If the air cooler tubes are lined with rubber, the corrosion problem can be eliminated." "If a particular value for x is chosen, a straight line will be defined on the plot." After those revisions, there is nothing grammatically wrong with either sentence, although in each case the sentences do remain passive.

Just remember, introductory participial phrases and passive verbs don't mix well. Here are some further examples, all taken from one small set of reports handed in by seniors and graduate students in a technical writing class.

Negative Examples

After taking off and climbing to the test altitude, a 6-step procedure was used to obtain the test data. (The procedure took off and climbed?)

Looking at the permeabilities, the gravel has a value of approximately 10,000 ft / day. (The gravel looks at permeabilities?)

When analyzing the current method, it became apparent that reducing the walking time could significantly improve performance. ("It" analyzed the current method?)

By raising the austenitizing temperature, the solubility of the carbon in the austensite is increased. (The solubility raises the temperature?)

By using a specialized shaft sleeve and movable housing seal, three basic tests may be run. (The tests use a shaft sleeve and housing seal?)

By simply changing the attachments, nearly all the desired testing parameters can be achieved. (The testing parameters change attachments?)

In executing a safe, low-level turn, preparation should be made by visually cross-checking the nose track with respect to the horizon. (The preparation executes a safe, low-level turn and visually checks the nose track?)

By having employees evaluated by more than one person, these biases will be reduced. (The biases have employees evaluated?)

❑ 10.2 ❑ Clarity

All of the correctness issues we have discussed (and many others) can cause problems of clarity. Yet there are other sources of difficulty that you should also recognize. Specifically, we note three particular problem areas: noun strings, unclear pronouns, and unclear synonyms.

Noun Strings

Technical people often string nouns together to make names for new things. Example:

Laser
Laser Printer
Laser Printer Paper
Laser Printer Paper Tray
Laser Printer Paper Tray Retainer
Laser Printer Paper Tray Retainer Spring
Laser Printer Paper Tray Retainer Spring Clip
Laser Printer Paper Tray Retainer Spring Clip Fastener
Laser Printer Paper Tray Retainer Spring Clip Fastener Screw

As you see, one can just keep adding nouns. In each case, however, the last noun in the string is the thing really being talked about; all of the other nouns effectively become adjectives modifying that noun.

You might regard this manufactured example as an absurd one, but it is surprising how many difficult noun strings one finds in technical text. For example, the following strings were found in a few reports in one company:

Main chain bushing torsional strain

Field test injection molded crescent seals

Tree harvesting operations primary piling

Torque divider planet carrier spline wear

In the same company we found one writer with a special flair for noun strings, as this sentence suggests:

> The XXX piston ring damped lube oil pump relief valve and improved rubber endsheet baffle retention system cooler field test plan is presented on Attached Sheet A.

The rule of thumb here is simple: "unstring noun strings." Put the real noun, the last in the string, first. ("The plan for . . . is presented on attached sheet A.") This revision will get the main noun into an initial position, whether as subject or object. That will help readers immediately identify the essential topic of discussion.

Realize that this suggested revision is not merely a matter of stylistic preference. From research, we know that noun strings require more cognitive processing than "unstrung" nouns.[2] That is, noun strings require more time of readers and produce less comprehension. Hence, to improve the comprehensibility of your technical text, unstring noun strings.

Ambiguous Pronouns

In any text, you should be able to draw an arrow from every personal, relative, and demonstrative pronoun to its specific antecedent. However, the first draft of almost any technical document has numerous "its," "this's," "that's," and "which'es" floating around unclearly, attaching themselves to unclear antecedents or wrong antecedents.

As negative examples, here are sentences in which the underlined word could logically attach to several antecedents:

> For simplicity, I will use the A procedure for analysis. *This* will easily provide an indication of the required current when the system is used with mean rated wind velocity. (Does "this" refer to "procedure"? Or does it refer to the "use of the procedure"?)
>
> The starting torque of the cylinder remained at 3.3 oz, before and after running the experiment for 30 minutes. *It* was repeated a third time after several days, yielding the same results. (What does "it" refer to?)
>
> Further tests are needed to confirm fuel consumption prediction capabilities. *This* is not expected to be a problem, though, due to the excellent correlation shown for engine speed, torque, and manifold pressure. (What does "this" refer to? Does it mean that testing should be no problem, that confirmation should be no problem, or that prediction should be no problem? One just can't tell, although no doubt the writer knew exactly what he or she meant.)
>
> The output shaft bore was found to be undersized, *which* may have decreased the running clearance with the input shaft. *This*, plus the decreased lube flow, contributed to the bearing seizure. (What does "which" refer to, and does "this" refer to the fact that something was found, to the fact that something was undersized, or to the fact that something was decreased?)

A good strategy is to look closely at the pronouns in your text to assure that they connect directly to only one antecedent. If you are at all in doubt, either replace the pronoun by repeating the noun or add the noun to the demonstrative pronoun. ("This decreased running clearance . . .")

An additional comment about the relative pronouns *which* and *that*. Many writers assume that they can improve their text by hunting down and removing the "which'es" and "that's." (Sometimes called a "which hunt.") On the surface, removing these words often seems to make sense. For example, in the sentence, "We opened the cannister which was recovered from the crash site," it is tempting to simply remove "which was." The sentence would then read, "We opened the cannister recovered from the crash site." The "which" clause is reduced to a phrase, which seems to make the sentence more concise and direct. Actually, however, researchers tell us that these relatives help readers to process the information more efficiently. Although it is not intuitively obvious, your better editorial strategy is to leave in some "which'es" and "that's".[3]

Unclear Synonyms

Writers often fear that they bore the reader by repeating the same terms, so they prefer instead to provide "variety" by switching to synonyms. Unfortunately, in difficult technical text this strategy works only if the readers recognize the terms to be synonyms. Otherwise, they may think that a new topic is being introduced.

Negative Example

After 2072 hours, the tracks were removed from the machine for a bushing-turn operation. This inspection was reported in Letter 41, File 12345. The analysis revealed that the track groups performed satisfactorily in a low-impact, high-abrasive environment but that the seals were flattened.

In the example sentence, the first key term is "tracks." Later the term "track groups" is introduced. Are they the same thing or two different things? Also note that the first sentence uses the term "operation," the second sentence uses the term "inspection," and the third sentence uses the term "analysis." Are these three different activities, two activities, or one activity?

In the mind of the writer, of course, there is no confusion here. He or she knows precisely that "tracks" and "track groups" either are or are not synonyms. But readers do not have that same background knowledge, so they have to guess. If they guess wrong, the writer could be in trouble. (Only 4 of 18 engineers in an in-house technical writing course in the corporation knew whether or not "tracks" and "track groups" were synonymous.)

For clarity, then, repeat the same terms. Also, you may find it helpful to use appositives, which use two terms but put them together ("My cousin, a lawyer, . . ."). Appositives allow you to introduce synonyms in a way that clearly connects them in the reader's mind. Once the connection is established, you can use either term in subsequent text.

❑ 10.3 ❑ Directness and Economy

Good technical prose is direct and economical as well as correct and clear. Of course, numerous factors can interfere with any one of these characteristics. Yet, as we did with correctness problems and clarity problems, again we can focus on a small number of frequent difficulties, and again for each we can give some specific suggestions for making changes in your rough draft texts. Specifically, four common problems cause indirectness and wordiness.

Needless Passivity

You have heard a lot about not writing in the passive voice, and it is certainly true that the active voice is more economical and more direct than the passive most of the time. Yet we won't tell you that the passive voice is evil. It isn't. It performs the helpful function of shifting attention from the actor to the effect or what is affected. That is, you could write "A causes B" or "B is caused by A." Which choice is right depends entirely upon what you want to put into focus by putting it into the subject slot of the sentence. Moreover, the appropriate choice depends upon what the surrounding sentences do; you will confuse your readers if you switch from the active voice in one sentence to the passive voice in the next.

If the passive voice is often useful and necessary, nonetheless many sentences in documents are needlessly passive. This is a problem in several senses. First, research indicates that, in many cases, active-verb sentences are easier to understand and can be recalled better than passive-verb sentences.[4] Second, the active voice has the advantage of putting the conceptual subject of a sentence near the beginning of the sentence, not near the end. Third, the active voice usually requires fewer words than the passive voice. In short, if there is a good reason to use the passive, by all means do it. But if there is not a good reason, the active voice is much the better choice for most of your sentences.

Unfortunately, many professionals have developed a heavily passive style that causes their prose to be indirect, wordy, and often awkward. A major source of this needless passivity is the use of verbs describing human behavior in sentences which have no humans in them. Writers incorrectly equate the active voice with personal prose, and passive voice with "objective" prose. They keep humans in their sentences by using human-behavior verbs, but "hide" the humans by making those verbs passive. Objectivity, however, is in your analytical method, not in the fact that no human performed the analysis.

A better solution is either to put the humans entirely into the sentence ("We expect") or to take them entirely out of the sentence ("Published information indicates"). For example, consider this sentence: "From published information it is expected that the 4140 steel will maintain its hardness throughout the entire specimen." Note the passive verb, "is expected." Note also that there are no human beings in the sentence to "expect" anything. Humans "expect" things, but no human is in sight.

Of course, you could fix the sentence by putting a human being into it, saying "I expect" or "the staff expects" or "Charlotte expects," but is there really any reason to use the verb "expect" at all? Do you need to introduce the "expecter," a human being, into the sentence? Why not simply express the idea directly by saying, "Published information indicates that the 4140 steel will maintain its hardness throughout the entire specimen." That makes the verb, "indicates," a perfectly meaningful active verb. Or turn the sentence around: "The 4140 steel should maintain its hardness throughout the entire specimen, as published literature indicates." Now the verb is "should maintain," another good, active verb. Furthermore, the subject has now shifted to the apparent real topic of the sentence, steel.

As you can see from this example, you can shift the verb to the active voice without introducing a human being into the sentence.

Negative Examples

It is expected that the X-groove drag will be reduced in the recovery region. (Humans "expect," but no humans are in the sentence. Note that the writer probably chose "expect" to soften his or her prediction that the X-groove drag will be reduced. Accordingly, you could revise the sentence to say "The drag should be reduced . . ." or, if you are more certain, you could say "the drag will be reduced." Notice that the verbs in both revised sentences are passive, but they are deliberate passives which focus on effect, not on cause. In this case, as in many sentences, it is the *needless* passive that causes problems—not the passive itself.

It was found that the fatigue life of the part would be exceeded in a relatively short time under these conditions. (Someone found something, but who? Either say "we found" or just eliminate the human-behavior verb and start the sentence with "the fatigue life . . .")

Agreement was reached to include only composition and measurable processing variables in the field of potential variables. [Someone agreed to something, but who? Either put the humans in ("The group agreed . . .") or take the humans out entirely ("The field of potential variables will include only composition and measurable processing variables.").]

It is felt that no further complete cooling tests should be required on the 220 series. (Someone "felt"—that is, "thought" or "concluded" something. But who? Either say "Smith concluded" or just start the sentence with "No further cooling tests . . ." By the way, most of the time professionals don't "feel," they "conclude.")

The output shaft bore was found to be undersized. (Someone found something, but who? Just drop the phrase, "found to be.")

It was noted that a dynamic timing band of from three to four degrees was experienced while operating in Notch 2 conditions. (Who noted? Was experienced by whom or what? While what or who operated? Note that the participial phrase, "operating in Notch 2 conditions," dangles because it has no noun to modify.)

Learn to recognize passive verbs. Then when you edit, simply look at each passive to determine if it is a deliberate passive, intended to focus on the effect of an action rather than on the cause of an action. If it is, keep the passive (assuming that it fits the surrounding context). If not, and if the verb is a "human action" verb, change the passive to an active. Either introduce the human actor or eliminate the passive "human action" verb.

Redundancy

Redundancy is not repetition; it is saying essentially the same thing more than once but in different ways.

A common problem is that first-draft texts often include redundant expressions. These redundant expressions are caused by the writer's working simultaneously on two levels of generality—an abstract level and a concrete level. For example, the phrase "a duration of two minutes" really says the same thing twice: "duration" is a measure of time (an abstract measure); "two minutes" is a measure of time (a specific measure). Hence, the same idea is expressed, but on different levels of generality. And think how often we do say things like "a period of time," "a weight of 50 pounds," "a distance of 10 miles," "a speed of 200 miles per hour," "a volume of 900 cubic feet," and so on. In these cases, the simple solution is to cut the abstract expression from the idea.

Negative Examples

More testing was done, five additional test runs. (Cut "more testing" and move the verb.)

Analysis indicates a cost of $33 per foot length to construct the proposed design. ("$33" is "a cost" and a "foot" is a measure of "length." Two ideas, each repeated.)

After a period of two months of data collection . . . ("two months" is a "period.")

Indirectness

You should make your sentence structures directly reflect the subjects and actions of the idea you are trying to convey. That is, the conceptual subject of a sentence should also be the grammatical subject. Similarly, the conceptual verb of a sentence should be the grammatical verb. However, in rough drafts we often find sentences in which the grammatical subject is different from the conceptual subject, or the grammatical verb is different from the verbal concept. If you cannot immediately put your finger on the conceptual subjects and actions of rough draft sentences, they need to be edited. When you edit, identify the subjects and verbs of your rough draft sentences, and change them if they are not the conceptual subjects and actions.

In the following example, you cannot immediately identify the core subject: "It is the responsibility of the Department of Highways and Traffic to investigate all fatal automobile accident sites in the city."

Neither the first word, "it," nor the fourth word, "responsibility," is the conceptual subject of the sentence. The grammatical subject should therefore be changed so that the sentence reads, "The Department of Highways and Traffic is responsible for investigating all fatal automobile accident sites in the city." As you can see, in its simplest form, the idea behind the sentence can be expressed in this way: "The department is responsible for investigating accident sites." In the original sentence, that simple idea was somewhat obscured because the real subject (department) and the real verb (is responsible for investigating) were not in the subject slot and verb slot, as they should have been—a tension between the idea of the sentence and the form of the sentence.

The same basic problem is evident in another example, but in this case the grammatical verb is not the conceptual verb: "The filtering technique involves separation of signals of different frequencies."

Here you cannot immediately identify the conceptual action. The verb, "involves," fills the verb slot, but that word suggests a focus on the stages of the filtering process. Actually the conceptual verb is "separates," which focuses on what filtering does. The grammatical verb should therefore be changed. The sentence should be edited to read: "The filtering technique separates signals of different frequencies." Or, "Filtering separates signals of different frequencies." Or, "Signals of different frequencies are separated by filtering."

Any of these revisions improves the sentence. Which is most effective depends on the context.

In this group of additional negative examples (all of which come from actual reports), we illustrate how to achieve directness. We have edited the first four sentences to focus on the conceptual subjects:

> After a brief meeting with Jim Blackman it was decided that four tests would be performed on the samples. (In this version of the sentence, the reader cannot determine who made the decision—who the subject is. This sentence may have resulted from the writer's unwillingness to say, "I decided.")
>
>> Edited version: Jim Blackman and I decided to perform four tests on the samples. Or: After meeting with Jim Blackman, I decided to perform four tests on the samples.
>
> If the antenna is to be used in an urban area, where many different UHF stations exist, it is imperative that the field pattern be very narrow. (The draft version is weak because the conceptual subject occurs after four nouns and pronouns. Three of these occupy subject slots.)
>
>> Edited version: The field pattern must be very narrow if the antenna is to be used in an area with many UHF stations.
>
> An important consideration which is almost always overlooked is that engines in a laboratory receive far more expert attention than the typical engine in a private owner's vehicle. (The conceptual subject, "engines," is obscured because the writer has placed a subordinate idea first.)
>
>> Edited version: Engines in a laboratory receive far more expert attention than engines in private cars. This fact is usually overlooked.
>
> In this report an efficiency test applied to a small steam turbine is described.
>
>> Edited version: This report describes an efficiency test applied to a small steam turbine.

An important characteristic of sentence subjects is that they should be "old" information, not "new" information.[5] That is, when a reader encounters a term in a subject slot, that term should be both the real subject and a familiar idea. "New" information will cause the reader to grope for the meaning. If the reader does not understand the subject term because (1) it was not familiar before the reader approached the text, or (2) it was not made familiar by the immediately prior text, the term will still be unclear to the reader even if it is the real subject.

To assure that your text is readable, then, first make sure the subject of every sentence is in fact the conceptual subject. Second, make sure that the reader can be assumed to know what that subject means. It is a simple task. Merely underline the subject of each sentence and then ask yourself these two questions: (1) Is this the concept I really want to stress? (2) Does the reader already know what this thing or idea is? If the answer is "no" to either question, recast the sentence.

In addition, you need to make sure that the verb of every sentence is the conceptual verb. Here is another group of negative examples which need to be edited to focus on the conceptual action—putting the verb idea in the verb slot.

> Selection of the circulation system will need to be carefully watched to ensure absorption of the proper wave length of incoming radiation to excite the methyl-viologen. (In the draft version, the grammatical verb, "will need to be carefully watched," has nothing to do with the meaning of the sentence.)
>
> > Edited version: The circulation system must be carefully selected to ensure absorption. . . . Or: We must carefully select the circulation system to ensure absorption. . . .
>
> The column is still operable when there is a small amount of flooding on the surface. (This draft version illustrates a very frequent indirect structure, in which the core action, "operates," is in noun or adjective form, "operable.")
>
> > Edited version: The column still operates with a small amount of flooding on the surface.
>
> This ion "cloud" is responsible for the corona effects.
>
> > Edited version: This ion "cloud" causes the corona effects.
>
> Data definition languages were finally accepted as a component of information processing systems equal in importance to programming languages with the advent of generalized data base management systems.
>
> > Edited version: With the advent of generalized data base management systems, data definition languages became as important as programming languages in information processing systems.

Directness is primarily a matter of keeping the grammatical form of a sentence consistent with the thought behind the sentence. If the conceptual subject of a sentence is "engines," the grammatical subject should not be, "an important consideration which is almost always overlooked is . . ." If the conceptual action of a sentence is "select," the grammatical verb should not be, "[a selection] must be made." Whenever possible, then, edit the form of a sentence to directly reflect the idea.

A corollary on directness. As you make the grammatical form of the sentence consistent with the thought behind the sentence, make sure to keep your subjects as near to the beginning of the sentence as possible. And make sure to keep your verbs as close to those subjects as you can get them. As research clearly indicates,[6] sentences which excessively separate the verb from the subject are harder to understand than sentences with the subject and verb near to each other and near the beginning of the sentence.

Also, watch that you do not fall into a stylistic habit common among professionals, the habit of turning a verb idea into a noun and then having to depend entirely upon weak verbs such as *make, have, do, use, get,* and *be.* This habit is a sure-fire way to turn direct prose into indirect, bureaucratic prose.

Here are some negative examples of common "weak verb" sentences:

The discovery of a problem *is made.* (Someone discovers.)

A decision to undertake an investigation *is reached.* (Someone decides to investigate.)

Consideration of alternative approaches *is given.* (Someone considers alternatives.)

The choice of a particular strategy *is made.* (Someone chooses a strategy.)

A test *is undertaken* (fancy way of saying "done"). (Someone tests something.)

The evaluation of results *is performed* (fancy way of saying "done"). (Someone evaluates the results.)

Identification of a feasible solution *is made.* (Someone identifies.)

Conclusions *are formulated* (fancy way of saying "made"). (Someone concludes.)

Recommendations *are made.* (Someone recommends.)

Actions *are taken.* (Someone acts.)

A resolution of the difficulty *is achieved.* (Something resolves the problem.)

In this little "history" of an investigation, you can see that the real verb often masquerades as an abstract noun.[7] ("Specification" rather than "specify," "recognition" rather than "recognize," "conclusion" rather than "conclude.") When that happens, there is no verb until a "scotch tape" verb such as *make, have, do, use, get,* or *be* is stuck in. Writers often intuitively recognize how uninteresting these verbs are and try to "fancy them up" by using synonyms: "make" becomes "generate" or "formulate," "take" becomes "undertake" or "proceed to," "get" becomes "obtain," and "use" becomes "utilize." But such "fancying up" of a sentence really does not improve it. Quite the reverse. To improve the sentence, merely find the real verb idea and use it.

By the way, in the example above, we changed all of the passive constructions to active constructions: "The discovery of a problem is made" becomes "Someone discovers." If your subject focus requires a passive construction, you can keep the passive construction but still put the real verb in the verb slot: "A problem is discovered"; "Alternatives are considered"; "A strategy is chosen"; "The problem is resolved."

Inefficiency

You should make your sentences as efficient as possible. That is, think of each word as a device used to do work. If you invest more words than necessary to do the work, you are wasting your reader's time and reducing the clarity of the message.

When you edit, shorten and simplify sentences as much as contextual considerations permit. Make every word count, but don't assume that "short" necessarily equals "good." Some texts (and most computer programs for text-analysis) suggest that you count words to assure that your sentences do not exceed some theoretically acceptable maximum number of words. But short sentences are not necessarily efficient; nor are long sentences necessarily inefficient. Sentence length is a function of both the complexity of the ideas expressed and the "density" of the text (density is the ratio of content words to articles, prepositions, and other nonsubstantive words).[8]

To tighten your sentences, first make sure that each sentence expresses only one important idea. Next, make sure that the idea merits the grammatical weight you have assigned to it.

Rough-draft sentences can be inefficient because clauses containing important ideas are thoughtlessly spliced together when they should stand by themselves. Here are several examples:

> Most voids are pores which exist on the surface and some voids have no outlet to the surface and are known as dead pores. (The writer of this draft version has put two quite distinct ideas into one compound sentence. In doing that, he or she has lost sense of the meaning.)
>
> > Edited version: Most voids are pores which exist on the surface. However, some voids, known as dead pores, have no outlet to the surface.
>
> Unless the polymer chemically bonds with the carbon a crack may propagate along the carbon-polymer interface and thus negate most of the strengthening effect of the polymer. (The draft sentence describes a causal process but puts two cause-effect ideas into one sentence. This is inefficient because the causal sequence is obscured.)
>
> > Edited version: Unless the polymer chemically bonds with the carbon, a crack may propagate along the carbon-polymer interface. The crack negates most of the strengthening effect of the polymer.
>
> But since the operating temperature of this process is between 153°C and 100°C, formic acid is going to form and accumulate at the bottom of the column, which presents a highly undesirable situation. (The draft sentence has a double focus; it is both cause/effect and evaluative. The edited version presents each focus in a separate sentence.)

Edited version: Since the operating temperature of this process is between 153°C and 100°C, formic acid will accumulate at the bottom of the column. This accumulation is highly undesirable.

Information concerning volume and tempo changes is extracted from the graphic display, but identification of individual notes and instruments requires the isolation of a large number of fundamental frequencies (some of them being the same for different instruments) and the even more arduous task of associating the hundreds of harmonics with their corresponding fundamentals. (The draft sentence has a double focus. The edited version uses two sentences to differentiate between the two different topics, one a contrast, the other a parallel series.)

Edited version: Information concerning volume and tempo changes is extracted from the graphic display. However, identification of individual notes and instruments requires isolating a large number of fundamental frequencies (some the same for different instruments) as well as associating hundreds of harmonics with their corresponding fundamentals.

A rough draft sentence can also be inefficient because excess verbiage and indirect constructions lend an idea more weight than it deserves. Often modifying clauses should be reduced to modifying phrases; modifying phrases should be reduced to single words; and single words should be eliminated. Indirect constructions usually should be eliminated.

Here are four examples in which editing has reduced modifying clauses to phrases:

While this situation does not constitute a problem at this time, it should be explored further.

Edited version: Although not a problem at this time, this situation should be explored further.

We have determined that we need a simple tension bar in our structure which will support a maximum expected tensile load of 40,000 lbs. (40 kips). (In this example, the phrase, "which will support," easily reduces to, "to support," and the meaning of the clause, "we have determined," is made implicit in the verb, "need.")

Edited version: We need a simple tension bar to support a maximum expected tensile load of 40,000 lbs. (40 kips).

Since this voyage is longer than 24 hours, overnight accommodations in the form of staterooms are required for all of the estimated 800 passengers who will be aboard the craft. (In this example, the phrase, "in the form of staterooms," becomes a word, and the phrase, "for all of the estimated," becomes "for the estimated." Notice also that a redundancy, "overnight accommodations . . . staterooms," can be eliminated, and that the redundant clause, "who will be aboard the craft," can be eliminated.)

Edited version: Since this voyage is longer than 24 hours, staterooms are required for the estimated 800 passengers.

A strength of 8500 pounds per square inch (psi) is required of the concrete.

Edited version: 8500 psi concrete is required.

Here are two examples in which editing has eliminated unnecessary "it-that" constructions (expletives):

It is the opinion of this department that a heat problem will not occur on production vehicles.

Edited version: This department concludes that no heat problem will occur on production vehicles.

Consequently, it is one that deserves the attention of all those who are in a position to improve the staff's ability to tackle this problem. (This draft sentence, like the preceding one, is weakened considerably by an unnecessary "it-that" construction. Notice that the "it-that" construction forces the core subject and action into subordinate slots.)

Edited version: Consequently, the problem deserves the attention of those in a position to improve the staff's ability to tackle it.

These examples are typical of the first-draft efforts of most of us. We grope for the idea as we write; in the process, we use more words, phrases, and sentences than necessary to do the work of conveying the idea. Thus, unless we spend the necessary time on editing, we force our readers to spend unnecessary time in deciphering the message. Little increments of inefficiency pile up; little bits of time accumulate.

To give you an idea of how serious the problem is, when consulting we find surprising inefficiency in corporate communication. We can usually chop 30% out of most of the documents we examine (managers and writers invariably agree with us) without changing the meaning. Think about that. If your gas mileage were 20 MPG when it should be 30 MPG, you would think it time for a tune-up. The same applies to written text. If it takes 30 words to express a 20-word idea, it is time for some editorial tuning.

In summary, as you look at your prose on the sentence level, look for the following problem areas:

Correctness

Diction
Spelling
Punctuation with commas
Punctuation with semicolons
Punctuation with colons
Punctuation with hyphens
Parallelism
Subject/Verb agreement
Dangling modifiers

Clarity
- Noun strings
- Ambiguous pronouns
- Unclear synonyms

Directness and Economy
- Needless passivity
- Redundancy
- Indirectness in choice of subjects
- Indirectness in choice of verbs
- Excessive separation of subject and verb
- Wordiness

As we said at the outset, you cannot develop a good style merely by avoiding problem areas. However, many writers are so troubled by several of these basic problems that they never really "get to" the more substantive issues of style. So perhaps this list of frequent problem areas might be a good starting point as you begin to examine your style.

If you have significant difficulty with several of these problem areas, we suggest you review your drafts for one specific problem at a time. (You may find it easier to see problems if you look for them one at a time.) If you do not have much difficulty in these areas, examine your drafts for one general category of problem at a time. Whatever your approach, though, try to provide your readers with documents written in correct, clear, direct, and economical prose. That is not a "nicety" for you as a professional; it is a responsibility which goes with your role.

Notes

1. There are numerous excellent books on issues of correctness and style; however, we recommend these sources as particularly helpful:

   For issues of correctness and usage: J.C. Hodges et al., *The Harbrace College Handbook*, 11th ed. (New York: Harcourt).

   For issues of style: Joseph M.Williams, *Style: Ten Lessons in Clarity and Grace*, 2nd ed. (Glenview, IL and London: Scott, Foresman 1985).

   We should point out that some suggestions for report writing style are inconsistent with stylistic conventions in other types of prose. Report writing style, for example, often is more concise and direct than essay or journalistic style. That doesn't mean that report writing style necessarily is a better style. It does mean that one function of a report is to get a message across as clearly and quickly as possible. In a report your task is not to engage a reader in the way you do in an essay; your task is to enable that reader to make a decision or to do something efficiently.
2. For an excellent synopsis of research relating to issues of style, see Daniel B. Felker et al., *Document Design: A Review of Relevant Research* (Washington, DC: American Institutes for Research, 1980). See also, Daniel B. Felker et al. *Guidelines for Document Designers* (Washington, DC: American Institutes for Research, 1981). For a discussion of noun strings, see *Guidelines for Document Designers*, 63–65.

3. See Felker et al., *Guidelines for Document Designers*, 39–40.
4. See Felker et al., *Guidelines for Document Designers*, 27–29.
5. Our colleagues, Thomas Huckin and Leslie Olsen, discuss the importance of putting "given" information before "new" information in chapter 21 of their excellent text, *Principles of Communication for Science and Technology* (New York: McGraw, 1990) 395 ff.
6. See Felker et al., *Guidelines for Document Designers*, 45–48.
7. See Felker et al., *Guidelines for Document Designers*, 35–40.
8. See Thomas N. Huckin, "A Cognitive Approach to Readability," in *New Essays in Technical and Scientific Communication*, ed. Paul V. Anderson, R. John Brockmann, and Carolyn R. Miller (Farmingdale, N.Y.: Baywood, 1983), 90–108, and Felker et al., *Guidelines for Document Designers*, 41–48. The complexity of issues involved in "readability" is such that it is impossible to equate "short and simple" with good or "long and complex" with bad. Accordingly, the "readability formulae" which many writers and companies rely upon to evaluate text are of very questionable usefulness at their present level of development. In fact, the consensus of researchers is that readability measures are useful only as gross indicators of problems. You should be very skeptical about assuming that such readability measures are accurate predictors of true readability. After all, such measures can take account of only a few sentence features such as number of words, number of syllables, number of letters, frequency of repetition, and so forth. They cannot examine more substantive issues such as appropriateness of vocabulary, grammar (except on an elementary level), content in relation to audience, prior knowledge, or a host of syntactic, grammatical, and lexical features. The result is that writers and companies often believe that they are using good measures when the research shows they are not. (Go ahead and run your document through the text analyzer because you might learn something from it, but don't expect much.)

CHAPTER 11

Using Essential Visuals in Reports

Throughout this book, we have been talking about a process by which you compose technical and professional documents. We have mentioned the visual elements of those documents only infrequently (for example, when we discussed description), and our emphasis has been almost entirely upon the document's verbal elements. Yet almost no technical or professional document is entirely verbal. Most of them combine at least two of the three languages of science and the professions: words, pictures, and symbols (mathematics). Therefore, in a very real sense, the process of planning and writing a document—everything we have talked about—inevitably involves both making choices among these three languages and finding ways to combine them effectively.

These choices, strategies, and design decisions cannot occur at the end of the composing process. Rather, they must be distributed throughout the entire process. In other words, you can't plan and write a document and then start considering its visual or mathematical elements. Instead, from the moment you begin to think about your audience's needs and your own purposes, you must be thinking about the possible use of visual and mathematical means—as well as verbal means—to address those needs and to accomplish those purposes.

Although it is beyond the scope of this book to discuss the language of mathematics, we do want to talk in this chapter about the use of visual aids in technical and professional reports, or "essential visuals" as we prefer to call them.[1] Our preference for the term is based on the fact that most visuals in technical and professional reports are not just complementary or supplementary "aids" to a text. Certainly they are not text ornaments, such as one finds on the sports page or business page of pop journalism (elaborate graphics, but little information). They are not even like the presentational graphics used as attention grabbers and summary tools in oral presentations. Rather, they are "working" graphics—integral, essential means for presenting substantive information that would be difficult if not impossible to conceptualize and present otherwise.

We should emphasize that in this chapter we discuss the use of visual materials in reports. In doing so, we are limiting the field of visual communication considerably. That is, we are distinguishing between the use of visual aids and visual communication, where information is communicated by images rather than by words.

Technical communication can rely considerably upon the use of visuals. In oral presentations, this is immediately apparent. Oral presentations, such as those in organizational meetings or those by a consultant to a client, often seem to be entirely visual. A common practice in business and industry is to put the entire presentation on slides or transparencies consisting, for example, of photographs of a production facility, diagrams of a control system, drawings, graphs, and tables of financial statistics. These slides and transparencies can also include some verbal information such as an outline of the presentation and a listing of the conclusions and recommendations. But in a very real sense the core of many presentations is visual, with most of the verbal communication consisting of commentary about the slides and transparencies.

Organizational reports can also rely primarily or even entirely upon visual communication. For example, certain instructions can be entirely visual, such as procedures for assembly, repair, and operation of equipment. Similarly, status reports can be predominately visual. For example, the monthly *Licensed Operating Reactors: Status Summary Report*, which is over 500 pages long, consists almost entirely of data and graphs such as those shown in Figure 11.1. These status reports indicate the operating status of every nuclear plant in the United States; yet, as you can see, they contain almost no words other than those designating the categories of reporting.

With written documentation, the use of visuals depends primarily upon the purpose of the communication. Problem-solving reports require verbal communication to request an action, make a recommendation, communicate a decision, or effect organizational action. The primary objective of the communication requires verbal communication. Even if the majority of such a report consists of visual communication, this serves to support the information that is presented verbally. With administrative reports, such as status reports or manuals, which are primarily informational, the visual commu-

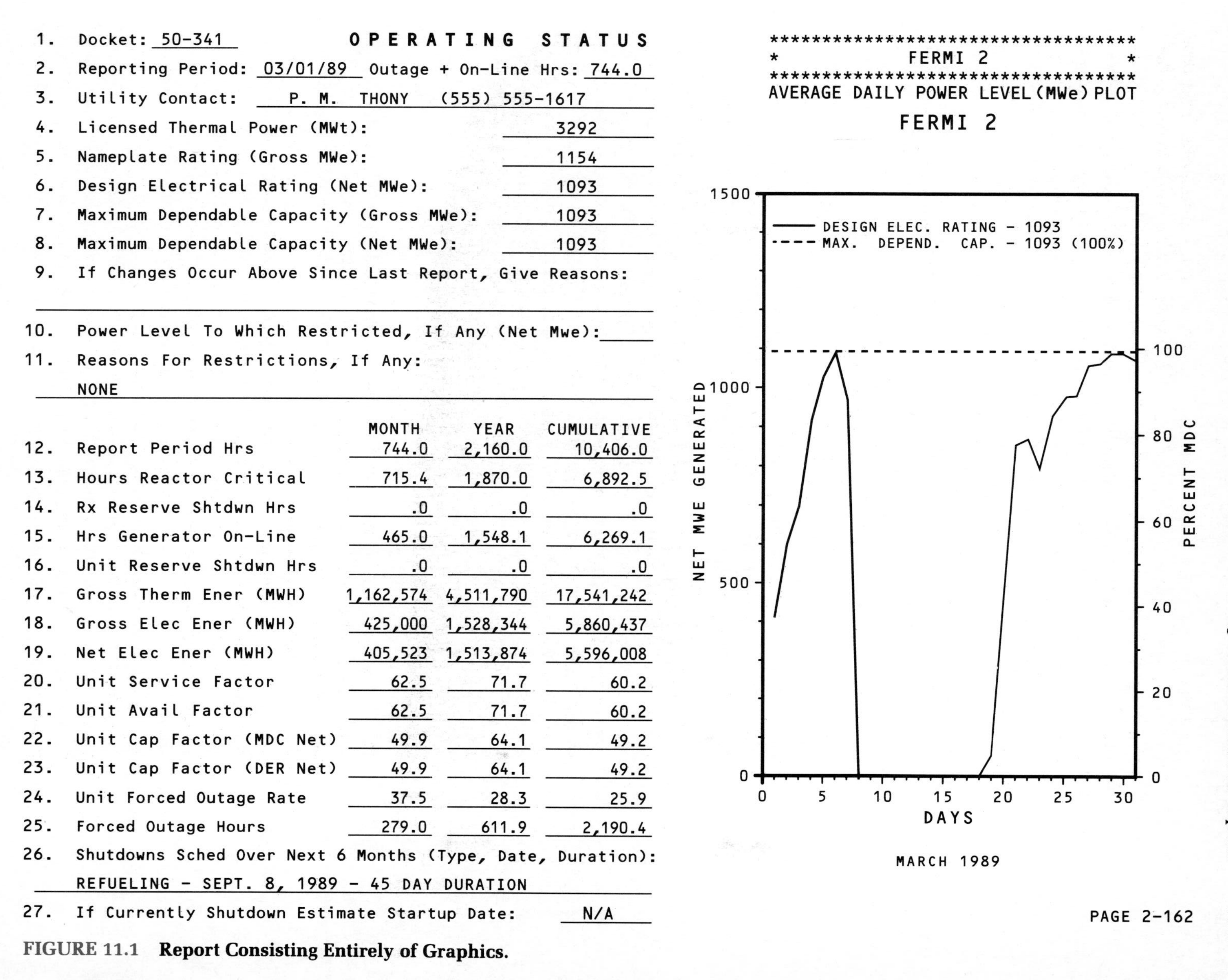

1. Docket: 50-341 **OPERATING STATUS**
2. Reporting Period: 03/01/89 Outage + On-Line Hrs: 744.0
3. Utility Contact: P. M. THONY (555) 555-1617
4. Licensed Thermal Power (MWt): 3292
5. Nameplate Rating (Gross MWe): 1154
6. Design Electrical Rating (Net MWe): 1093
7. Maximum Dependable Capacity (Gross MWe): 1093
8. Maximum Dependable Capacity (Net MWe): 1093
9. If Changes Occur Above Since Last Report, Give Reasons:
10. Power Level To Which Restricted, If Any (Net Mwe):
11. Reasons For Restrictions, If Any:
    NONE

		MONTH	YEAR	CUMULATIVE
12.	Report Period Hrs	744.0	2,160.0	10,406.0
13.	Hours Reactor Critical	715.4	1,870.0	6,892.5
14.	Rx Reserve Shtdwn Hrs	.0	.0	.0
15.	Hrs Generator On-Line	465.0	1,548.1	6,269.1
16.	Unit Reserve Shtdwn Hrs	.0	.0	.0
17.	Gross Therm Ener (MWH)	1,162,574	4,511,790	17,541,242
18.	Gross Elec Ener (MWH)	425,000	1,528,344	5,860,437
19.	Net Elec Ener (MWH)	405,523	1,513,874	5,596,008
20.	Unit Service Factor	62.5	71.7	60.2
21.	Unit Avail Factor	62.5	71.7	60.2
22.	Unit Cap Factor (MDC Net)	49.9	64.1	49.2
23.	Unit Cap Factor (DER Net)	49.9	64.1	49.2
24.	Unit Forced Outage Rate	37.5	28.3	25.9
25.	Forced Outage Hours	279.0	611.9	2,190.4

26. Shutdowns Sched Over Next 6 Months (Type, Date, Duration):
    REFUELING - SEPT. 8, 1989 - 45 DAY DURATION
27. If Currently Shutdown Estimate Startup Date: N/A

FIGURE 11.1 **Report Consisting Entirely of Graphics.**

nication can be more important than the verbal communication. Thus, the use of visual communication is not one of degree, but one of importance and function.

In our discussion of the use of visuals in reports, we explore the process of composing visuals, the functions they serve, the principles and conventions governing the use of some frequent types of visuals, and some practical matters related to the actual mechanics of integrating visuals into your documents.

❑ 11.1 ❑ The Process of Composing Visuals

Effective visuals are the products of an effective graphic composing process; they are not the products of merely following a set of specifications or using software that helps you to deal with the mechanics of drafting. In this respect, graphics and writing are alike. That is, the process of producing effective visuals is essentially like the process of producing effective writing. This point is nicely demonstrated in recent research by Lee Brasseur in a pioneering study of the graphic composing process.[2] Brasseur's research examined the individual composing process of over thirty test subjects who were given the task of composing a set of graphics based upon the same set of numerical data. The test subjects all had the same tools, the same amount of time, the same audience, and the same stated purpose for the work.

As her test subjects worked, Brasseur documented the process by which they went about their tasks. Then she sent the resulting graphics to a panel of experts drawn from industry and academia. She asked the experts to evaluate the graphs, recording their evaluations both numerically and in open-form comments. Then, analyzing both the processes used by her test subjects and the evaluations by the experts, she looked for meaningful connections between procedure and results.

Her findings clearly indicate that those graph designers who had the most effective composing process produced the most successful graphics. Specifically, she found five ways in which the performance of the successful designers was distinctive (Table 11.1).

First, successful designers spent more time planning than did those who were rated unsuccessful. Successful designers studied both the instructions and the data, and they spent more time experimenting with different ways of representing the data than did other test subjects. Unsuccessful designers, on the other hand, tended to choose a method of representation quickly and to dive into its execution with little planning or analysis. Second, successful designers revised more frequently than unsuccessful designers, who tended to stick with their preliminary choices even when they knew that they had made mistakes. Third, successful designers focused much of their attention on audience considerations, while unsuccessful designers tended to focus

TABLE 11.1: Typical Behaviors of Both Successful and Unsuccessful Graphic Designers

Behavior	Successful Designers	Unsuccessful Designers
1. Planning	Analyze both instructions and data, experimenting with alternative designs.	Use little analysis or planning, quickly choosing a method of representation.
2. Revision	Revise frequently. Abandon unsuccessful attempts.	Revise infrequently, sticking with choices even when designers know they have made mistakes.
3. Focus	Emphasize audience needs and interests.	Emphasize details of data and mechanics of drafting.
4. Verbal design	Devote attention to accuracy and appropriateness of titles, legends, callouts.	Omit or make mistakes in verbal aspects of the graphics.
5. Visual design	Know and effectively use principles of visual design.	Create eccentric and aesthetically unsuccessful designs.

their attention on either the details of the data or the mechanics of the drawing. Fourth, successful designers devoted particular attention to the accuracy and appropriateness of titles, labels, and legends of their graphics, while unsuccessful designers tended to leave out or make mistakes in verbal portions of the graphics. And finally, successful designers made decisions which were consistent with visual design principles, while unsuccessful designers sometimes created eccentric visual devices.

Although Brasseur's research is the first to demonstrate a clear link between process and product in visuals, these findings should not be a surprise. That is, we have known for some time that an effective composing process produces effective writing.[3] Therefore, we might have anticipated that the change from verbal language to visual language should not change the outcome: good process produces good product.

Yet, if these research findings are not surprising, nonetheless they are important for you to consider. In particular, these findings make clear that you must take your time to analyze the audience and purpose of your communication before choosing a mode of graphic representation. Furthermore, you must be willing to experiment and to revise—even if it means trying out several visual forms and then pitching out the unsuccessful ones. You must also be willing to look for and to correct problems, especially problems that

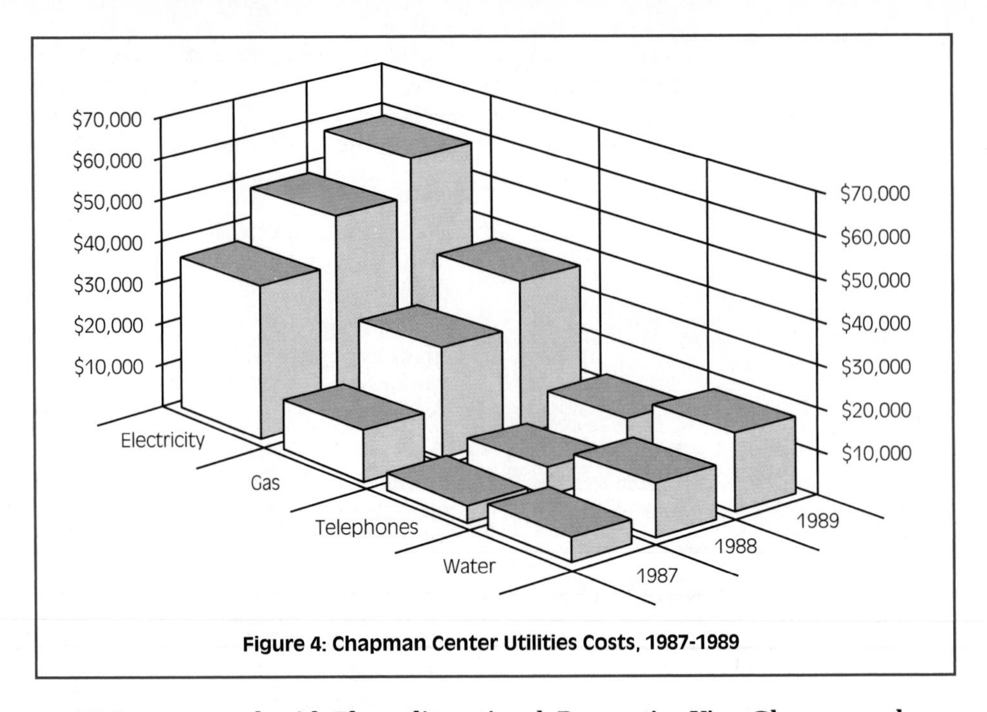

FIGURE 11.2 Graph with Three-dimensional, Perspective View Obscures and Distorts Data. (*Note*: This bar chart, similar to one drawn by a test subject in the Brasseur study, is based on a format provided by one software package for charting.)

are likely to occur in the verbal portions of graphic devices. And finally, these findings make clear that your success as a designer of graphics requires you to be sensitive to graphic principles and conventions—not an easy matter in a society that scholars have characterized as being visually illiterate.[4]

Just as the successful designers point up some approaches that you must take seriously if you are to design effective graphics, the failures of the unsuccessful designers point up problem areas to avoid. Particularly—and somewhat surprisingly—the experts found the largest degree of failure to be in the verbal aspects of the graphics. That is, titles, legends, and labels in graphics were areas where unsuccessful designers made a lot of mistakes. Indeed, almost half of all of the negative evaluations by the experts focused on some failing in the verbal elements of the graphics designed by the test subjects. That tells you something important to look for in your own visuals. Another significant source of difficulty was violation of conventions and design principles in the visuals themselves. For example, the experts agreed that the use of the third dimension in bar charts (Figure 11.2) is an ineffective technique that adds nothing conceptually and that might actually obscure comparisons of data. Some of the unsuccessful designers, however, used the 3-D effect because they thought it just looked good. (Unfortunately, computer graphic programs increasingly provide a 3-D effect as a flashy graphic technique for bar charts and pie charts, but just because it can be done does not mean it should be done.) And finally, the unsuccessful designers tended to have a number of problems with issues related to scales, axes, and accuracy.

Among these findings, three points are critically important aspects of the visual composing process: audience consideration, willingness to experiment and revise, and attention to verbal elements.

Audience Considerations

Designers who think about the implications of their readers—both in terms of who they are and what they need—tend to get better results than designers who focus on the content of their graphics or the mechanical aspects of the artwork. That is no surprise, but it is an important realization that should affect your approach to the design of visuals.

The identity and the needs of the audience of course mean that the choice of graphic mode can be quite different for each audience and situation. For example, a water treatment system can be shown as a detailed schematic diagram (Figure 11.3) or as a simplified block diagram (Figure 11.4). The first version comes from the appendix of a report. Its readers were the city engineers, who are both familiar with schematic drawings and interested in their details, such as sizes of pipes and locations of valves. The second version comes from the body of the same report. It addresses the primary audience for that report, the Planning Commission, an audience that needs only general information about the stages and capacities of the water treatment system being discussed, concepts obscured by the detail of the first drawing.

Both versions are appropriate to their own audiences; the writer made a wise decision to design two versions of the same graphic, one for each of his two types of audiences. Most writers would not have taken the effort to do that, instead using one version for both audiences. Yet, either version alone would not have communicated with some of the readers of the report. Probably, in fact, if only one version had been used, it would have been the detailed schematic, because that version was closer to the writer's own understanding of the system. Hence, had there been only one graphic, it probably would have been the wrong one to address the primary audience of the report.

As this example suggests, you must suit your design to your readers. No figure is good in the abstract. Like a piece of text, a visual is good only in relation to its context of audience and purpose.

Willingness to Experiment and Revise

Particularly important in the process of composing graphics is your willingness to evaluate your graphic choices and to revise or even discard them if they are inappropriate or ineffective. Perhaps because visuals are time-consuming to produce, we tend to resist revising or eliminating them once they have come into our heads and have been translated to our paper or our screens.

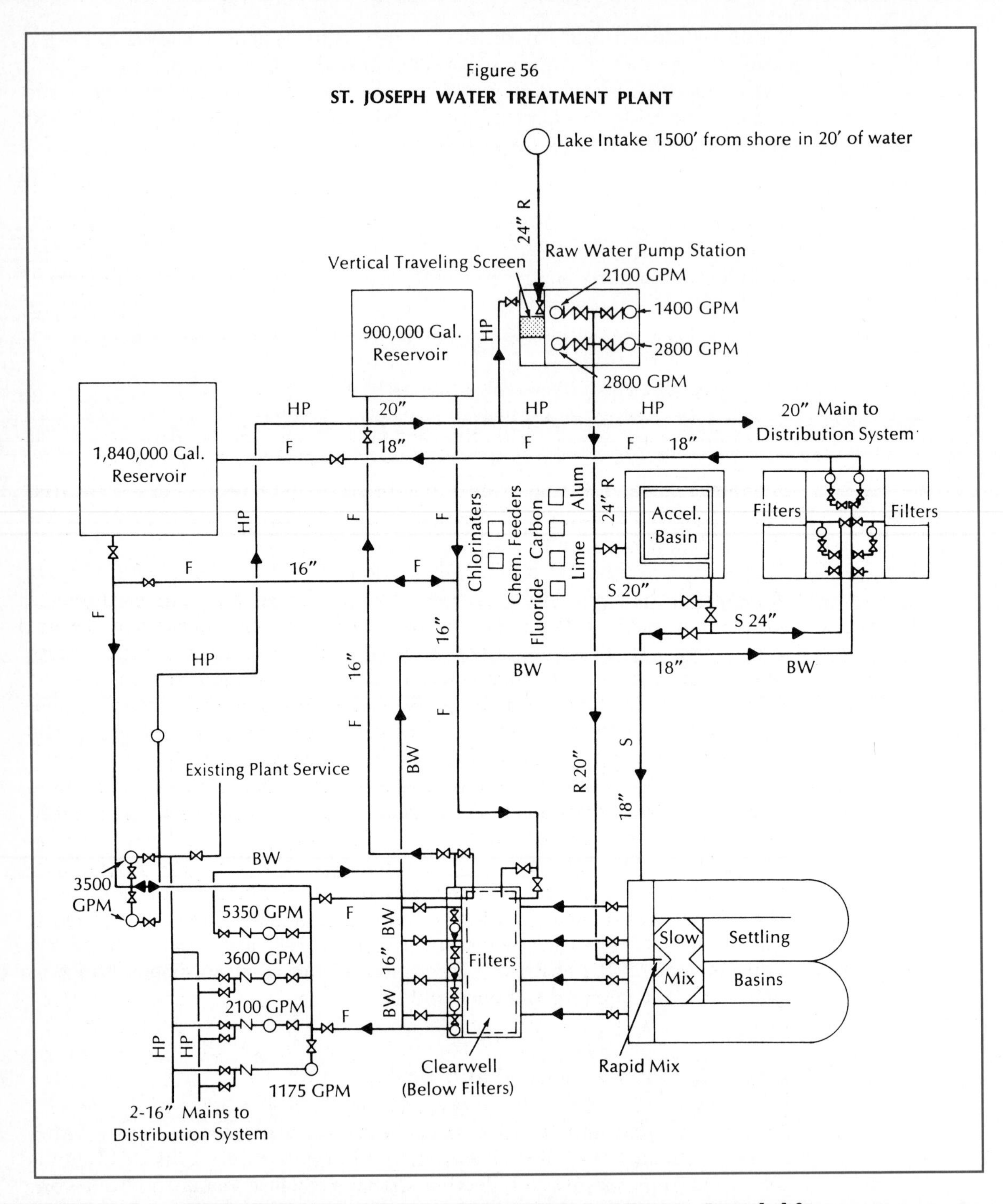

FIGURE 11.3 Detailed Schematic Diagram of Water Treatment System, Intended for an Audience of Technical Specialists.

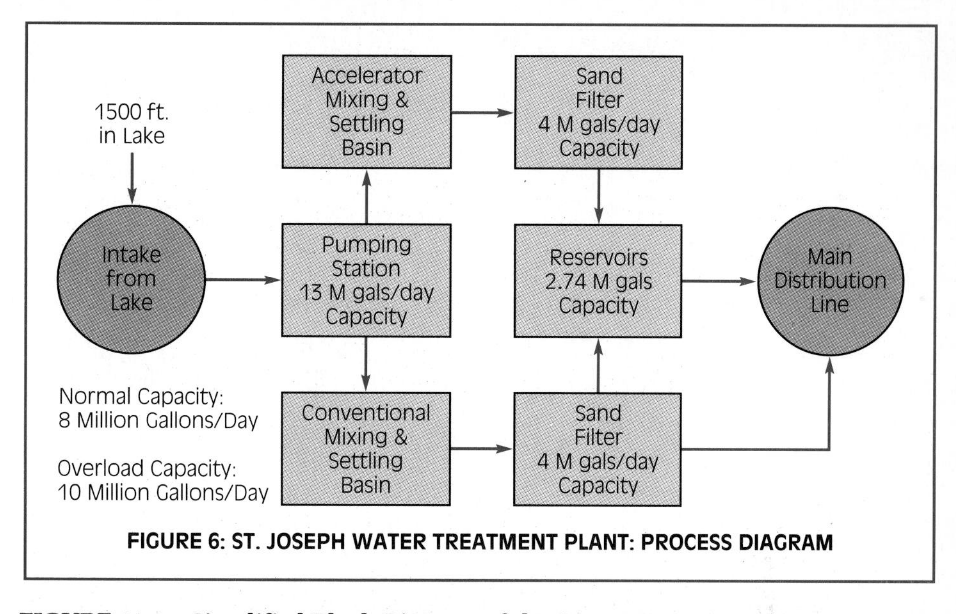

FIGURE 11.4 Simplified Block Diagram of the Same Water Treatment System, Intended for an Audience of Municipal Administrators.

We go with what we first "see." (Poor writers do the same thing, but words are easier to change than pictures.) We simply urge you to be aware that you will tend to stick with the first concept that pops into your head. Force yourself both to experiment and to discard what does not work well.

As a nice example of the need to revise, consider the pair of figures showing precipitation and runoff into the Green River (Figure 11.5). (Unfortunately, these are final versions from an actual report, not planning versions.) The writer who first thought of those figures saw them in his mind's eye as separate: precipitation is one thing; runoff is another. Accordingly, he graphed the two separately. Yet the whole intent of the graphic was to show the relationship between runoff and precipitation, and separate graphs do not show relationship well. Hence, the first mistake of the designer was to think in terms of separate treatment when he should have been thinking in terms of combined treatment. In retrospect that seems obvious.

The second problem of the original pair of figures is that the left axis scales of the two graphs are not the same. Accordingly, the visual effect is to mislead readers: runoff seems about equal to precipitation, and in some years it even seems to exceed precipitation! Yet if you notice that the scales are not the same, it is clear that the runoff is actually only a small percentage of the precipitation. The designer made a basic conceptual mistake in the original design, but did not see it and did not correct it. An alternative version of the same graphic (Figure 11.6) eliminates the misleading impression by superimposing the runoff data on the precipitation data. This alternative version perhaps has the problem of being so detailed that year-by-year matching is difficult, but at least the basic relationship between runoff and precipitation

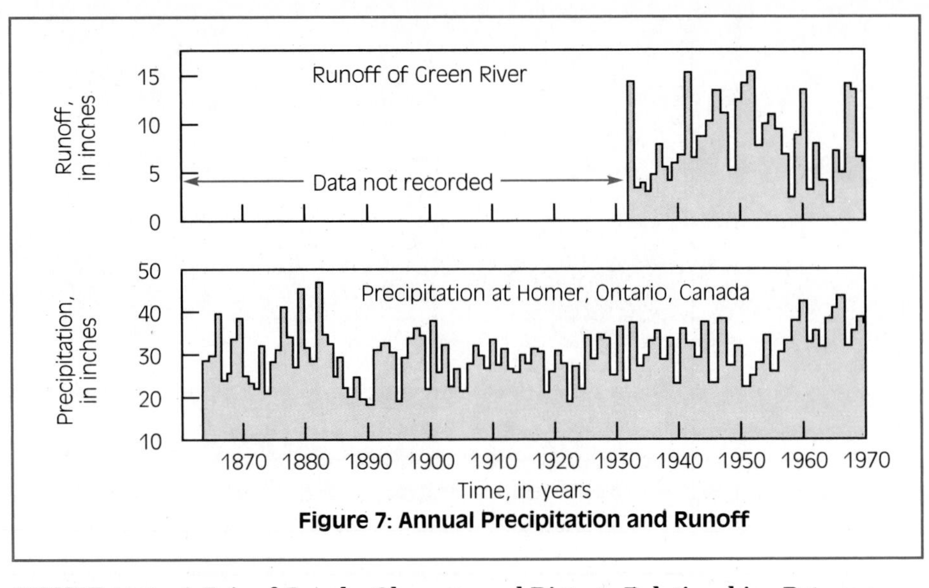

FIGURE 11.5 A Pair of Graphs Obscures and Distorts Relationships Between Two Sets of Data.

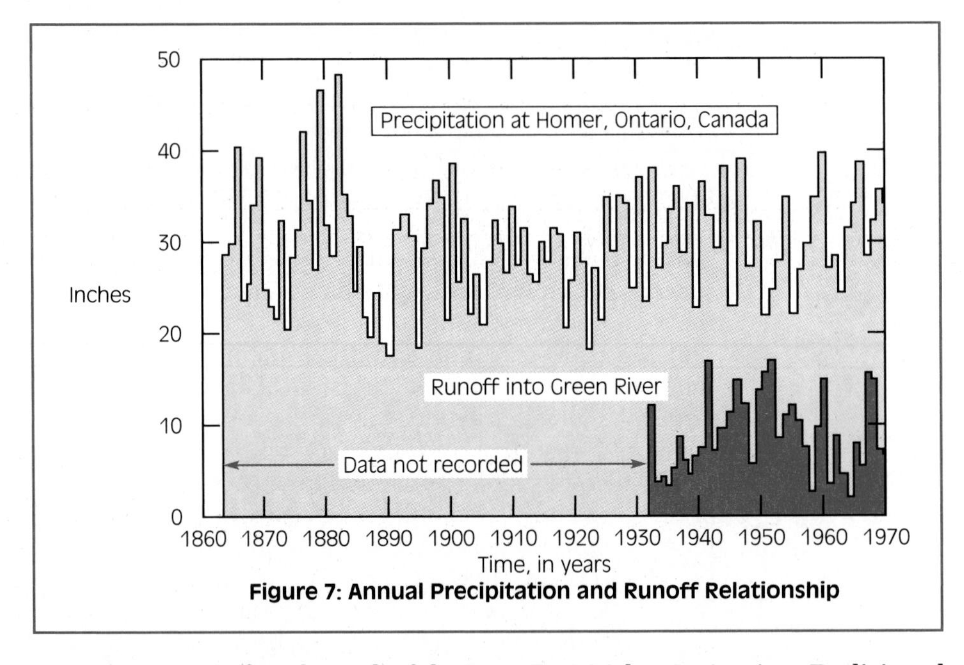

FIGURE 11.6 Combined Graph of the Same Data Makes Comparison Explicit and Clarifies Relationship Between Two Sets of Data.

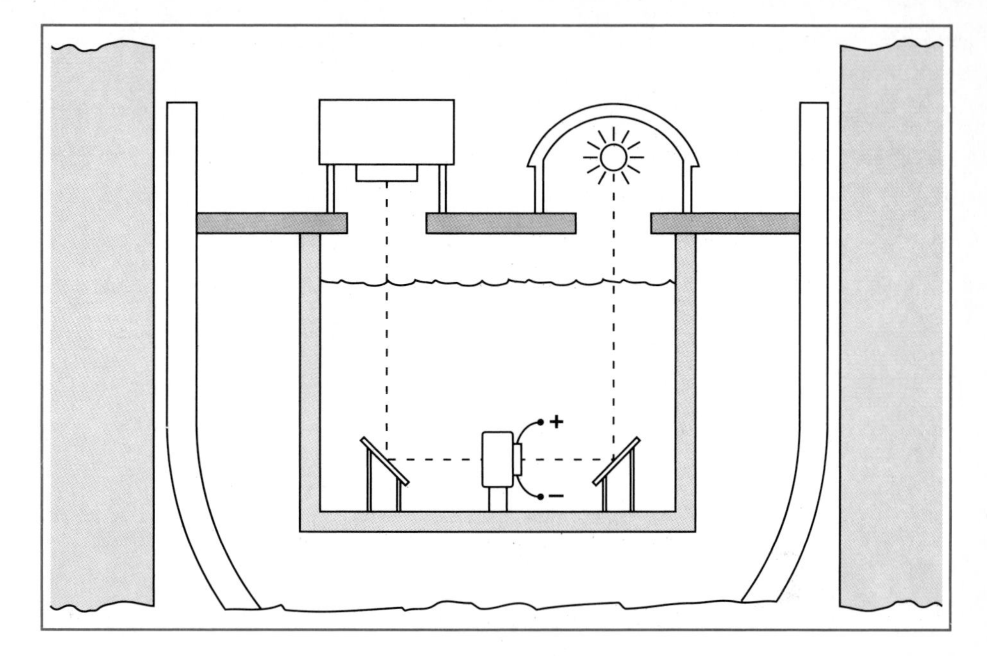

FIGURE 11.7 **Visual Without Heading or Labels Must Depend Excessively on the Text.**

is made clear, whereas the original drawings both obscured and distorted that relationship.

As this example suggests, you must be prepared to challenge your own design concepts and to discard those which do not hold up. Unfortunately, drawing a graph, sometimes even with software, is a time-consuming process that tends to make you satisfied with your first products. Hence you will have to push yourself.

Attention to the Verbal Elements

You must also remember that designers tend to be inattentive to the verbal elements of their visuals. Either the legends and callouts are missing entirely (as shown in Figure 11.7), or they are present but meaningless (as in Figure 11.8).

Why this problem exists is unclear. However, experienced professional proofreaders and editors tell us that routinely they find errors and omissions in the verbal portions of graphics and in the type associated with titles and headings in text. Perhaps the explanation is what one editor called "the authority of print." That is, the larger or more imposing the type, the easier it is to overlook an error in it—especially in standardized elements such as titles for figures or tables. In any case, be aware that you make mistakes in the verbal elements of your graphics. (We are referring to the verbal elements of

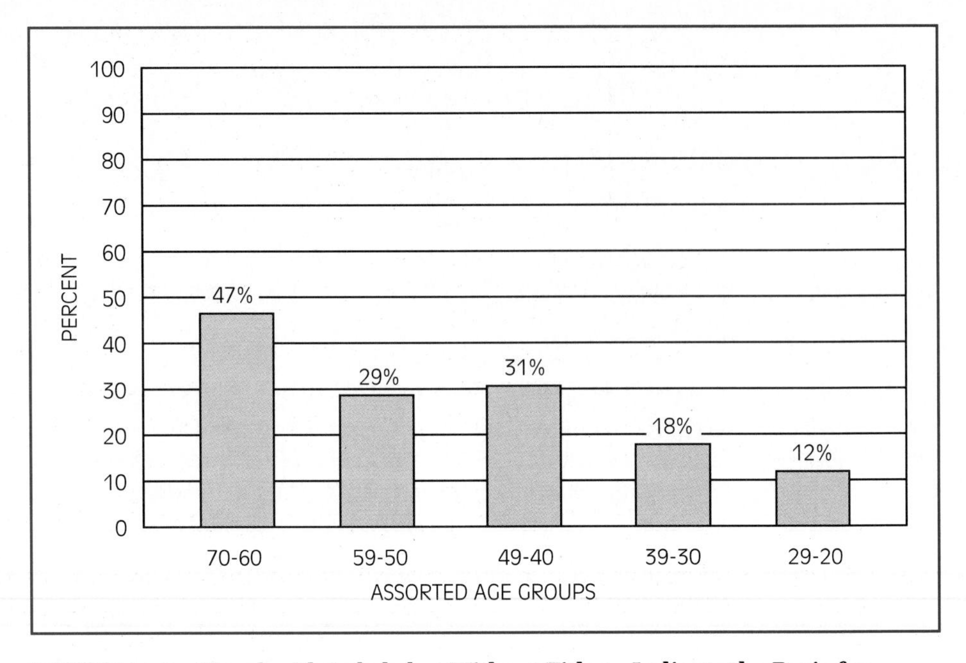

FIGURE 11.8 Visual with Labels but Without Title to Indicate the Basis for Comparison.

the visuals themselves. In addition, you must look closely at the verbal elements of the text as it introduces and discusses a visual. Carelessness with words may not be limited to those on the visual itself.)

But merely proofreading the verbal elements is not enough. You must also design the verbal elements so that your graphics communicate with maximum efficiency. Figure titles, like the subject lines of reports, should communicate the topic and purpose of the communication. Similarly, callouts and legends (rather like the headings and subheadings in a report) should convey specific and clear information about what is in the drawing. Like subject lines and headings, the verbal elements in visuals need to be correct, informative, and succinct.

A Suggested Process for Designing Visuals

You can approach the visual design process much as you do the verbal design process. Our recommended approach consists of the following six steps:

1. Start by thinking consciously about the implications of your audiences' needs and knowledge. What kinds of information will they understand, like, and need?

2. Define your motive for using the visual. That is, decide what the function of the visual is (more about this in the next section).

3. Sketch several different approaches, attempting to find one that best meets your considerations of audience and purpose and that functions effectively on both informational and aesthetic grounds.

4. Select a design concept and develop a first draft of the visual, paying attention as you do to the verbal detail as well as the visual detail.

5. Carefully evaluate and edit the visual in terms of design principles and conventions, accuracy and correctness, and effectiveness of verbal elements. If you find mistakes, be prepared to take the time to rethink or redo the entire visual. If the concept is correct and appropriate, polish the execution of the visual just as you would edit the details of a written text. If the concept is defective, redesign the visual.

6. Integrate the visual into the text both verbally and physically.

We believe that if you follow a design process such as this with the visuals, as you do with the words, your visuals will be more effective than if you follow the "ad hoc" approach of unsuccessful designers.

We might add that the availability of software for charting and drafting does not change the fundamental fact that a conscious, process approach to the design of visuals is essential. If anything, the availability of these tools makes the need for a deliberate approach even more important. That is, spreadsheet programs will take a page full of numbers and turn it into any of numerous types of graphic representation. A click on the menu is all that is required. Similarly, drafting programs enable everyone to draw quickly and well (meaning neatly). Yet behind the successful use of these charting and graphic tools must be intelligent decision making about what is conceptually effective and appropriate for the audience and purpose. Especially this is true because many of these graphics tools are designed with a tempting array of colorful fill patterns, flashy effects, and sometimes eccentric but unforgiving templates. Thus, designers can easily be seduced into confusing neat drafting with good design—just as word processing programs can trick writers into confusing good format with good writing. Of course the tools are useful, but don't assume that they are going to do your thinking for you.

❑ 11.2 ❑ The Functions of Visuals

You need to think consciously about the functions of your visuals before you can begin to decide which is the best type of visual or what specifically it should contain. Several ways of thinking about the functions of visuals have been proposed.[5] Of these, one of the most useful is a 4-part analysis of functions defined by Michael Macdonald-Ross.[6] He says that graphics in texts are there for any of four fundamental reasons:

1. To show what an object looks like and to identify and label key parts (an "iconic" function).
2. To display the results of empirical observations (a data display function).
3. To show the logical relationships among key ideas (an explanatory function).
4. To help readers perform some well specified task (an operational function).

Although it is possible to think of visuals that do not fit into one or another of these basic categories of function, Macdonald-Ross's list clearly does encompass most of the major uses of visuals in texts. Consequently, the list can provide a starting point for two kinds of decisions: first, should you use a visual at all? And second, if you do use a visual, what are the appropriate choices? In other words, the list can help you to think analytically about your choices and means for presenting information visually.

Of course, these four basic functions are not entirely separate; any particular visual can perform more than one of these functions simultaneously. For example, a photograph or sketch might serve an essentially iconic function—to show what something looks like. However, if labels are imposed on key parts of an object in a photograph, the reader's attention will be drawn not just to the object's appearance but also to logical relationships and key ideas that the writer/designer has imposed upon the object (the object itself has no labels). Furthermore, a photograph or sketch can be used to show relationships. (For example, A connects to B in this way, or here is how big A is in relation to B.) Photos and sketches can also explain how to perform a task. (Operation manuals that are used in international contexts often try to escape the translation problem by using entirely visual instructions, for example.) Similarly, a flow chart may be a principal tool for showing relationships among the stages and components of mechanical, electrical, or chemical systems as they operate (logical relationships). However, a flow chart can also be used to explain how a human activity is to be done: who does what and when. For example, Figure 11.9 is a flow chart that assigns times, tasks, and responsibilities to individuals who are performing a collaborative proposal-writing activity in which the longest time required to perform the activity must be determined and in which the contributions of the individuals must be coordinated in a sequential, time-sensitive manner.

Each of the basic functions previously identified can be further elaborated. One useful way to do that is simply to ask questions that might be answered for your readers if you choose to use a visual. For example, here is a list of

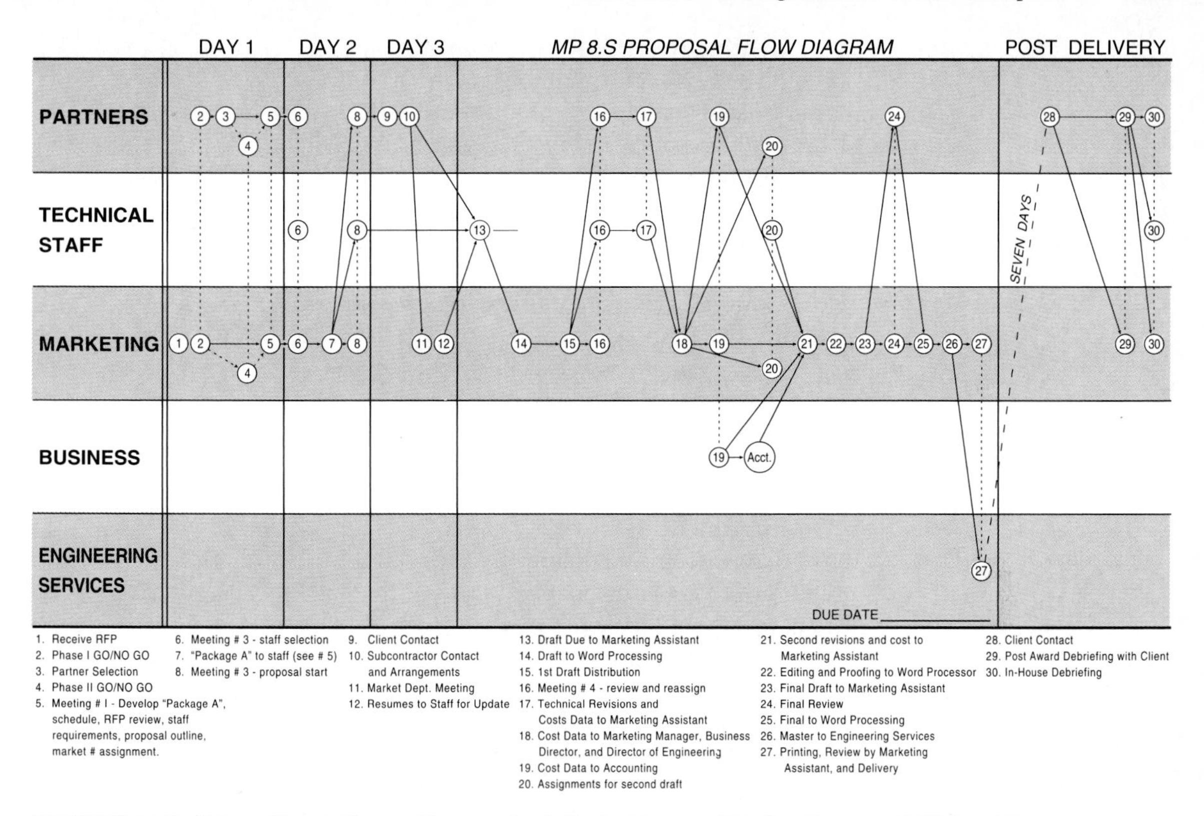

FIGURE 11.9 Flow Chart Shows Human Activity in Terms of Tasks, Responsibilities, Times, Relationships.

functions to which we have added a number of more specific questions that you might have behind your decision to use a visual (you can add others, no doubt):

1. **Iconic function:**
   What does it really look like?
   In general terms what does it look like?
   If we enlarge it, what does it look like?
   If we reduce it, what does it look like?
   What are the parts?
   How do the parts fit together?
   How does the thing relate to other things around it?
   What does it look like in action?
   What changes are observable over time?
   What effects are observable?
   How do similar things differ in appearance?
   How are different things similar in appearance?
   What is used to do something?

2. **Data display function:**
   What are the specific numbers?
   What are the aggregated numbers?
   How are the numbers logically structured?
   What do the numbers mean?
   What causes the numbers?
   What effect does time have upon the numbers?
   What are the effects of other variables on the numbers?
   How do the numbers compare to some standard?
   How do the numbers compare to still other numbers?

3. **Explanatory function:**
   What are its basic parts or components?
   What are the relative quantities of the constituent elements?
   Are the parts organized in some logical way?
   What are the differences?
   What are the similarities?
   What are the relative sizes?
   What are the relative quantities?
   How are things connected?
   How do the parts interact?
   What things are distinct, unconnected?
   What things are partly connected?
   How do things change over time?
   How does the thing relate to other things around it?
   What does it look like in action?
   What effects are observable?

4. **Operational function:**
   Who does what?
   What is used to do it?
   When is it done?
   Where is it done?
   What tasks are involved in doing it?
   What is the sequence of steps in doing it?
   How should it not be done?
   How does something happen?
   What are the stages of its happening?
   What are the inputs?
   What are the outputs?
   What are the conditions of choice?
   What are the alternatives?
   When do the alternatives apply?
   What are the consequences of doing it?
   What are the problems of doing it?
   What are the dangers of doing it?

As you can see, many of these questions are similar to those that could be asked in choosing between, let's say, a descriptive pattern and a process pattern for text or between a cause/effect pattern and a comparative pattern. Just as text blocks have specific functions, visuals also have functions. By probing deliberately into your purposes, you can clarify and specify the functions of a particular visual. This is not to say that the message of a visual can always be reduced to a question or simple verbal statement of purpose. Indeed, precisely because visuals are inherently nonverbal, you convey different kinds of meaning than with words. Nonetheless, you need to clarify the function of a visual in order to choose and design it with maximum efficiency.

Once you have thought about the functions of your visuals in some detail, then you need to think about the various alternatives available to you to accomplish those functions. That is, several different types of visuals can be employed to accomplish any of the basic functions or to answer any of the specific questions. For example, here is the list of functions with suggestions of some likely candidates for accomplishing the functions:

1. **Iconic function:** photo, representational sketch, map.

2. **Data display function:** table, line graph, bar graph, scatter plot, pie graph, histogram.

3. **Explanatory function:** schematic drawing, flow chart, table or matrix, decision tree, organizational chart, Venn diagram, photo with superimposed interpretation, representational sketches such as exploded drawings or simplified functional drawings that either eliminate or "ghost" out part of the detail.

4. **Operational function:** photo, sketch, flow chart, decision tree.

The choice among alternative means for accomplishing a function must be based upon your appraisal of the interplay between your purposes and the needs and interests of your audiences. You need to know enough about the various modes to recognize what their relative strengths are and what conventions and design principles govern their use.

❑ 11.3 ❑ Principles and Conventions for Design of Visuals

Having chosen to use a visual, you need to decide which type of visual might be effective, given your audience and purpose. That entails weighing the relative merits of the alternatives and choosing one that is appropriate and practical within the limits of your technology, budget, and time. In the following discussion, we look at some strengths, limitations, and design considerations of each of several frequent forms of visuals.

Iconic Function

If your purpose in choosing to use a visual is iconic—to show what something looks like—your choice is between photographs and representational sketches. Photos have the virtue of being quick to create and inexpensive to produce, especially with Polaroid cameras and digital scanners. Sketches, in contrast, are relatively time-consuming to produce, even with graphics programs. On the surface, then, photos are an especially attractive alternative. However, the choice between photographs and sketches should not be made solely upon the issue of how time-consuming it is to produce them. Rather, the choice should be based on an analysis of the relative capabilities of the two tools.

Photographs Photographs are good for showing surface detail and external appearance, especially with details that are too small to be seen with the naked eye (Figure 11.10). They are ineffective, however, for showing internal and functional details that cannot be seen on the surface. That is, they can show the outside of the black box, but they cannot easily show what goes on inside. Similarly, they are good at showing relative size, context, and use, but

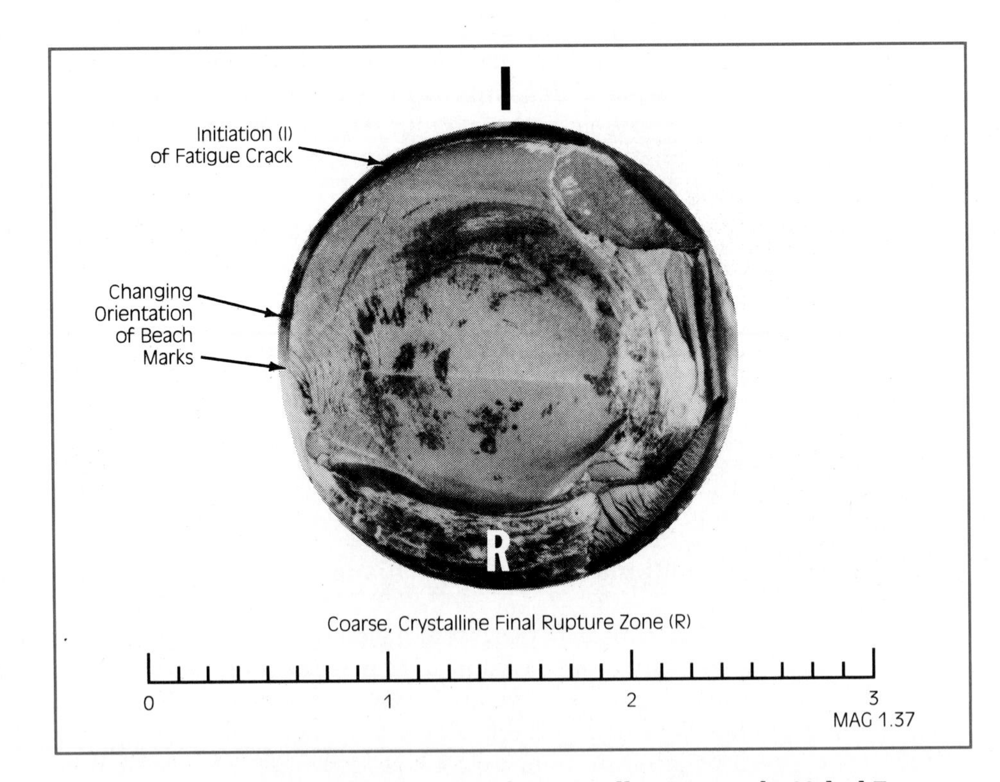

FIGURE 11.10 Photograph Shows Detail Too Small to Be Seen by Naked Eye.

they are poor at showing composition. Photos have the capacity to capture rich detail, including the detail of three-dimensional representation.

But precisely because they are detailed, photographs require careful composition and craftsmanship. Just as there is a significant difference between the work of a studio photographer and that of an amateur, effective photos for use in reports may require special lighting, special composition, and even special processing. Merely scanning a candid shot from a Polaroid camera may not be effective at all because the resulting illustration is likely to be poorly lit, badly composed, and excessively full of detail.

Representational Sketches Sketches have special capabilities that set them apart from photographs. They are good at showing general shapes and relationships, both on the surface and on the interior of objects, without getting overwhelmed by the detail that might be caught in a photograph. As inherently selective creations, sketches can simplify and clarify reality by leaving out details and by deliberately featuring other details. More importantly, however, sketches can look beneath the surface, unlike photographs, and thereby show things that cannot be captured easily in the lens of a camera (Figure 11.11). Although sketches do not show surface appearance or detailed and

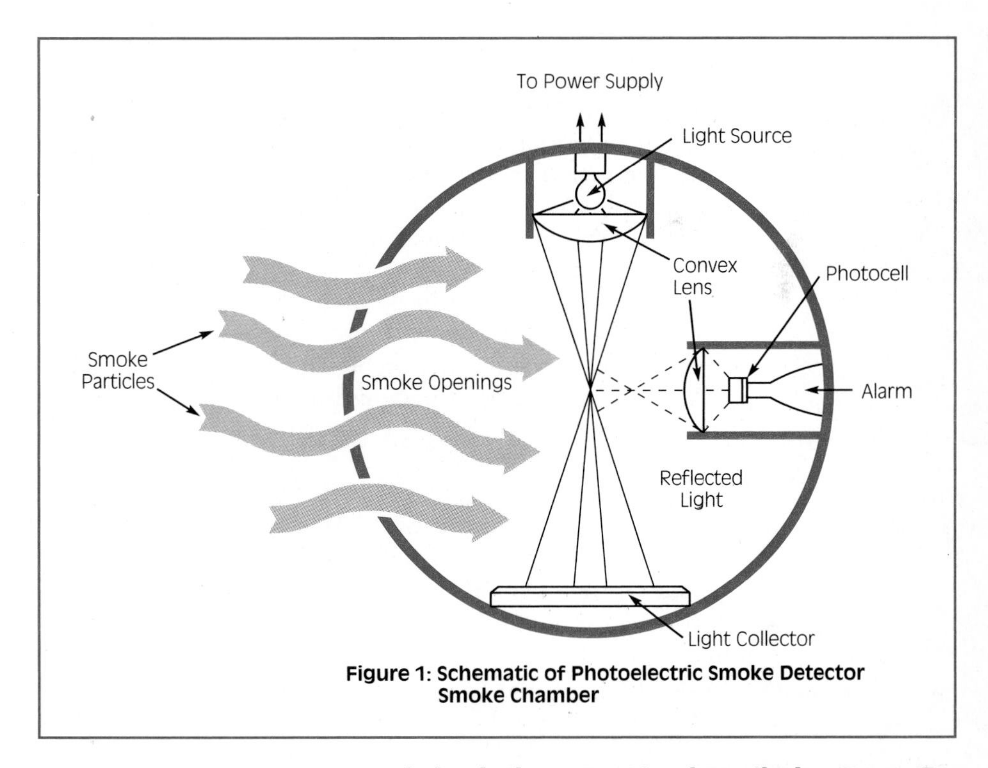

FIGURE 11.11 Representational Sketch Shows Functional Details that Cannot Be Photographed.

complex relationships as well as photographs, they nonetheless are more powerful tools for showing simplified, quasi-representational views of insides of things. Exploded drawings, ghost drawings, and cross-sectional views, for example, all allow you to represent reality in ways that cannot be seen but which are no less real than what is shown in a photograph. Obviously, all of these representations require substantial time and skill to develop. Yet, if your objectives are not primarily to show the surface appearance of something, the investment of time may be well worthwhile.

Data Display Function

Showing data requires you to decide whether to use a table or a graph of some kind.[7] You choose the table if you wish to convey precise information; you choose a graph when you wish to show relationships among data or sets of data rather than precise quantities. That is, tables allow precision but are less strong at showing relationship; graphs are strong on showing relationship but less strong at showing precision. A further basis of choice can be how much data there is to be displayed: tables can handle massive amounts of data; graphs—at least most kinds of graphs—can handle lesser amounts of data.

Spreadsheet programs give you the option to represent data in either tabular form or graphic form (or to use both forms) merely by choosing an option from a menu. They allow you to experiment and to plot alternative graphs and charts from tabular data in a matter of seconds. Indeed, these programs provide an increasingly wide range of alternative graphic types for representing numerical data. Yet behind these types, there are basic design principles that you can keep in mind as you consider how to display data.

Tables Tables are good for showing the large amounts of numerical detail and discrete bits of data that professionals often need from a report. Furthermore, because they consist of columns and rows, tables are powerful ways of showing groupings of data and organization within data. In addition, the columns and rows of tables allow readers to access the data randomly, not linearly, so they can examine the data efficiently even though the table itself might contain a great deal of detail that is not of direct interest. A limitation of tables is that they are not very effective for showing trends in data or relationships among data points. Although tables allow side-by-side comparisons of numbers, they are not strong devices for showing the big picture within a mass of numbers. Inherently, they focus on the abstract detail of cell-level information, not upon the overall interpretive level of chart-level information.

If you choose to design tables, you should be aware of design principles that have been demonstrated to improve their functioning.[8] The following suggestions are appropriate to keep in mind in your design of tables. (Figure 11.12 shows both good and bad table designs):

TABLE A

Table from "Facts in Focus" (London: Penguin, 1974). As published, it breaks most of Ehrenberg's rules.

Unemployment in Great Britain

	1966	1968	1970	1973
Total unemployed (thousands)	330.9	549.4	582.2	597.9
Males	259.6	460.7	495.3	499.4
Females	71.3	88.8	86.9	98.5

Source: *Facts in Focus*

TABLE B

The same data as Table A, but with rows and columns interchanged and numbers rounded and placed in decreasing order of size. Visual comparisons and mental arithmetic are now straightforward. The effects of such rearrangements are even more marked with complex tables.

Unemployment in Great Britain

GB	Unemployed (000's)		
	Total	Male	Female
1973 (D)	600	500	99
'70 (B)	580	500	87
'68 (A)	550	460	89
'66 (C)	330	260	71
Average	520	430	86

TABLE C

The same data arranged to show the deleterious effect of artificial spacing-out to fit the page.

Unemployment in Great Britain

	Unemployed		
	Total	Male	Female
1973	600	500	99
'70	580	500	87
'68	550	460	89
'66	330	260	71

Figure 1. Examples of Ehrenberg's rules applied to a simple table.

FIGURE 11.12 **Table B Illustrates Table Design Principles, in Contrast to Tables A and C. [From Michael Macdonald-Ross, "Graphics in Texts," in Lee S. Shulman, ed., *Review of Research in Education,* 5 (Itasca, IL: F.E. Peacock, Publishers, Inc., 1977), pp. 49–85. Copyright American Educational Research Association.]**

1. Round numbers to two significant digits unless greater detail is essential. (You might have greater detail, but if it is not necessary to present it, don't.) The rounding of numbers to two digits will help readers to do the mental arithmetic required to relate numbers. Unnecessary detail makes their task harder.

2. Provide row and/or column averages to give some reference point against which data can be interpreted. It is hard to get a sense of numbers without some yardstick by which to view them.

3. Design the table so that columns, rather than rows, are used to present the most important comparisons. (Side-by-side comparison is easier for readers than up-down comparison.)

4. Order the rows and columns by some logical principle such as size of the numbers rather than by some arbitrary principle such as alphabet. (Again, to impose order on detail is to impose meaning also.)

5. Design the table to be compact but not cramped. Don't artificially space out the table to fit onto the page. Although cramped tables are difficult to read, tables with too much white space are especially difficult to read. Use space (and lines) functionally to show groupings and logical connections between blocks of numbers.

As Macdonald-Ross indicates, complex tables are not well handled by what he calls the "general readership." Therefore, you may want to restrict your use of tables to smaller amounts of data and perhaps to use with more technically knowledgeable audiences. However, tables are a familiar and useful tool for sharing information with professional users. As he says, they are the preferred format for professional users. They permit meaningful presentation of large bodies of numerical data on one page—something that words cannot do.

Graphs Graphs of various types can also be used to present numerical data. Their use tends to focus the attention of viewers upon the big picture—upon relationships and trends—rather than on the discrete details of the data. They are effective tools for giving a meaningful overview of massive amounts of data that would be hard to comprehend otherwise. (Remember, of course, that you can put a table into an appendix and a graph into the text of a report if you wish to take advantage of the strengths of both types of tools. This is especially easy now that spreadsheet programs are available, but it has always been an option.) Line graphs, bar graphs, and pie graphs are all effective tools, each with its own strengths and weaknesses.

Line graphs are good for showing direction and trend in large bodies of data. They can show the relationships among data points and the importance of certain significant or "nodal" points within the data (a nodal point is a point at which significant change occurs). For example, a graph of stock market activity could reflect the influence of external factors on market trend: the president of a large corporation resigns, and stock sales react in a way that can be dramatically graphed. Important but non-nodal data points within the data, however, are not well reflected. Also, line graphs are not especially effective tools for showing relationships among numerous sets of data that have complex relationships. For example, a line graph with about four lines on it is easy to interpret unless the lines begin to cross one another. At that point, or if there are many more lines, the graph becomes difficult to read.

Some suggestions for line graphs are as follows:

1. Use them only if there will be a small number of lines, especially if the lines must cross.

2. If possible, put labels directly upon the lines themselves. If you use a legend to identify and distinguish between lines and symbols, readers will have more difficulty in keeping track of the distinctions between lines than if you put the interpretation directly onto the lines themselves. (Note: many graphics programs do not give you the option of labeling the lines directly—a clear weakness in the programs.)

3. Keep axes proportionate, clearly label them, and clearly designate the units of measure along them. If possible, avoid turning the type to run up and down rather than across the page.

4. Identify the significance of nodal points by means of "callouts" superimposed upon the line graph.

(Figure 11.13 shows a line graph that incorporates most of these principles.)

Bar graphs are good for showing volume rather than trend. (A filled line chart can also show volume as well as trend, but it can handle only small sets of data.) Like line graphs, bar graphs are also limited in their ability to handle large numbers of data sets. For example, a graph representing ten years of population data could work quite effectively as a bar graph. You might even use double bars or stacked bars to represent the men and women within the population over those ten years. However, if you tried to use a bar chart to show the proportions of ethnic groups within the population over a 10-year period, your graph would rapidly become so complex as to be worthless. You might manage three or four relationships at a time, but not more than that.

Despite their inherent limitations, bar graphs can very effectively show relationships as well as trends and are especially effective at showing the contrast between large and small numbers. They also show the similarities between similar numbers well. Yet bar graphs are not very effective for showing discrete data points or specific numbers.

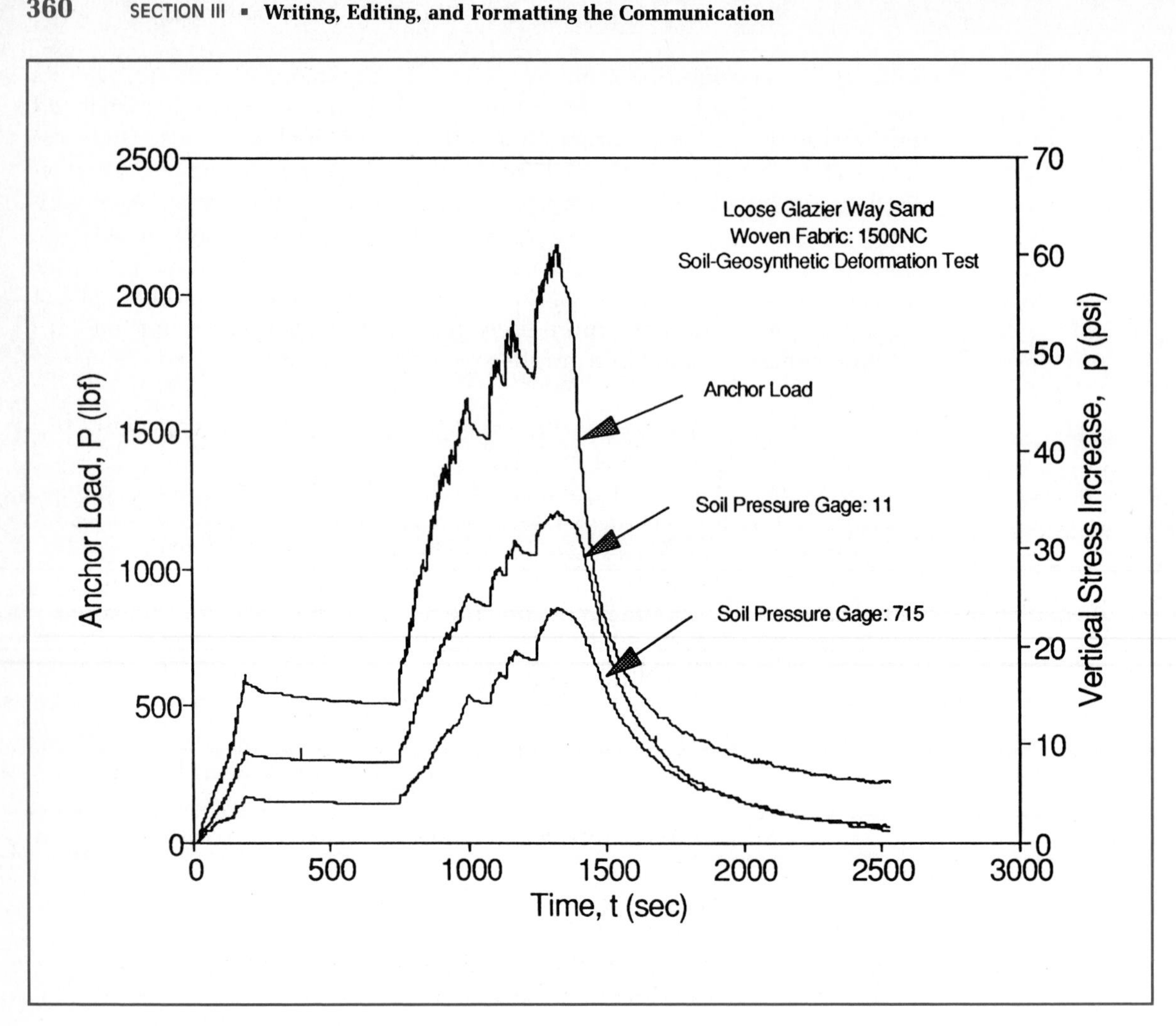

FIGURE 11.13 Sample Line Graph Incorporates Most of the Suggestions on Design of Line Graphs.

Bar charts can be either vertical or horizontal. The vertical format is the more frequent, but the horizontal format has some attractive features that recommend it to your consideration. Notably, the labels that identify the bars on a horizontal chart can be put directly upon the bars themselves, not into a legend. Furthermore, even large numbers can be added at the tops of bars to provide specific numerical data in a horizontal format, whereas in a vertical format often neither the legend nor the numbers will fit.

Some suggestions for using bar graphs include the following:

1. Present only a small number of bars.

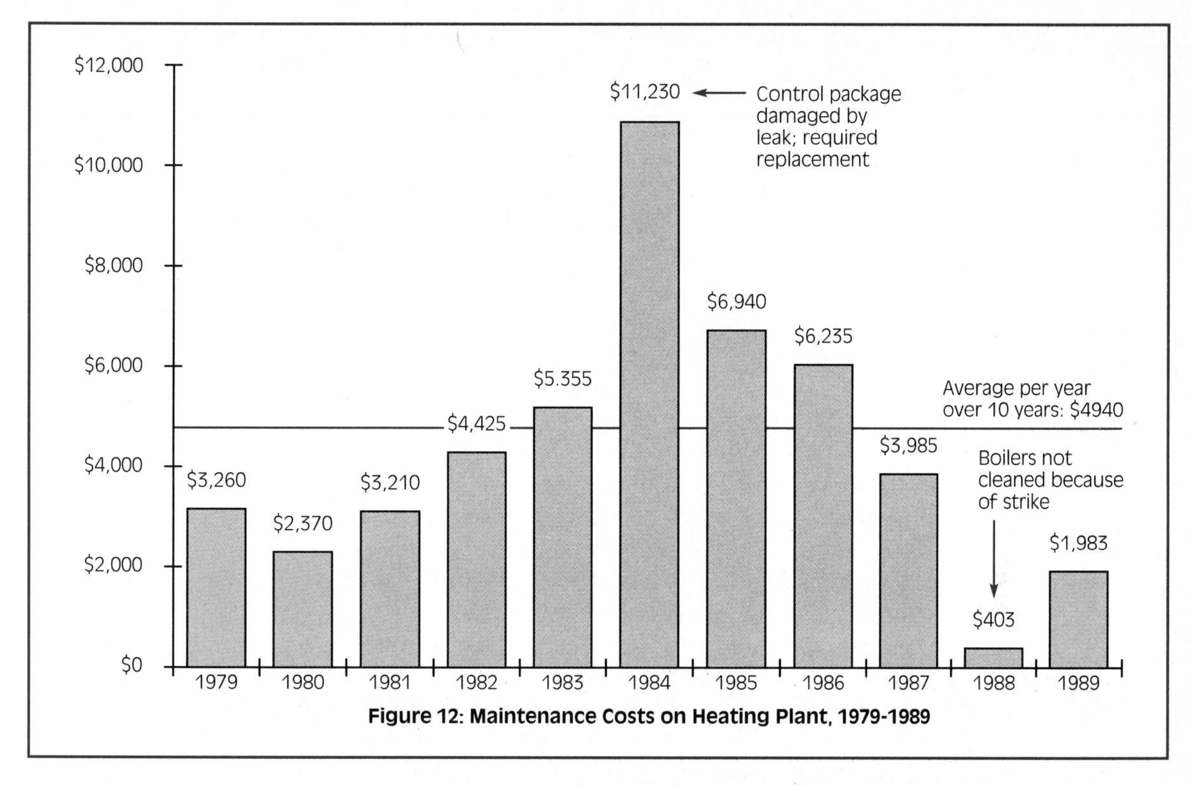

FIGURE 11.14 Sample Bar Graph Incorporates Most of the Suggestions on Design of Bar Graphs.

2. Restrain the urge to use elaborate fill patterns (you can quickly create a confusing and altogether useless visual effect).

3. Do not use three-dimensional bars. The addition of the third dimension adds absolutely nothing informative to a bar chart (there are two axes, not three), and although it might "look good" in some instances, the technique will confuse viewers.

4. Provide informative labels, if possible on the bars themselves.

5. If possible, provide specific numerical quantities at the heads of bars.

6. Keep axes proportionate, clearly label them, and clearly designate the units of measure along them. If possible, avoid turning the type to run up and down rather than across the page.

(Figure 11.14 shows a bar graph that incorporates most of these suggestions.)

Pie graphs are good for showing relationships among percentages that make up a whole. They are also effective for showing the contrast among small and large percentages and the relative similarity of similar percentages. However, pie graphs are very poor tools for showing specific quantities or percentages, small differences between similar percentages, or relationships among more than six or seven parts.

In a research project of our own, we found that in general pie graphs work well with two kinds of data: population data and economic data. However, we also found that there were severe limitations of which you need to be aware. First, engineering audiences did not like them very well even though engineers were more successful at interpreting pie charts than general audiences. (General audiences liked the charts, but engineers could accurately estimate the percentage of unlabeled wedges of pies, whereas general audiences could not. In fact, when general audiences were asked to assign percentages to unlabeled wedges, often they assigned percentages that added up to over 100%.) Second, pie graphs are effective for showing relationships among components only if there are more than about three or four yet fewer than six or seven. In other words, the range of appropriate use for pies is extremely narrow. A two-wedge pie looks ridiculous and gives almost no information. Conversely, an eight- or nine-wedge pie becomes excessively detailed and difficult to assimilate (as well as impossible to create effectively in the sense that you need room on the wedges for the percentages, quantities, and labels if the pie is to be easily readable).

Suggestions on pie graphs are as follows:

1. Use them primarily for general audiences, not for technical peer audiences.

2. Don't use them if you have fewer than three or four components or more than seven or eight.

3. Put the percentages and actual numerical values directly onto the wedges of the pie (or immediately outside) so that the size of the wedge is not the only device to convey percentage and quantity.

4. Start the wedges at the 12-o'clock point and move in a clockwise fashion around the circle, arranging the wedges according to some logical principle. Size of numbers is a reasonable principle by which to order the wedges, but the information may lend itself to much more meaningful sequencing if you look for a logic. (There is dispute about starting at "12 o'clock"; however, among the experts in the Brasseur study cited earlier, there was a consensus that pies should start at 12 o'clock and that the sequence of wedges was an issue.)

5. Restrain the urge to use elaborate fill patterns, three-dimensional effects, and exploded pies such as those provided by graphics pro-

grams. These are clutter ("chartjunk" as one expert calls it) that add no meaning. Simple contrast will suffice, if you use clear labels on the pie itself to identify the significance of the different wedges.

(Figure 11.15 shows a sample pie graph that incorporates most of these suggestions.)

Explanatory Function

The overall objective of explanatory graphics is to explain logical relationships. Unlike photos and sketches, which represent the world of visible things, explanatory graphics represent a wide variety of physical states, functional relationships, and activities that cannot be captured in a representational form. Also, unlike tables and graphs, explanatory graphics do not deal so much with quantitative information as with other kinds of qualitative information, such as composition, process, quality, relationship, and function.

For example, a meeting, such as the one we mentioned in our discussion of minutes of meetings, could be photographed or sketched, but what could be learned from a picture about the function of the meeting, its internal dynamic, or its output? (A picture would convey meaning about the energy

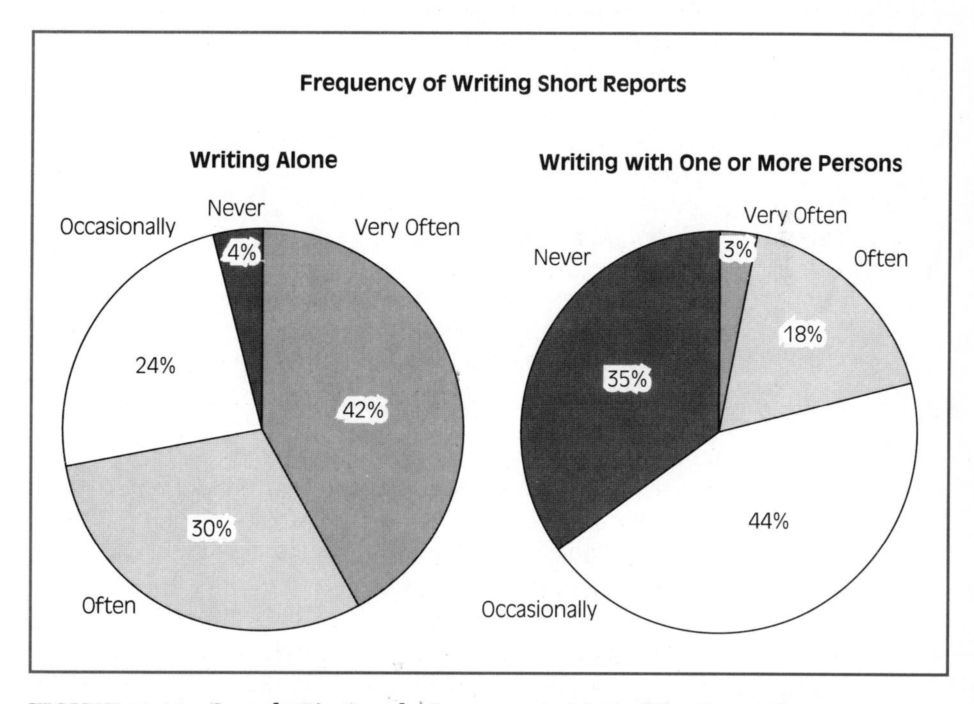

FIGURE 11.15 Sample Pie Graphs Incorporate Most of the Suggestions on Design of Pie Graphs. (Based on data from L. Ede and A. Lunsford, *Singular Text/Plural Authors: Perspectives on Collaborative Writing*. Carbondale, Ill.: Southern Illinois University Press, 1990.)

level of the meeting, perhaps, but not of its content or result.) Of course, words could be used (minutes) to interpret the meeting effectively, but non-verbal, visual means might also be used, perhaps in isolation, but probably in combination with words. You may recall that the meeting we discussed in Chapter 6 was one involving choice of a manager for a branch office. As a decision process, the choice could be modeled by either a flow diagram or a decision tree—both of which are devices that transform the invisible aspects of a process into a logical explanation of the process. Also, you may remember, the choice was based upon several criteria and involved evaluations of several candidates. Matching candidates against criteria, we could easily produce a matrix or table that would interpret the relative evaluations of the individuals and identify the chosen candidate. Both techniques would provide a perspective on the procedure and outcome of the meeting, yet both would rely upon predominantly visual rather than verbal information to do it. In doing so, both techniques would add meanings. A flow diagram or a decision tree would focus on the process of the decision makers. Conversely, a matrix would focus on the topics under discussion—the candidates and the criteria. In each case, the effect would be significantly different, but each would be appropriate and meaningful in certain audience and purpose situations.

Given the limitless variety of possible explanatory graphics, it is impossible to itemize and explain the strong and weak points of each. Accordingly, we restrict our comments to basic principles that can be applied to a wide variety of visual types: schematic drawings, flow charts, decision trees, organizational charts, Venn diagrams, photos with superimposed interpretations, representational sketches such as exploded drawings, or simplified functional drawings that either eliminate or "ghost" out part of the detail. In particular, we would like to mention four basic design principles.

Visual Efficiency One of the standard bits of advice that you will encounter is to keep your visuals simple. On one level that is good advice, for indeed needless complexity and clutter do make for problems in visual language as well as in verbal language. However, we would like to refine this advice just a bit because people often assume that "simple" means "undetailed." It need not.

We would advise you to make your visuals as efficient as possible—not just at simple as possible. That is, if you draw an arrow, draw it because there is a function for the arrow, not just because it looks good. If you use a fill pattern, put it there for a substantive reason. Put a third dimension into your drawings only if you really need it, and edit out any graphic detail (or print detail on the graphics) that does not contribute. We say "edit out" because often your visual materials are ones that were not expressly created for your report but are instead the byproducts of your investigation and of your professional work. With these especially, you need to evaluate their communication efficiency to make sure that they can be used in the document. Frequently you will find that these drawings, printouts, flow diagrams, and graphs contain superfluous detail that is nonessential for the communication purpose. Some of the visuals you have prepared during your investigation, however, will be appropriate for your report without modification. For example, you generate a graph of the temperature stratification of the air in an atrium (Figure 11.16) during an analysis of the HVAC (heating-ventilation-air conditioning) system. In this architectural analysis, the graph is used to determine whether or not stratification occurs and, if so, when. The data themselves do not provide the answers to these questions, as thousands of data have to be statistically averaged. Instead, the data are entered into a spreadsheet program so that the answers can be determined by graphical analysis. This particular graph then can be used in the report because it is appropriate for communication purposes as well as analytical purposes. (On Figure 11.16, the graph is supplemented with a schematic of a cross-section of the atrium to indicate the location of the ten sensors that collected the data for the graph.)

Other visuals are not appropriate for communication purposes without modification. They are working documents or images on the screen used for

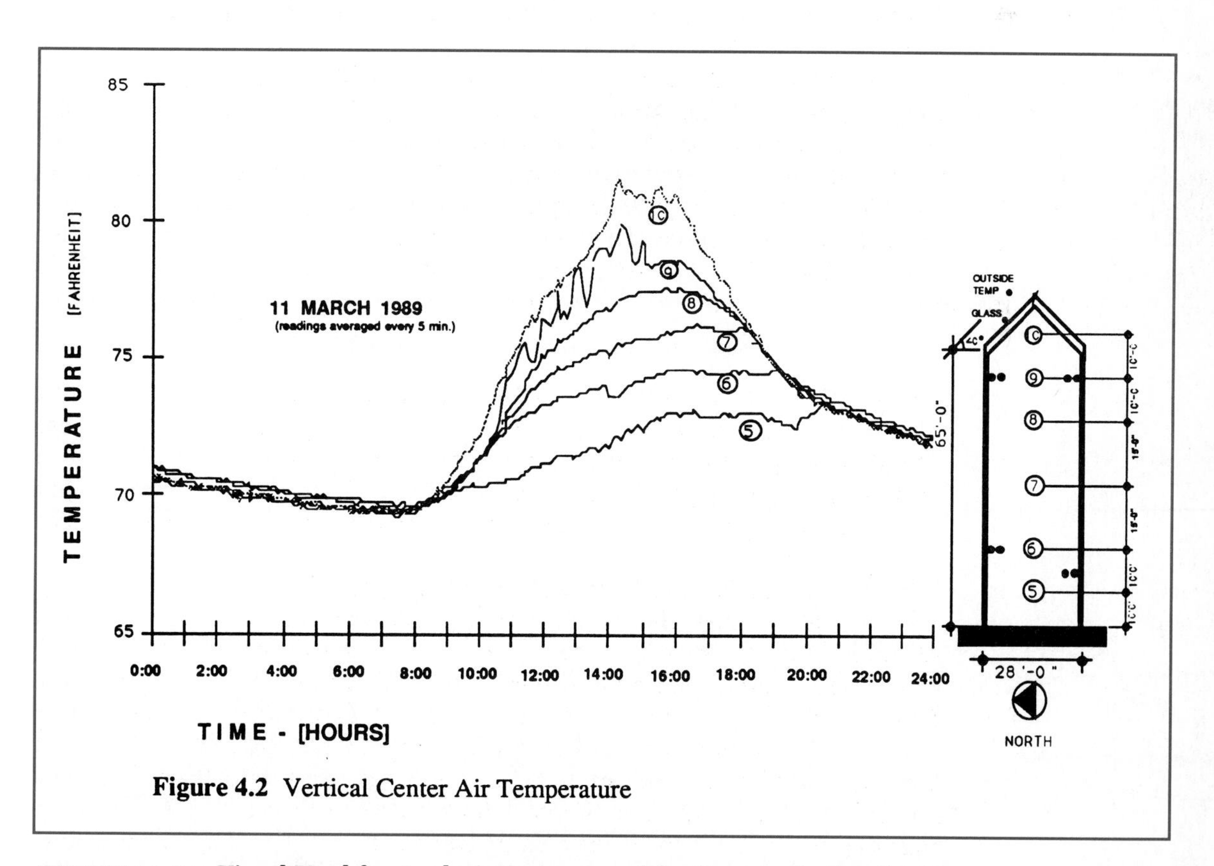

Figure 4.2 Vertical Center Air Temperature

FIGURE 11.16 Visual Used for Analysis Purposes and for Communication Purposes.

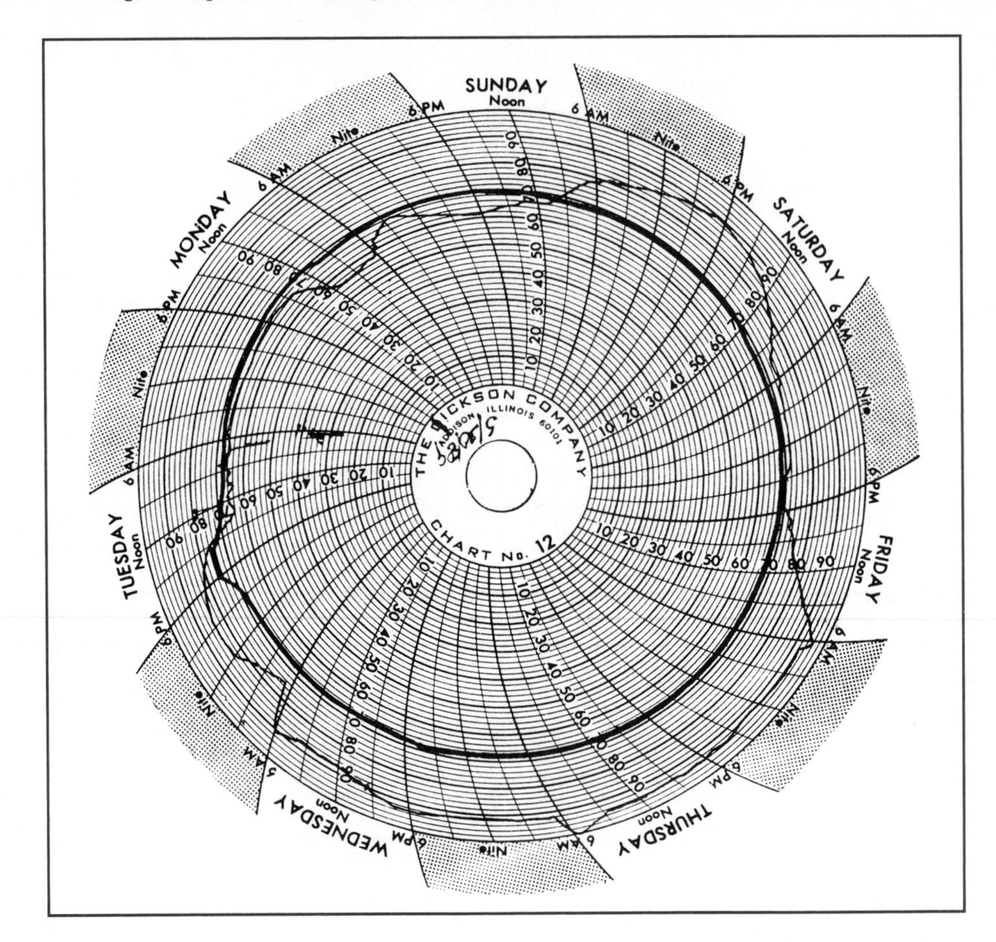

FIGURE 11.17 Visual Used for Analysis Purposes but Ineffective for Communication Purposes.

design or analysis purposes but have little utility for communication purposes (Figure 11.17). The detailed information on daily temperature and humidity is not germane to the purpose of the document; the overall trend is germane, but is not immediately clear from the detail.

But remember, the issue is not how much information there is. Rather, the issue is how germane and efficient the information is.[9] For example, a process flow diagram in a design report for a chemical process (Figure 11.18) is highly detailed. The diagram includes considerable technical detail, and the stream tables contain considerable data. Yet, even though a design report is distributed throughout the corporation, the detailed process diagram is expected and necessary. As detailed as it seems, all of the information is essential to an explanation of the process, intelligible to most of the readers, and welcome. Indeed, the combination of the flow diagram and the table is welcomed

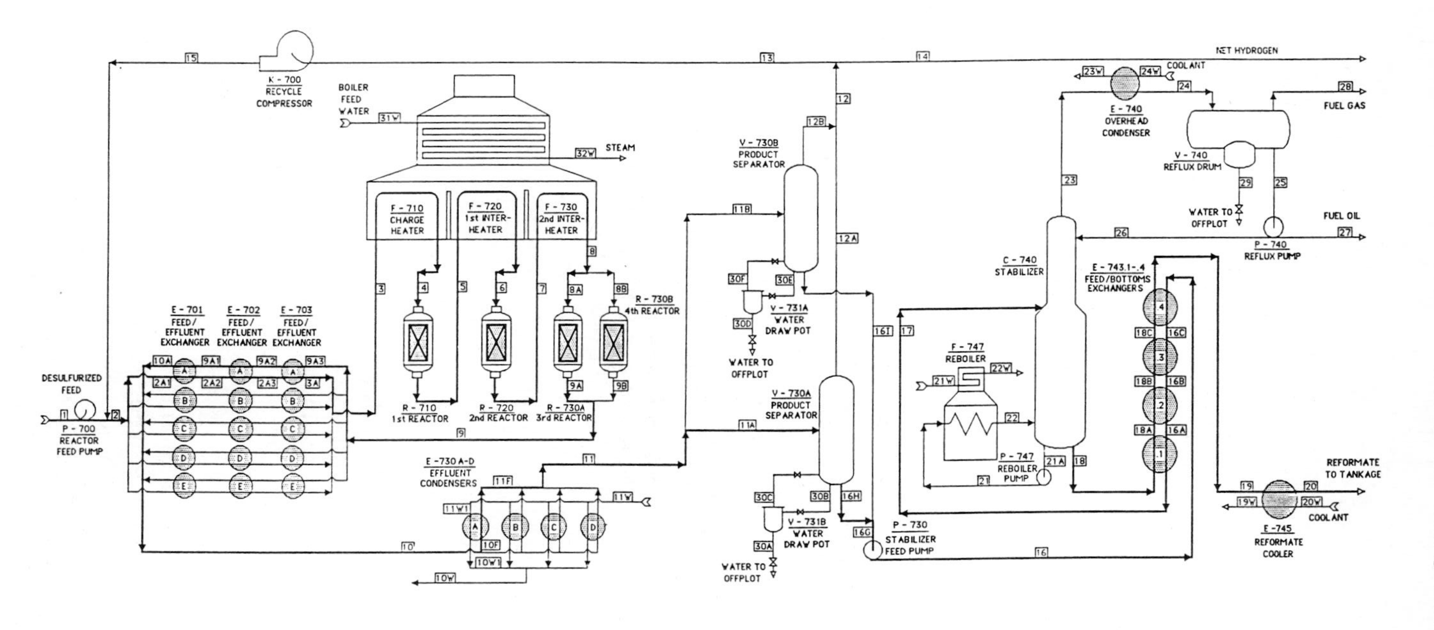

Figure 4 Process Flowsheet

G.S.H. 4/11/86

STREAM NUMBER	FLOW RATE (LB/HR)	TEMPERATURE (°F)	PRESSURE (PSIA)
1	596,979	200	115
2	983,379	220	199
2A1-2E1	196,677	220	199
2A2-2E2	196,677	310	199
2A3-2E3	196,677	555	199
3A-3E	196,677	870	199
3	983,379	870	199
4	983,379	980	194
5	983,379	830	193
6	983,379	980	186
7	983,379	930	185
8	983,379	980	182

STREAM NUMBER	FLOW RATE (LB/HR)	TEMPERATURE (°F)	PRESSURE (PSIA)
8A	491,690	980	182
8B	491,690	980	182
9A	491,690	952	181
9B	491,690	952	181
9	983,379	952	181
9A1-9E1	196,677	418	159
9A2-9E2	196,677	650	167
9A3-9E3	196,677	952	181
10A-10E	196,677	250	149
10	983,379	250	149
10F-I	245,845	250	149
10W	3,034,068	120	90

STREAM NUMBER	FLOW RATE (LB/HR)	TEMPERATURE (°F)	PRESSURE (PSIA)
10W1-4	758,517	120	90
11	983,379	140	140
11A-B	491,690	140	140
11F-I	245,845	140	140
11W	3,034,068	80	90
11W1-4	758,517	80	90
12	484,972	140	140
12A-B	242,486	140	140
13	386,037	140	140
14	98,934	140	140
15	386,037	201	199
16	498,408	140	223

STREAM NUMBER	FLOW RATE (LB/HR)	TEMPERATURE (°F)	PRESSURE (PSIA)
16A	498,408	350	221
16B	498,408	280	221
16C	498,408	210	221
16G	498,408	140	140
16H	249,204	140	140
16I	249,204	140	140
17	498,408	425	220
18	484,533	471	222
18A	484,533	396	221
18B	484,533	326	220
18C	484,533	257	219
19	484,533	187	213

STREAM NUMBER	FLOW RATE (LB/HR)	TEMPERATURE (°F)	PRESSURE (PSIA)
19W	415,275	120	90
20	484,533	120	212
20W	415,275	80	90
21	103,575	448	222
21A	103,575	448	227
21W	18,973	399	155
22	103,575	471	222
22W	18,973	364	150
23	61,600	202	216
23W	75,000	120	90
24	61,600	118	214
24W	75,000	80	90

STREAM NUMBER	FLOW RATE (LB/HR)	TEMPERATURE (°F)	PRESSURE (PSIA)
25	58,861	118	214
26	47,725	118	217
27	11,136	118	214
28	2,739	118	214
29	-----	-----	-----
30A-B	-----	-----	-----
30C-D	-----	-----	-----
30 E-F	-----	-----	-----
31W	195,143	339	610
32W	195,143	900	600

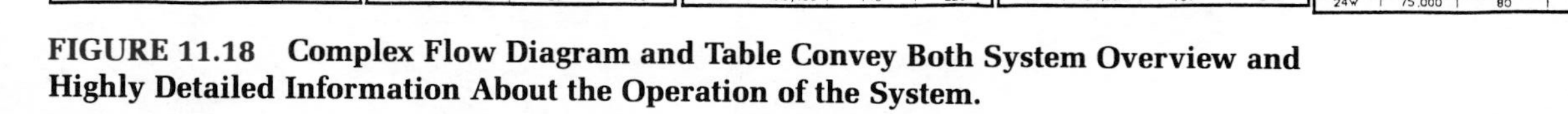
FIGURE 11.18 Complex Flow Diagram and Table Convey Both System Overview and Highly Detailed Information About the Operation of the System.

because readers regard it as an efficient combination that makes an enormous amount of different types of information immediately accessible. Readers appreciate that. The entire process is displayed on one diagram so that the reader can interpret any one stage of the process (because of the interrelationships of all stages of the process). Of course, the process diagram can be broken down into simplified subsections for explanations of stages of the process; however, the entire diagram is needed for an overview of the process and of the interrelationships among the stages and streams.

Visual efficiency, then, is economy of visual elements. Don't draw what is not needed, but feel free to make your visual creations detailed if they need to be detailed. They are not just visual relief from a printed page; they are functional drawings. Just make them do their work efficiently.

Verbal Clarity Visuals should be self-sufficient; that is, they should have labels and a title. Although a visual is interpreted in the text, the visual must stand alone. When a reader refers to a visual, he or she looks at it as an independent communication unit. It needs to make sense by itself. In addition, visuals often are read separately from the text. They are copied and distributed independently. (Also, some readers only look at the "pictures" in a document.)

To label visuals is important so that the reader does not have to reread the text in order to understand the visual. The visual on the squeeze casting process (Figure 11.19) is not uncommon. Notice that on the drawing neither the stages of the process nor the parts of the casting equipment are identified. Thus, to understand the squeeze casting process, the readers must either refer to the text or take the time to construct their own interpretations. Even if they might be able to do so, you should not force them to do so. The visual communicates inefficiently.

In addition, a visual needs a heading or title to inform readers how to interpret it. Every visual should have an explanatory heading placed directly underneath along with a figure number. (Table titles and numbers conventionally are placed directly above a table.) The heading should explain the aid sufficiently for a reader to be able to understand it without referring to the text.

Visual Accuracy Your visuals must be accurate. With professionals "Graphical Integrity"[10] is assumed. Accurate representation is more than a matter of being objective. It is a matter of the significance or interpretation that the reader will make of your visual.

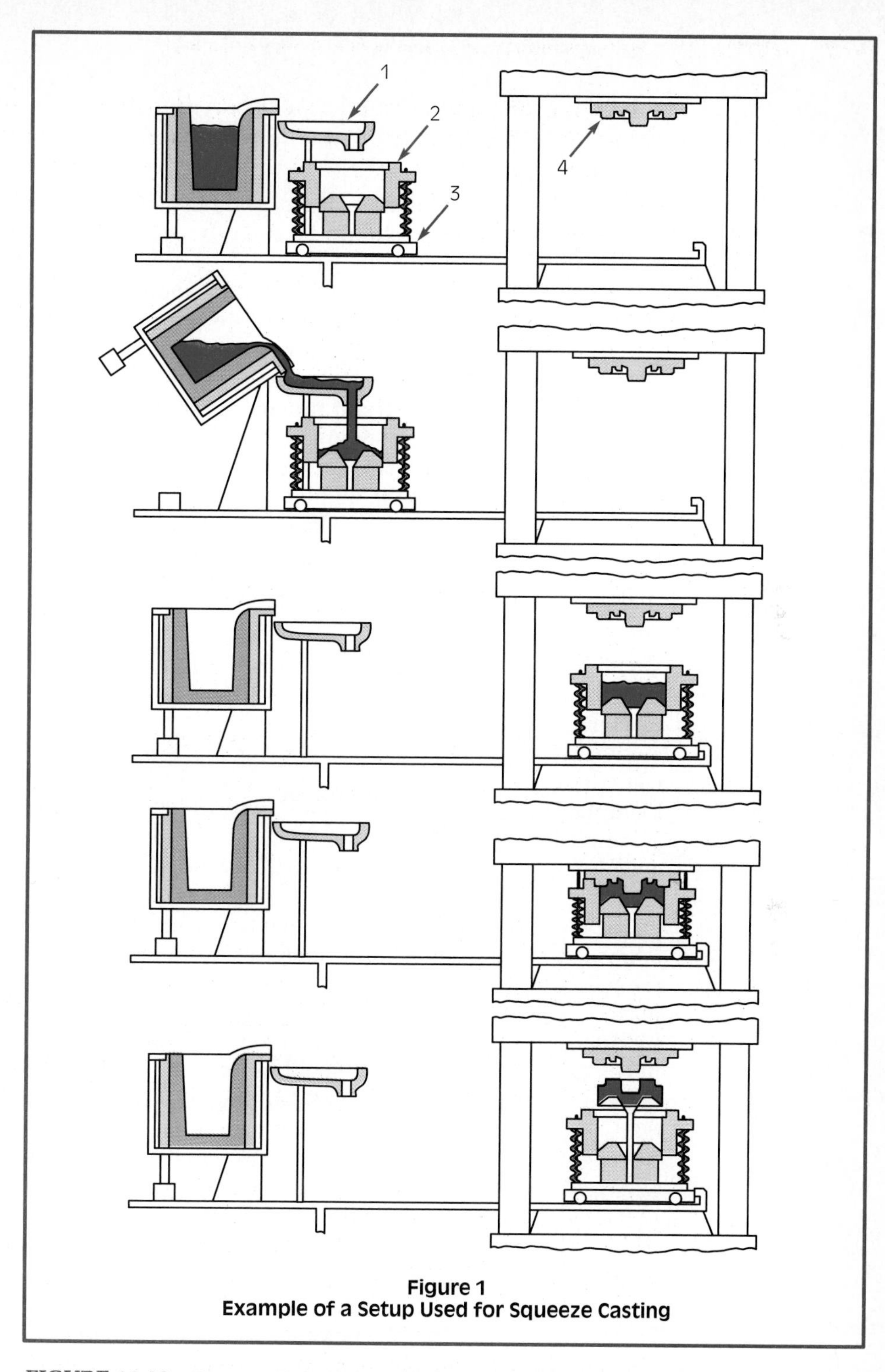

FIGURE 11.19 Process Description Without Words to Clarify the Stages of Process and the Parts of the Equipment Used.

Because misrepresentation can occur unintentionally, these principles enunciated by Edward Tufte should be followed:

- The representation of numbers, as physically measured on the surface of the graph itself, should be directly proportional to the numerical quantities represented.
- Clear, detailed, and thorough labeling should be used to defeat graphical distortion and ambiguity. Write out explanations of the data on the graphic itself. Label important events in the data.
- Show data variation, not design variation.
- In time-series displays of money, deflated and standardized units of monetary measurement are nearly always better than nominal units.
- The number of information-carrying (variable) dimensions depicted should not exceed the number of dimensions in the data.
- Graphics must not quote data out of context.

Accuracy is the badge of a professional.

Visual Focus and Balance Visuals are inherently artistic creations that involve both the concepts of balance and focus just as do other art forms. That is, the placement of the elements of a drawing is not a matter related to content alone. It is a matter of balance and focus, the same principles that might allow us to distinguish between an amateurish snapshot and a well composed photograph. The picture should fit well within its frame, and the elements of the picture should relate well to one another so as to create a harmonious, balanced, focused visual image that draws the eye to what we want viewers to see.

For example, an organization chart is a schematic representation of the administrative structure of an organization. As such, it obviously needs to reflect accurately both the levels and kinds of substructures that combine to make up the overall structure of the organization; the form of the drawing has its basic origin in the organization itself. But the organization chart also needs to convey a sense of the harmony of the organization and the solidity of its structure as well as its literal configuration. Although the organization's configuration could be drawn in an infinite number of ways, a good drawing will convey a sense of the functional equality of individual roles and the partnership relationships of parallel unit structures as well as of the hierarchy of the organization. The drawing needs to be constructed carefully to achieve visual effect that reinforces and reflects the conceptual meaning of the organization itself.

Unfortunately, the first version of a drawing seldom achieves this sense of balance and focus. Yet, the fact that you seldom achieve a well-balanced and focused drawing on the first attempt should not deter you. It just means that you should sketch your design concepts several times, looking not only at their overt meaning but also at their aesthetic qualities of balance and focus.

Operational Function

Explanatory graphics convey high-level, abstract concepts, promoting understanding and long-term memory. In contrast, operational functions involve giving someone enough task-oriented information to do something quickly and correctly, perhaps with little conceptual information accompanying it. For example, if you want to show someone how to install a new board in a computer, perhaps your best choice is to provide a cookbook approach that says in effect, "to install the board, do this, this, and this." Little teaching is going on, and concepts about how the board functions or even what it does may be omitted (which makes these instructions less useful in anomalous situations or during unanticipated faults). Conversely, if you wanted to explain the function and operation of the board, you might provide almost no information about how to install it (which makes the information being imparted of less task-oriented value than might be required to install a board successfully). The limitations of each of the two modes of explanation perhaps suggest that you would be wise to combine explanatory information with your task-oriented information, but it is clear in any case that the operational function can be well served by visual information.

Photos, sketches, flow charts, and decision trees can all provide operational information. Although all of these same tools can be used to enhance understanding and learning (an explanatory function), they are all effective as tools for giving information that is strongly action-oriented or task-oriented and that focuses upon short-term memory.

From experience and research, we know that strongly task-oriented and heavily visual instructions do work. For example, in one experimental evaluation, such aids were shown to reduce training time.[11] Effective aids, which were strongly graphic, allowed high school students with only 12 hours of training to perform with surprisingly good results when compared to experienced Air Force technicians with an average of 7 years of experience and 7 hours of training. Similarly, in another study, such aids were tested in terms of performance. The results showed that task-oriented, heavily visual instructions produced no errors among either of two groups who used them (a group of experienced technicians and a group of inexperienced ones). Conversely, another group who used conventional maintenance information over a 4-month period always had some proportion of error. Finally, assessing the reactions of test subjects to task-oriented and heavily visual instructions, researchers found that 78% of the experienced Air Force technicians questioned liked the task-oriented and heavily visual instructions better than the conventional instructions that these replaced.

As you can see, you must decide what your purpose is. Do you want to add to the long-term knowledge of your readers/viewers? Or do you want simply to give them short-term knowledge that will get them through a task? Or both? If you wish to do either of the latter two of these, photos, sketches, flow diagrams, and decision trees are appropriate choices. The photos and sketches can be used for actions that involve physical objects and physical movement; the flow diagrams and decision trees for actions that involve cognitive activity or actions that are not possible to photograph.

Photos and Sketches for Operational Information In the case of photos and sketches, the trick is to keep them highly controlled so that the only variables are significant, action variables. If, for example, a right-handed person is shown in one panel of a sequence, switching to a left-handed person in the next panel could cause awkward misunderstanding upon the part of readers and viewers. Viewers might ask, are we supposed to switch to the other hand? Are we supposed to switch people? Or are those changes unimportant? Similarly, making even subtle arbitrary changes in such things as color or fill pattern can cause confusion. We remember, for example, a graphic that we encountered in an electronics company. The instructions had to do with installing a piece of equipment. At one point in the instructions, a sketch showed a hand holding a plug. The plug was the black, 3-prong type of precisely the sort attached to the actual equipment. In the next panel, inexplicably, the plug was drawn in an unfilled pattern (white). In the writer's mind, the plugs were the same. In the viewer's eye, clearly they were not. And that caused confusion.

Other important points to consider in photos and sketches are scales, callouts, and legends. Scales should not be switched from panel to panel. That is, if one panel shows a 4-foot-high object and the next shows a 6-inch-high object, there must be some point of reference (human figures are good) to indicate the scale difference between the two photos or sketches. Otherwise, viewers can be confused. Similarly, callouts and legends focus attention on specific details of a photo or sketch, allowing the viewer to know what is being shown without relying entirely upon the visual information. Without words in callouts and legends, there is no guarantee that viewers will interpret what is being represented as the writer intends.

Flow Diagrams and Decision Trees With actions that cannot be shown by photos or representational sketches, both flow diagrams and decision trees are effective devices, although decision trees tend to be somewhat unfamiliar to nontechnical people. In both cases, effective design requires consistency. That is, in flow diagrams and in decision trees, different shapes (circle, cylinder, square, etc.) have meanings. For example, a circle in a flow diagram describes a condition, while a square or rectangle describes an action. Preserving the distinction between these requires consistency. Inconsistency can lead users to think that a difference in shape or volume is somehow a meaningful, deliberate change in pattern, when in fact it may be an accidental inconsistency in drafting.

Another concern in flow diagrams is feedback loops. Flows are sometimes completely linear; however, other flows often allow (or even require) feedback loops in which the effects of one stage of a process can be used to affect not only subsequent stages but prior stages as well. In this case the flow is an iterative process involving "forward" movement and also "backward" movement. Progress in such flow situations comes about through multiple repeated steps. Omitting the feedback loops transforms such a process into a linear, one-time-only process—an obvious misrepresentation. Unfortunately, often these feedback functions are neglected in flow diagrams.

In decision trees, error recovery procedures need to be considered. That is, if you set up a series of choices that somehow lead users into a dead end, they need a recovery procedure or escape route to go either forward or backward in the instructions. Finding none, users will quickly become frustrated.

Whether your purposes in using visuals are dominantly for iconic functions, data display functions, explanatory functions, or operational functions, visuals are tools for liberating your own thinking and the thinking of your readers (viewers) to look at information in a nonlinear yet functional way. As one design expert says, "Many people limit their thinking to the verbal, sequential level even when its effectiveness becomes increasingly strained by the amount of information that must be absorbed and expressed."[12]

❑ 11.4 ❑ Integrating Visuals and Printed Text

Once you have decided to use visuals and have gone through the process of considering alternatives and designing particular types of visuals, you still must use your visuals effectively by combining them with the printed text. This integration is partly a conceptual matter and partly a mechanical matter.

Conceptual Integration

Conceptual integration requires you to introduce and interpret your visuals verbally. You must make the visuals a part of the text, not ornaments on it. To interpret a visual, you need to explain to the readers basically why it is there, what you want them to see in it, or essentially how to read it. Without your interpretation, different readers may perceive different details and interpret the visuals differently from each other—and from you. Of course, the amount of interpretation necessary for a particular visual may vary considerably according to its complexity and communication value. Some with high communication value will need little explanation; the text instead can merely provide a brief focusing comment. With some visuals, however, the text must explain or interpret the significance of the data. This is especially true with visuals that present numerical information, but it can be true of any type of visual. The important point is that the visual does not necessarily "speak for itself" and will not function with maximum efficiency if you merely drop it in without verbal integration.

As you see, how you integrate a visual depends on both the type of information in the visual and the communication efficiency of the visual. Where you integrate it is another matter, depending upon the immediate importance of the visual to readers as they read your document.

Not all visuals should be put in the text itself, and certainly not all of them should be grouped in an appendix unless you have no other alternative. This is because some of the visuals will be essential to readers as they read; others are not essential during the act of reading but do provide important follow-up documentation. You must therefore determine which visuals to put in the text and which to put in the appendix. Those visuals likely to be necessary or even very useful to readers as they read should be introduced, presented, and interpreted directly in the text. Those not likely to be of immediate value to readers still need to be referenced in the text. Although visuals in the appendix generally are not interpreted within the text, you still may need to interpret at least some of your appendix visuals. You can put that interpretation directly upon the appendix visuals themselves so that they can be read in isolation from the text. (A good rule of thumb is that the appendix visuals should be able to fall out of the report, be picked up by a passerby, and still make sense. That often means that verbal interpretation must be added to visuals put into an appendix.)

Mechanical Integration

When drawings were hand-done, mechanically integrating them with the printed text was a difficult matter. The drawing itself had to be of sufficient size to be "drawable," yet once that size was chosen for the benefit of the graphic artist, it was difficult to change the size for publication purposes, at least in informal documents. (Printed documents could have photo-typeset reductions.) Now, however, computers and copy machines with enlargement and reduction features have made it easy to scale a drawing to almost any size and to "paste" it into a printed text at whatever point would be most useful to readers. But the questions remain, what are the best ways to put a visual into a document and how big does it have to be?

On one level, the obvious answer is that artwork needs to be big enough to be easily legible, but small enough to fit into the document with a minimum of difficulty at the precise point at which it is being discussed—or as near to that point as possible. There is also an aesthetic dimension to this matter: a simple pie diagram looks absurd in a full-page size, for example, while a complex diagram of the parts of a machine might look overcrowded even if legible in a small scale. In other words, you must please your own inner eye as well as your reader's literal eye.

Yet, the answer can be a little more specific if you consider that there are really only three basic sizes for drawings inserted into an informally designed text.

1. Full-width page element of less than one page in height

2. Page size

3. Double page, foldout size

When we allude to "informally designed text" we are referring to standard word-processed or typewritten text in single column format, not to "desktop published or printed" text in multiple column format. (We choose to limit our discussion here because most reports are word processed informally in the single column format. If you do use multi-column formats, you have additional sizes of drawings to consider—ones the same size as a column width and ones the same size as combined column widths.)

Although it might be theoretically possible to go above a double-page size in a drawing, we would not recommend that as a technique that readers will find very manageable: a 3-page foldout, for example, can become quite cumbersome. Similarly, going below a full width of the page in single column, informal text would be unconventional. Admittedly, desktop publishing tools would allow you to wrap the text around small photos even in a single column of text, but that is not frequently done in informal, "unpublished" reports, at least to this point.

Whatever the final size of the visual is to be, we suggest that you draw it in whatever scale seems necessary to permit you to draw effectively. But then don't stop by merely reproducing your drawing in its original scale, as most report writers do. Instead, choose a scale appropriate to the content of your artwork, basing your choice on three factors: legibility, aesthetic appeal, and the three basic sizes listed previously. You can then place the drawing physically into the document by using a grid.

Create a Grid The grid is a planning device used by page designers and graphic artists. It merely divides the page into a matrix of evenly spaced lines that can be used for alignment purposes as text and pictures are combined. Nonphotocopiable blue ink used to be used for grids, but today the planning lines are electronic. The outer margins of the text itself are defined by the grid, as are subdivisions of the page. These subdivisions indicate, for example, the page header, the page footer, and the columns within the text, usually with a maximum of three columns.

For Small Figures For many of your figures, you will not need a full page to present them (as in the graph of temperature stratification, Figure 11.16). If that is the case, size the figure and insert it directly at the point at which it is being discussed. The grid will allow you to design a box of appropriate size into which the figure will fit. The width of the box is a function of the outer margins of the text. Its height is a function of the drawing itself (proportionally, once the width has been chosen, the height follows).

If possible, place the figure at that point in the text where it has just been introduced. If that is not possible—for example, if you are near the bottom of the page—then put the figure on the next page, at the top.

Drawing a box (or shaded background) around a figure is not difficult on most computing systems and it gives the page a more polished appearance. If you choose to do it, place the figure within the box just as you would place a photograph or a picture in a frame to be hung on your wall. That is, position the picture so that it balances within the frame, a uniform "margin" all around and the center of the artwork at the center of the frame.

Page-sized Figures If your artwork is too detailed to fit onto part of a page, your next choice is to go to a full-page figure. In this case, you could still draw a box around your figure, setting its size to correspond to the outer margins of your text pages; if you need still more size, you can just eliminate the box and let the drawing itself go out as far as the margins on a printed page from the same text.

You have a choice on how to place the figure: vertically (corresponding to the orientation of the text) or horizontally (turning the page on its side). The best choice for your readers is to keep the vertical page, if the content of the drawing allows that. If not, turn the page so that the bottom of the drawing is on the margin that would be the normal right margin.

Double-page and Foldout Figures Sometimes a full-page format is still too small to handle the detail necessary in a figure, and you may need to go to a 2-page or foldout format. That does happen sometimes, although you probably should reconsider your design concepts if you find that it happens regularly. In any case, if you find that you need a 2-page figure, again you have the option to make it a "vertical" format with the box corresponding to the margins of the two pages combined, or to turn the figure to a "horizontal" format. As before, you don't want to force the reader to turn the text unless it is necessary to do so.

These double-page foldout figures are capable of handling a great deal of detail, especially of the sort found in the appendix for very technical reports that are intended for technically expert audiences. They are the types of drawings that readers pore over to learn details, not the type they quickly consult to get the general drift of something. The combined flow diagram and table from the chemical plant (Figure 11.18) is an excellent example of a double-page foldout, one which by convention is in the text of this type of design report rather than in an appendix.

Conclusion

We have said that the reality of your technical and professional writing situation will inevitably demand that you use visual materials. In that sense, to be a skillful writer, you have no choice but to understand the process, functions, principles, and mechanical requirements of the effective use of visuals.

However, do not think of visuals as merely a professional necessity for technical and professional writers. Visual thinking can be liberating for you and for your readers. Visual thinking can be a means of exploring ideas and translating complex thought into concrete form. As one expert says, "Graphic thinking is a tactile experience that brings you directly into contact with your ideas and allows emotional involvement with them. The sketches on paper are products of ideas, concrete rather than transitory."[13] In short, visuals are indeed essential, but their essential nature is not limited to their being products of technical and professional work. They are also helpful parts of the process of that work both for the writer and for the reader. For that reason we believe that visuals deserve a good deal of your attention and concern.

Notes

1. We are indebted for the term "visual essentials" to our colleagues, Ben and Lisa Barton. We would also like to acknowledge our considerable debt to them for their many articles on graphics, especially Ben F. Barton and Marthalee S. Barton, "Toward a Rhetoric of Visuals for the Computer Era," *The Technical Writing Teacher*, 12: 2 (Fall 1985): 126–45, and "Trends in Visual Representation," in *Technical and Business Communication*, ed. Charles H. Sides (Washington, DC: Society for Technical Communication, 1989) 95–135.
2. Lee Ellen Brasseur, *The Visual Composing Process of Graph, Chart and Table Designers*, diss., U of Michigan, Ann Arbor, 1990.
3. The most helpful understanding of the composing process has been provided by Linda S. Flower and John R. Hayes, "A Cognitive Theory of Writing," *College Composition and Communication* 32: 4 (Dec. 1981).
4. Carter J. Brown, "Excellence and the Problem of Visual Literacy," *Design* 85: 2 (1983): 12. Also see Rudolph Arnheim, *Visual Thinking* (Berkeley: U of California P, 1969).
5. Phillipe C. Duchastel, "Research on Illustrations in Text: Issues and Perspectives," *Educational Communication and Technology Journal* 28: 4 (Winter 1980): 286.
6. Michael Macdonald-Ross, "Graphics in Texts," *Review of Research in Education*, ed. L. S. Shulman, 1977, vol. 5, 70.
7. Given the central importance of graphs in much technical and professional work, a much fuller treatment of the topic than is possible here is appropriate. We recommend a book-length treatment of the topic by AT&T Bell Laboratories researcher William S. Cleveland. His most recent book on the subject is *The Elements of Graphing Data* (Monterey, CA: Wadsworth Advanced Books and Software, 1985). He also has another book, *Graphical Methods for Data Analysis*, Wadsworth, 1983.
8. See Michael Macdonald-Ross,"Graphics in Texts," pages 61 and following. These principles were first presented by A. S. C. Ehrenberg, "Rudiments of Numeracy," *Journal of Royal Statistical Society*, ser. A, 40 (1977). See also Patricia Wright, "A User-Oriented Approach to the Design of Tables and Flowcharts," in *The Technology of Text: Principles for Structuring, Designing, and Displaying Text*, ed. David H. Jonassen (Englewood Cliffs, NJ:

Educational Technology Publications, 1982) 317–19.

9. Ben F. Barton and Marthalee S. Barton, "Simplicity in Visual Representation: A Semiotic Approach," *Iowa State Journal of Business and Technical Communication* 1:1 (1987): 9–26.
10. Edward R. Tufte, *The Visual Display of Quantitative Information* (Cheshire, CN: Graphics Press, 1983) 13.
11. Robert J. Smillie, "Design Strategies for Job Performance Aids," in Thomas M. Duffy and Robert Waller, *Designing Usable Texts* (Orlando: Academic Press, 1985) 213–43. See also, G. Richard Hatterick and Harold E. Price, *Technical Order Management and Acquisition*, Air Force Systems Command, Brooks Air Force Base, Texas 78235, Report AFHRL-TR-80-50, May 1981. The 400+ page manual reviews and summarizes guidelines for presenting technical data for maintenance. It is relevant here, besides being a necessary reference for persons in the aerospace industry, because the guidelines are devoted almost exclusively to preparation of graphics. See also Richard Kammann, "The Comprehensibility of Printed Instructions and the Flowchart Alternative," *Human Factors* 17:2 (1975): 183–91. He concludes "that the algorithmic or flowchart format can produce a higher level of direct comprehension than is obtained with a standard prose format" (190). Patricia Wright and Fraser Reid, "Written Information: Some Alternatives to Prose for Expressing the Outcomes of Complex Contingencies," *Journal of Applied Psychology* 57:2 (1973): 160–66, however, conclude that the effectiveness of prose or alternative means of presenting information "depends on conditions of use."
12. Paul Laseau, *Graphic Problem Solving for Architects and Designers* (New York: Van Nostrand Reinhold, 1986) 13. This is an interesting book devoted entirely to four basic types of graphic and visual devices that architects and designers commonly use: bubble diagrams, area diagrams, matrices, and networks. Although different, these are, as Laseau says, the common graphic tools of problem solving, allowing you to deal with size, location, identity, relationship, and process. Although the book is for architects, it is useful reading for others.
13. Paul Laseau, 15.

CHAPTER 12

Formatting the Document

At several points in earlier chapters, we have briefly mentioned formatting considerations. Also, we have presented numerous sample documents that illustrate formatting possibilities and conventions. Now it is time to talk more explicitly about formatting and to present some suggestions to help you choose appropriate physical design features for your professional documents. In particular, we will look at issues of how your documents should be word processed and how they can be formatted to increase readers' speed and comprehension.

The basic point is that your professional documents are physical things as well as words and ideas. No matter how elegant the theoretical conception, if the physical document is not effective you risk discouraging your readers by forcing them to work overly hard to grasp your message. This is not merely a matter of superficial "packaging" to influence readers' attitudes. It is a much more substantive matter of human performance. That is, reading is work that requires both perceptual skills and cognitive skills. If you do not take these demands into account, you jeopardize the success of your documents. Although some professionals dismiss formatting as relatively unimportant, we hope to convince you that you have important design decisions to make on this level as well as on all of the other levels discussed so far.

Of course, many of your physical design decisions may be made for you by the conventions or guidelines of your organization or by the constraints of your word-processing packages and hardware. Seldom will you start with full freedom in document design because every organization has conventions, many have specific format guidelines, and all organizations have specific hardware and software. But at least you should be aware of the need to adapt conventions and to manipulate fixed formats so that they will represent your documents in the most favorable light possible. Trade-offs might be necessary, but make them informed trade-offs.

❑ 12.1 ❑ Basic Issues: How Should the Document Be Typed?

To type a document, you have to make decisions about typefont, character size, margins, use of justification (that is, the squaring off of the right margins), and line spacing. Even in these very basic decisions you can make deliberate choices based upon research in reading behavior.

Choose a Character Size

Although readers do differ in their ability to read type of different sizes, research indicates that type smaller than 8-point type (newspaper size) is difficult for average readers to read in sustained text. (Type is measured in "points." One point equals 1/72 of an inch.) Conversely, we know that very large type—although visible—does not get read efficiently and accurately. It is interesting, for example, to see how many typographical errors show up in large-print headings—suggesting that they are not being carefully proofread. Accordingly, you should choose type in the optimum range between 8-point and 12-point for the major type of your document. We would even narrow that to a safer range between 10-point and 12-point type. You can use larger type for headings and subheadings, and you can use smaller type occasionally (for example, in figures). For optimum reading performance, however, the basic type of non-typeset text probably should be in the 10- or 12-point range.[1]

Choose a Line Length

You need to decide how many characters will appear on a line of type in your document. Although you probably think in terms of "setting the margins," actually, setting the margins is determining the number of characters (and thus words) on a line of typed text. From research we know that this decision has a significant impact on reading efficiency. Excessively long lines tend to cause readers to lose their places easily and therefore to slow their reading in order to reduce the risk. Excessively short lines, on the other hand, force excessive eye motion and again slow readers down. Accordingly, you should choose a line length that is within, or not far above, the optimum range for human reading performance. Somewhat surprisingly, that range is between 50 and 70 characters per line, generally about ten or twelve words per line.[2]

We say "somewhat surprisingly" because if you look at a 70-character line in a typical word-processed text, you will find that it looks quite short in terms of the conventions operating in schools and in many companies. For example, Figure 12.1 shows lines of typewritten text of different lengths. At what point do they begin to look "normal" to you?

Probably the longest line illustrated in Figure 12.1 looks normal to you, even though it is well above optimal length for reading. It is even a relatively "short" line by some company standards. For example, we recall one company for which we consulted. The company traditionally set margins at one-half inch from each edge of the paper, producing about 130 characters per line, and put far too many lines on a page. No one in the company thought this particularly undesirable. In fact, the choice, which was implemented by default in the steno pool, was "justified" to us on the grounds that it saved paper. Our response was what you might guess: yes, it saved paper, but what was the cost in terms of reader speed and efficiency?

We suggest that you set your line length at not much more than 70 or 80 characters per line in conventionally word-processed or typed text. (If you have desktop publishing software or a word processor that permits double-column text, you could have short lines in double columns for some types of documents, such as manuals, formal reports, technical papers, proposals, and other "formal" documents, although these "printed" formats are still relatively unusual for daily use in most companies.) For conventionally typed documents, provide generous margins that balance your line lengths comfortably between company convention and optimum performance. Certainly resist setting line lengths that push much beyond 80 characters per line. If you let your lines extend beyond that, you are going to pay a price in terms of readability.

Decide Whether to Justify the Text or Not

"Right-justified" text is squared off on the right margin. We suggest that you not right-justify the text. That is, leave the right margin "ragged." This suggestion is based on research showing that at least some readers have difficulty in dealing with the unevenly distributed spacing along lines which is caused by right-justification on most word-processing systems.[3] To achieve the squared-off right margin, most word processors merely introduce looser or tighter spacing between letters and words, thus stretching or squeezing the lines arbitrarily to provide a neat margin. That causes spacing along word-processed lines to be quite inconsistent. In some word processors, the right-justification

These are 30-character lines.
Below optimum reading limits,
they require excess eye motions.

These are 40-character lines. Also below
optimum reading limits, they too require
excessive eye motions and encourage
skimming.

These are 50-character lines. They are at the bottom
limit of the optimal line length. As you can see, the
lines are still very short, and they require quite a bit
of eye motion. Readers may think of these lines as
too short, but there is a difference between reader
preference and reader performance. With these lines,
research shows that readers can read efficiently.

Here is a 60-character-per-line passage. As you can see, this line
is midway within the range that reading researchers tell us
to be the most easily readable. Perhaps this line length may
begin to look OK to some readers, but it still looks very short in
terms of conventions of typewritten documents such as term
papers in school and most business documents.

Here is a 70-character-per-line passage. As you can see, this line still
looks short to most readers. Interestingly, though, it is at the upper limit
of the line length that researchers tell us to be the most easily read and
comprehended by average readers.

Here is an example of an 80-character-per-line text. This exceeds the limit for
optimal reading, and it causes many readers to lose their places on the page when
switching back to the left margin from the right margin. Surely in many companies,
though, you will find documents which have been typed in lines much longer than
these.

Here is an example that demonstrates a 90-character-per-line passage. It is, as you can see,
a perfectly conventional line length for many technical and professional documents.
However, again, research suggests that this line length will tend to make reading tasks
and comprehension more difficult for readers. You are therefore faced with a dilemma:
will you go with convention or with optimal readability?

As a final example, here is a passage of text that is written on a conventional word-processing
system until the lines "wrap around" at the default margins provided by the software package. As
you can see, these lines are approximately 100 characters per line. They look rather conventional to
most of us, but reading research suggests that they are significantly beyond the optimum for reading
performance.

FIGURE 12.1 Typewritten Lines of Different Lengths. The Type is 12-point Times (here reduced). Which Line Length Looks Most "Normal" to You?

is achieved by hyphenating words that appear at the ends of lines, splitting words so that they appear on two lines. Neither technique is particularly desirable. The recommended solution for most report writers is to print the text with a ragged right margin. Even if you think that right-justified text looks "neat and tidy," and even if your word-processing package provides justification by default, probably your best choice is a ragged right margin.

Choose a Typefont

Having chosen a type size, line length, and right margin, you must decide upon a type font. Your choices are numerous on some word-processing systems, and almost nil on others. For example, Figure 12.2 shows the choices of type available by default on the computer which we are using to write the manuscript for this book. Many others are available for our system. Obviously, we would choose to ignore typestyle number 9 as a possible choice for professional documents. However, we probably could reasonably choose any of the remaining eight typefonts. Which to choose?

The answer is certainly a matter of personal preference—an aesthetic choice—because some fonts look more pleasing to you than others. However, you can also temper your choice in terms of a principle. Fonts 1, 2, 6, and 8 are all serif fonts. That is, they have little strokes crossing at an angle on the ends of letters. Fonts 3, 4, 5, 7, and 9 are all essentially sans serif fonts: no little strokes. While the research on the relative desirability of these two basic forms of type does not strongly favor either one or the other, the slight margin

1. Bookman type (plain) in 12-point size looks like this.
2. Palatino type (plain) in 12-point looks like this.
3. Geneva type (plain) in 12-point looks like this.
4. N Helvetica Narrow type (plain) in 12-point looks like this.
5. Helvetica type (plain) in 12-point looks like this.
6. Times type (plain) in 12-point looks like this.
7. Avant Garde type (plain) in 12-point looks like this.
8. New York type (plain) in 12-point looks like this.
9. **Chicago type (plain) in 12-point looks like this.**

FIGURE 12.2 Typestyles Available by Default on a Word Processing System.

of preference goes toward serif fonts. Moreover, if you look at the frequency with which serif fonts are used in a wide variety of typical publications, you will find that the serif outnumber the sans serif publications by a substantial margin. Most of this preference for serif fonts is a matter of aesthetics on the part of writers and experience on the part of readers. Yet some of it may have to do with readability.[4] Apparently, the little strokes on the ends of letters do help readers to differentiate similar letters: **b, h, p,** and **d,** for example. The horizontal strokes of the serifs evidently help readers to distinguish among the largely vertical strokes of our type. Thus, although subjectively many writers believe that the simplicity of the sans serif fonts gives these fonts the edge, in fact your better choice—by a slight margin—is a serif font.

Choose a Typestyle

Next, you need to choose a type style. That is, you need to choose among plain, italic, bold, underlined, and all-capital-letter text. Without question, you should avoid anything but infrequent uses of any of these but plain type. The others, useful for occasional highlighting, are difficult to read in heavy doses, as research has demonstrated.[5] Indeed, you don't really need the research to recognize how difficult it is to read the examples shown in Figure 12.3. We suspect that even these few examples will adequately convince you not to rely excessively on highlighted type styles in your professional documents. In any event, be aware that research suggests that highlighting in isolation works well, but that excessive highlighting actually reduces readability.

You may think it odd that we need to point out this obvious fact. Yet recent experience tells us that not everyone regards it as obvious. We found one university office sending fund-raising letters typed in all capital letters. And we found a manual written by a computer programmer typed in all-capital-letter text. Both documents were easy on the typist, but extremely hard on the readers.

Decide What Line Spacing to Use

Your remaining basic choice is line spacing or, in the language of printing, "leading" (LED-ING). (At one time strips of lead were inserted between lines of type to create the separations of lines.) That is, you have to decide whether to use single space, space-and-a-half, or double space. And on some word-processing systems (and with desktop publishing software) you can choose points of leading, just as you would choose points of type.

In most organizational letters, memoranda, and internal documents, the customary choice among professionals is single space. Students are often more accustomed to double space because most professors want it so they can annotate or comment between lines. However, the advantage of single space is that it allows readers to see a good deal of text at a time and so to

ALL-CAPITAL LETTER TEXT TENDS TO BE HARD FOR READERS TO READ IN SUSTAINED PASSAGES. AN OCCASIONAL HEADING OR ISOLATED WORD IN ALL-CAPITAL LETTERS DOES NOT SEEM TO CAUSE MUCH PROBLEM, BUT AS YOU CAN SEE, WHEN YOU READ A LONG PASSAGE OF IT, THE LETTERS ALL TEND TO BLUR TOGETHER. IT IS INTERESTING THAT YOU SOMETIMES SEE WARNING MESSAGES TYPED IN ALL-CAPITAL LETTERS, APPARENTLY UNDER THE BELIEF THAT THEY WILL STAND OUT. HOWEVER, THAT IS PROBABLY NOT A GOOD IDEA FROM A READABILITY STANDPOINT.

Underlined text in sustained passages also tends to be very difficult to read. The underlining just provides a great deal of distracting detail which the eye must sort out from the significant detail. We suspect you would not want to read more than a few words at a time in this type style.

Italic type is similarly very difficult to read in sustained passages. It may make an occasionally effective highlighter in text, but even one sentence would be a heavy dose of italics.

Even bold type can quickly lose its legibility. In isolation, the bold letters really stand out and can be read easily; however, in several lines of text, the boldness blurs together, and the stark contrast of the black print on the white page makes for a bit of eye strain.

FIGURE 12.3 **Examples of Highlighted Text in Sustained Passages. Note How Difficult It Is to Read all the Examples.**

gain an increased perspective on hierarchy and relationships within the text. Double spacing may obscure that. Admittedly, in professional contexts, external documents such as proposals and "full dress" formal reports are often presented in double space, perhaps because of the traditional use of double-spaced text in college writing. (We recommend space and a half in such instances.) Still, the usual choice among professionals is single space, and that is our recommendation for almost all of your documents.

With single-spaced text, we suggest you use a block indentation and a line of white space between paragraphs, as illustrated in Figure 12.4. Conventional 5-space indentation of paragraphs, with no additional line of white space between paragraphs, makes for a very crowded page in a single-spaced document (shown in Figure 12.5). Indentation is necessary in a double-spaced format because there is no additional line of white space between paragraphs.

❑ 12.2 ❑ Use Format Cues to Increase Readers' Speed and Comprehension

So far we have discussed choices of format that can be selected at the outset from the menu on the word processor. Our message is that you should make strategic choices, not merely accept what is given to you.

1020 H.H. Dow Building
Ann Arbor, Michigan 48109-2136
313/763-1233
FAX 313/936-3492

College of Engineering
The University of Michigan

1 November 1986

Mr. Jack L. Harris, Construction Manager
Webster Construction Company
1246 Hudson Mills Road
Scio Township, Michigan 49109

Dear Mr. Harris:

I am writing to request immediate corrective action for problems which have been identified in the Dow Building, which Webster Construction Company is now completing at The University of Michigan.

Specifically, I request your personal attention to three instances in which the contract specifications for the building have not been met in the television classroom, a room which of necessity has been in full-time use daily for three weeks. There we have found that improperly installed desks have come loose, improperly installed television monitors are blocking aisles, and improperly installed lighting fixtures have reduced the necessary lighting level in the facility. Especially because they include safety hazards, these three problems require both your prompt attention and special scheduling.

The three problems which must be corrected are in room 1023 Dow. They are as follows:

1. Inadequately secured desks: Desks (with attached seats) are secured by screws into lead anchors in the concrete floor. In only three weeks, the entire first row of desks has come loose and therefore cannot be safely used. This both causes a safety hazard and prohibits the room from handling classes of the size necessary. We request that the first row of desks be reinstalled and that all other desks be checked to assure that they are properly installed. (Checking the installation will be time-consuming because carpet covers the entire floor of the classroom.)

2. Missing ceiling mounts for television monitors: The four television monitors in the room should have been installed on ceiling mounts. The absence of mounting brackets has necessitated their temporary placement on carts in the aisles, a further safety hazard because fire exits are partially blocked. The appropriate mounting hardware should be installed immediately and the monitors should be hung from the ceiling, as specified in the contract.

3. Improperly positioned lighting fixtures: Strip lights have been improperly placed to run the length of the room rather than the width of the room, as the contract requires. This has reduced the lighting level in the room to just slightly more than half of the specified level, a fact with both limits the room's usefulness as a television facility and which

FIGURE 12.4 Technical Letter Typed in Conventional Single-spaced Format with Block Indentation (Additional Line of White Space Between Paragraphs).

2

causes students and instructor to work with less than adequate lighting at their desks. These lighting fixtures must be taken down and reinstalled in their specified positions above the desks.

In order to correct these problems, Webster Construction Company will have to arrange a special "after hours" schedule. The classroom is presently used from 8:30 a.m. until 4:30 p.m. daily during week days, and because of our TV broadcast schedules, we cannot interrupt the daily use of the room. However, I am happy to work with you or your on-site Supervisor, Jerry McCarty, to arrange a mutually satisfactory schedule for your work crews.

I am confident that you are eager to see the construction at the Dow Building completed in accordance with our contract. Please contact me at your earliest convenience to work out the details of the necessary corrective action.

Thank you for giving this matter your prompt personal attention.

Sincerely,

B. K. Burgess, Programming Manager

Copy: J. McCarty

Michigan Engineering Television Network

1020 H.H. Dow Building
Ann Arbor, Michigan 48109-2136
313/763-1233
FAX 313/936-3492

College of Engineering
The University of Michigan

1 November 1986

Mr. Jack L. Harris, Construction Manager
Webster Construction Company
1246 Hudson Mills Road
Scio Township, Michigan 49109

Dear Mr. Harris:

I am writing to request immediate corrective action for problems which have been identified in the Dow Building, which Webster Construction Company is now completing at The University of Michigan.

Specifically, I request your personal attention to three instances in which the contract specifications for the building have not been met in the television classroom, a room which of necessity has been in full-time use daily for three weeks. There we have found that improperly installed desks have come loose, improperly installed television monitors are blocking aisles, and improperly installed lighting fixtures have reduced the necessary lighting level in the facility. Especially because they include safety hazards, these three problems require both your prompt attention and special scheduling.

The three problems which must be corrected are in room 1023 Dow. They are as follows:

1. <u>Inadequately secured desks</u>: Desks (with attacthed seats) are secured by screws into lead anchors in the concrete floor. In only three weeks, the entire first row of desks has come loose and therefore cannot be safely used. This both causes a safety hazard and prohibits the room from handling classes of the size necessary. We request that the first row of desks be reinstalled and that all other desks be checked to assure that they are properly installed. (Checking the installation will be time-consuming because carpet covers the entire floor of the classroom.)
2. <u>Missing ceiling mounts for television monitors</u>: The four television monitors in the room should have been installed on ceiling mounts. The absence of mounting brackets has necessitated their temporary placement on carts in the aisles, a further safety hazard because fire exits are partially blocked. The appropriate mounting hardware should be installed immediately and the monitors should be hung from the ceiling, as specified in the contract.
3. <u>Improperly positioned lighting fixtures</u>: Strip lights have been improperly placed to run the length of the room rather than the width of the room, as the contract requires. This has reduced the lighting level in the room to just slightly more than half of the specified level, a fact with both limits the room's usefulness as a television facility and which causes students and instructor to work with less than adequate lighting at their desks. These lighting fixtures must be taken down and reinstalled in their specified positions above the desks.

In order to correct these problems, Webster Construction Company will have to arrange a special "after hours" schedule. The classroom is presently

FIGURE 12.5 The Same Technical Letter Typed in a Single-spaced Format, with Indented Paragraphs and No Additional Line of White Space Between Paragraphs.

2

used from 8:30 a.m. until 4:30 p.m. daily during week days, and because of our TV broadcast schedules, we cannot interrupt the daily use of the room. However, I am happy to work with you or your on-site Supervisor, Jerry McCarty, to arrange a mutually satisfactory schedule for your work crews.

I am confident that you are eager to see the construction at the Dow Building completed in accordance with our contract. Please contact me at your earliest convenience to work out the details of the necessary corrective action.

Thank you for giving this matter your prompt personal attention.

Sincerely,

B. K. Burgess, Programming Manager

Copy: J. McCarty

Now, however, we need to consider additional format devices; these require substantive design decisions that you must implement more directly. Specifically, you need to consider uses of white space, headings and subheadings, numbering and bulleting, highlighting techniques, and perhaps color. All of these are format devices that you can use to increase the readability of your documents if you are aware of principles behind their best use.

Use White Space to Show Hierarchy

White space is an effective cue for readers because it shows how elements of a document relate to one another. That is, coordinate elements of the document—those with equal and parallel relationship—are identified and set apart by white space set around them. Similarly, subordinate elements in the document—those with hierarchical or subordinate relationships—are identified and set apart by white space. Thus, even without headings and subheadings or numbering, you can show readers a physical cue of relationship that will help them to interpret your text as they read it quickly and selectively, as most readers will.[6]

White space is both "vertical" and "horizontal." If you use vertical white space, you use a consistent pattern of variation in line spacing to signal relationship among clusters of paragraphs. In single-spaced text, that pattern might be as follows:

▪ Between lines within a paragraph	no additional white space
▪ Between paragraphs	one line of white space
▪ Between clusters of related paragraphs	two lines of white space
▪ Between major sections	three lines of white space (or in formal reports, begin each major section on a new page)

Under this spacing scheme—or a variation that you consistently apply throughout the document—readers will see the hierarchy of relationships within the text merely by your vertical spacing. Instead of forcing them to read the text to discover the relationships, you can use variations in vertical spacing to give them a "map" by which to interpret the text. (Note that if you use this scheme you do not need to indent paragraphs to signal where they begin. That would be a redundant cue and would merely introduce a distracting unevenness in the left margin.)

"Horizontal" white space is an alternative and complementary cue. It uses varied block indentation to signal coordinate and subordinate relationships within the text. That is, it establishes a consistent variation of both the line lengths and beginning points for lines within your text. For example, the pattern might be varied as follows:

▪ To signal a paragraph	set left margin at 1.5 inches (75-character line)
▪ To signal a subordinated paragraph or cluster of subordinated paragraphs	set left margin at 1.5 inches + 5 spaces (70-character line)
▪ To signal sub-subparagraphs or clusters of sub-subparagraphs	set left margin at 1.5 inches + 10 spaces (65-character line)

Of course, this technique is practical only for a limited number of levels of subordination (perhaps three or four). After that, the indentation begins to produce an absurdly "skinny" column of text down the right side of the page. In an extended passage this will look rather odd.

These two patterns for using white space can be combined, of course, as is illustrated in Figure 12.6. If you use no other cues, such as numbering or headings and subheadings, you probably should consider combining them. Just be sure to be consistent with whatever design you choose throughout the document. You might try deciding upon the pattern early in the actual writing of the document, so that you can implement at least the broad outline of your plan as you compose. This can help to keep you alert to where you are in the hierarchy of the document, and it will help produce a document that readers can scan quickly without missing cues of coordinate and subordinate relationship. (Note: Don't get overly tied up with these matters if they are distracting to you as you compose. You can always put these cues in during the editing stages. Indeed, some writers prefer to impose these format cues after composing; others prefer to implement them as they go along. Find your own technique, but use the cues.)

Use Informative or "Scenario" Headings and Subheadings

For headings and subheadings two suggestions are in order. First, use them. Second, to the extent possible, make them specifically informative or "scenario" headings rather than generic or topical headings.

Our experience in many companies suggests that writers often tend not to introduce headings or subheadings into their documents. (Probably this is because they know where the logical units within the documents are, and therefore they just don't appreciate the need for headings and subheadings.) We also see documents in which there are headings, but no subheadings. In other words, the text is "chunked," but not into bite-sized bits. And finally, we see numerous documents in which the headings (and subheadings) are generic, essentially meaningless headings that could be transported unaltered to almost any report or document (for example, "Discussion"). Often involving Greg Turner and Bob Stovall in the completion of this PCP. I will update existing documentation for archival purposes if time is available, but no other action is warranted.

4. Merit Internal LANS

I am generally responsible for all elements of data connectivity to Merit staff within the Computing Center Building (CCB) and the Institute for Science and Technology Building (IST). I administer most of these networks in their entirety, including installing and repairing the hardware, designing the configuration, developing new installations, and resolving problems.

Networks internal to the CCB and IST require a different kind of attention than those elsewhere. Nearly all network use is by staff knowledgeable in the media. Continuous connectivity is often of great importance, and Network failure or poor performance requires immediate attention. This is particularly true of networks in the Network Operation Center (NOC). There, continued connectivity is of vital importance, yet redundancy is difficult due to the high concentration of people and machines. The NOC thus requires special attention to minimize difficulties.

Administering all of these networks is a time-consuming job. However, any single Merit LAN could easily be managed by any of a number of qualified technicians and engineers within Merit. My limited time is better spent on tasks no one else is qualified to do, and on teaching others to do those tasks. I would like to delegate these LAN responsibilities to others as soon as possible.

4.1. IST Networks

All data connections to Merit Staff in the IST are my responsibility. The Merit portion of the IST Building has no autonomous networks: All connectivity to Merit staff in IST is supplied via twisted-pair copper wire from the CCB. IST networks are merely extensions of CCB networks.

The cable between the IST and the CCB is currently connected in an ad hoc manner. A technician requires considerable prior experience with the cable in order to understand circuit connections. As a result, I will dismantle and rebuild all junctions in a modular scheme during December, 1989, after which connections should be straightforward and will not require any prior understanding of the cabling system. IST circuits needed after that date can thus be made by any technician and can be assigned on a per-circuit basis.

4.2. CCB Networks

Networks within the CCB are numerous and extensive. Much management time is spent on these LANs. There are six primary types of network in use.

4.2.1. Merit Asynchronous RS-232

I currently handle all aspects of Merit RS-232 connectivity for Merit staff. Staff needs change often, sometimes on a project-to-project basis, requiring frequent changes. I've replaced nearly all RS-232 wiring with modular

FIGURE 12.6 Page Format Shows Combined Use of Vertical and Horizontal White Space, as well as Numbering and Different Type Sizes and Styles to Signal Headings.

they are general or ambiguous. And routinely they are topic headings rather than informative, "scenario" headings. (The term "scenario" suggests a human being doing something in a context: a plot. To illustrate, "Bad Debts" is a topic heading; "How to Deduct a Bad Debt on Your 1992 Tax" and "Deducting Bad Debts" are both scenario headings that make the basic topic and purpose of the text clear.)

The research on uses of headings and subheadings is helpful.[7] It indicates that headings and subheadings are extremely useful devices that speed readers along, help them to see the global design of a text, and increase their understanding. Further, headings and subheadings serve useful reference purposes for readers seeking to find, or re-find, a specific passage of text within a larger text. In addition, we know that readers often form much of their impression of how difficult a text is merely by what they find in the headings. And finally, we know that scenario headings are better understood than the topic headings and subheadings.

For all of these reasons, putting good headings and subheadings into your reports is worth the time it takes to plan and implement them.

Use Numbering or Bulleting to Signal Coordinate and Subordinate Relationships

Numbering systems can also help readers to see both coordinate and subordinate relationships. You need only to adopt a consistent pattern that observes a few simple principles.

First, number pages. Pages without numbers are difficult to find for reference purposes, yet surprisingly many professionals forget about the necessity for page numbering. Probably this is because word-processing software often does not provide page numbering by default. After all, to your word processor, a text file is a "scroll" that can be unrolled, not a pile of discrete pages. Some systems do indicate where the page breaks will occur in the printed document, but many do not automatically number those pages for you. If this is the case, take the time to set the page numbering function. (This is a minor but vexing point for us; we find that fully half of our students hand in unnumbered pages on their assignments in our classes. Either they don't know how to set the numbering in action, or they don't bother. We find the same thing in industry.)

You may even want to go a step further for some documents, especially for documents to be transmitted by FAX machine or that serve explicit legal purposes, such as contracts or specifications. You may want to number pages as parts of the total document. That is, at the top or bottom of each page provide a number that identifies the page specifically in relation to the total document: "page 6 of 9" or "page 3 of 14," for example. This can be a reassuring cue to readers that they have received the whole document.

Having numbered the pages, next number the divisions of the document: major sections such as chapters, major headings, subheadings, and sub-subheadings. We suggest that you use the "all-Arabic" system (1.—1.1.—1.1.1., etc.) rather than the traditional Roman system (I.—A.—1.). This bit of advice is based upon three premises: the Roman system, although quite traditional in nontechnical prose, does not allow specific reference to an individual section of the document. That is, there are many "A" sections in the document, but there can be only one "3.1." Thus, all-Arabic numbering is useful for reference purposes because it is inherently more specific than the Roman system. Further, the all-Arabic system is increasingly standard in business and industry because it has been adopted as a specification by the military and by most federal agencies. Finally, the all-Arabic system is compatible with digital programming. You may have grown up with the Roman system and may even feel more comfortable with it, but its use in the professional world is quickly declining.

We do need to express one reservation. The all-Arabic system gets cumbersome after three or perhaps four levels of subordination. For example, the number 4.2.1.3. is accurate as an identifier for a section of a document, but you can see that it begins to require a bit too much space and it takes a reasonable amount of cognitive processing to interpret it. Hence, after three levels of numbering, you may want to switch to unnumbered units of text or to "bulleted" text, unless the nature of the document precludes that. (For example, specifications and contractual agreements require a complete numbering system to whatever level necessary.)

Finally, as you number coordinate units within the text, distinguish between those that actually require numbering and those that should be "bulleted." (A bullet is a darkened circle, ellipse, or square symbol usually available on most word processors. For example, "•." On a typewriter, a different symbol, such as an asterisk or hyphen, can be used.) If you use bulleting, however, you need to use it at the appropriate times. As a technique, bulleting is not interchangeable with numbering.

The principle of distinction is simple: numbers indicate order and they imply comprehensive treatment; bullets do not indicate order and their use implies illustrative treatment. A numbered list of five elements (or "ordered list," as it is sometimes called) has specific order: for example, element 4 occurs logically before element 5, and so on. Further, in a numbered list of five elements there are specifically five elements, not six. In contrast, a bulleted list of five elements is not specifically ordered, and the list includes only five elements, even though any number of elements might have been listed.

Use Highlighting Techniques

Highlighted text can be boldfaced type, underlined type, capitalized type, italic type, or type varied in size. As we have indicated, all of these are difficult to read in sustained text, and if overused any of them becomes confusing. So limit their use and be absolutely consistent in whatever pattern you establish. However, if you use these techniques sparingly and consistently, they can allow you to draw attention to key information and to signal consistent distinctions among different types of text within a document. A considerable body of research evidence supports the premise that both of these are worth doing.[8]

In a computer manual, for example, if you set all of the keystroke information in boldface type, you can easily help readers both to see it and to recognize its function. (For example, "Enter the following command to gain access to the message system: **$me**. After you type this command, to execute it, press the **Return** key.") In that same manual, you might signal all screen information as it actually will appear on the screen. (For example, "After you press the **Return** key twice, you will see these options on the screen: POST, DISPLAY, ABANDON? The computer is asking you what you want to do: to send the message to someone, to show you what the message will actually look like when it is sent, or to abandon the message.") Note that there are two types of highlighting in that last sentence, one for what the user does, and one for what the computer does. We know from research that making these distinct from one another helps readers. Using the same highlighting cue for both would confuse readers, as would using no highlighting cue at all.

Highlighting techniques can call attention to headings and subheadings, numbering, bulleting, examples, screen displays, notes, warnings, and cautions. They can also draw attention simply to information that you want to stress. (For example, "You **must** enter your account number correctly during either of the first two tries. If you do not, the teller machine will assume you are an unauthorized person attempting to gain access to a bank account and **it will not return your card**. If this should happen, you can recover the card only by visiting the bank during business hours.")

To establish effective highlighting, resist the urge to highlight everything. Establish a plan by which the highlighting will be introduced consistently as you write and edit the text. A simple highlighting plan can be part of a larger plan that establishes specifications of the type size and style as well as the margin sets for specific types of text. Such a plan might also include graphic factors such as use of boxes to set figures apart from the text. For example, Figure 12.7 presents a simple plan that you might use. Figure 12.8 presents part of the more elaborate plan that was used to typeset this book.

Whatever plan you establish for highlighting text, be consistent and acknowledge that there are different kinds of information included within many documents. By helping readers to spot the highlights and to recognize the differences in types of information, you will make their tasks easier and faster.

If Your Technology of Printing Permits It, Use Color

In a sense, color is nothing more than another highlighting technique. We treat it separately simply because, at present, most professionals in business

Format Plan

For headings:

Chapter titles	18-point bold, centered on the page
1st level heads:	plain, 18-point type, left margin, three lines of white space before, two lines after
2nd level heads:	bold face, 14-point type, left margin, two lines of white space before, two lines after
3rd level heads:	bold face, 12-point type, left margin, one line of white space before and after
4th level heads:	plain text, underlined, left margin, one line of white space before, none after

For types of text:

Basic text	plain text, 12-point type, double space
Danger, Warning, or Caution notes	bold face, 12-point type, single space margins at 1.5 and 6.5 inches. Boxed by bold line. Signal word underlined and followed by exclamation point (E.g., **Danger!**)
Examples	italics, 10-point type, single space, margins at 1.5 and 6.5 inches, boxed by plain line

FIGURE 12.7 **Format Plan for Office Procedures Manual Written by a Staff Member at the Direction of the Office Supervisor.**

Inserted figures and graphs	plain text, 8-point type, boxed by plain line
Key stroke or action information	bold face, 12-point type
Screen display information	facsimile of what the screen actually shows boxed by rounded-corner box to simulate screen shape. Box truncated if screen is not full.
For all other emphasis	underlined key words

(A) Part number: "SECTION" with roman numeral in 18 pt Antique Olive Bold caps, letter space to 29 pi wide, use less letterspace between numerals with word "Section" × 22 pi. wide; center height in height of drop-out panel in graduated color panel (see layout and see below). Insert 8 pt Zapf Dingbats (color) centered in # between each letter of "SECTION" and centered on height of cap. Set a 3/4 pt rule × 29, 2 pts optical # above and below (A)

(B) Part Title: Melior Bold c/lc, drop out (white) from color panel; cap height: 24 pts; max width: 25 pi, do not break words; center × 29; 33 pts b/b; sink baseline of last line 36 pi
Set a graduated color panel 29 pi wide × 39 pi deep, no sink, starting with 30% color at top and running to 90% color at bottom; drop out (white) a panel 29 pi wide × 2.5 pi deep, which sinks 2 pi, position (A) in this panel. Set a 20% black "shadow" 6.5 pi wide at left and bottom of color panel (see layout). Part Opener to come from Desktop.

(AA) Chapter number: "CHAPTER" with arabic no in 18 pt Antique Olive Bold caps (color), letter # "Chapter" to 24 pica wide with one digit, to 22 picas wide with two digits; center height in height of drop-out panel (see layout). Insert 8 pt Zapf Dingbats centered in # between each letter of "CHAPTER" and centered on height of cap. Set a 3/4 pt rule (color) × 29, 2 pts visual space above and below (AA)

(BB) Chapter Title: Melior Bold c/lc. drop out (white) from color panel; cap height: 18 pts; max. width: 25 pi, do not break words; center × 29; 27 pts b/b; sink baseline of last line 15.5 pi
Set a graduated color panel 29 pi wide × 18 pi deep, no sink, starting with 30% color at left and running to 90% color at right; drop out (white) a panel 29 pi wide × 2.5 pi deep, which sinks 2 pi, position (AA) in this panel. Set a 20% black "shadow" 6.5 pi wide at left and bottom of color panel (see layout)

Chapter Opener to come from Desktop

Opening text line (or subhead): If text, flush left × 29; if subhead, follow specs; sink baseline 29 pi

Additional information: Throughout: dingbats are ITC Zapf Dingbats 200, see xerox attached

SUBHEADS

(1) Head: Use a 7 pt. Zapf Dingbat flush left × 36 (centered on height of no.), 6 pts space to "0.0" in 12/12 Antique Olive Bold (color), 6 pts space to 7 pt. Dingbat; 48 pts. b/b above. On next line: 11/12 Melior Bold C/lc (black) × max width of 29 pi, do not break words; flush left × 36; 24 pts b/b below. Set a ¾ pt. rule (color) × 36, 2 pts. optical space above and below head.

(2) Head: 11/12 Melior Bold c/lc (color); flush left × 29; 36 pts. b/b above, 24 pts. b/b below.

(3) Head: 10/12 Melior Bold c/lc (color); flush left × 29; 24 pts. b/b above, text runs on after 1 em space.

(5) Head: 10 pt. Antique Olive Bold Caps (color); flush left × 36; 36 pts. b/b above, 24 pts. b/b below. Set a ¾ pt. rule × 36, 2 pts. optical space above and below.

When subheads follow each other without text between, use space below superior head between them.
First line of text following head aligns left with head.

FIGURE 12.8 Part of the Format Plan for this Book.

and industry do not have direct access to technologies that permit effective use of color in routine documentation. They can have color only if the document goes through a printing process. Admittedly, they can use different-colored paper to identify title pages for chapters or to distinguish a part of the document (Executive Summary sheets in green or beige, for example, as found in many companies) or colored divider sheets between sections of a formal report. They can introduce preprinted color tabs. And perhaps they could introduce some minor color cues by using multiple-color ribbons on dot-matrix printers or typewriters. But only gross distinctions can be signaled in a text by changing colors of paper or by using color tabs. And in the case of colored type, generally it is a crude technique that produces only semi-legible type unless the letters are exceptionally large.

Yet color is a powerful cue that probably will soon become a common part of your "toolbox" for producing routine corporate documents that are not typeset. Already we have color displays that allow us to show color on the computer screen. And already there are color printers which, although expensive and not widely available, will no doubt come down in cost and will soon become routinely available in the corporate workplace. What about use of color then?

Again, we do have a body of research to draw upon.[9] Hence we offer the following four suggestions.

First, like any other highlighting technique, color can be used to show consistent relationship. Thus, for example, putting a color wash behind a single type of information within a document (let's say screen examples in a computer manual) can help readers to differentiate information types. Publishers often use a light blue or a light green wash behind screen examples in computer documentation for this reason, as well as simply to perk up the appearance of the page. No question about it: readers do prefer these color treatments, and they do work. Publishers also provide color bars behind headings or behind headers at the tops of pages to help divide information into topics and types. They print headlines in color. And they use colored lines around figures to make them stand out plainly.

Second, limit the number of color cues that you use in any document to a very small number—no more than one, two, or perhaps three in a document. If you use more than that, the color can quickly become distracting rather than helpful. For example, probably you have seen software that uses too much of the enormous range of colors that can be displayed on many computers. The software industry went through that "phase" when color first became available, just as many of us went wild with type fonts when computers first gave us a range of choices. "Palettes" of up to 256 colors are already available on some computers, and a few even allow you to define your individual color palette by slightly varying the basic red, blue, and green content of any individual color. But these are tools for the graphic artist, not for most writers. Most viewers can distinguish only a very small range of colors. They cannot see fine distinctions within shades of blue, for example. (Indeed, 6 percent of the population is color blind.) Thus if you start introducing subtle color distinctions simply because they are available to you, you will be wasting your time and perhaps confusing your poor readers whose eyes do not allow them to see subtle differences in color. Use only a few colors, limiting yourself to clearly distinguishable ones—and be sure you clearly indicate to your readers what the different colors mean.

Third, recognize that some color is used according to already established "rules." For example, stop signs are red, yield signs are yellow, green lights mean "go." To arbitrarily change them would unquestionably confuse drivers. Similarly in most countries, the conventional color signals for danger, warning, and caution notes[10] are ones you should know and follow if you do use color: red is for danger (indicates imminent risk of serious injury or death); orange is for warning (indicates possible serious injury or death); and yellow is for caution (indicates potential for injury or damage). In short, danger, warning, and caution notes are specific instances in which you do not have free choice of colors if you choose to use color.

Finally, if you do use color, remember that color always carries emotional baggage with it. We have all learned to associate the red, orange, and yellow end of the spectrum with warmth, excitement, and energy. The green, blue, indigo, and violet end of the spectrum tends to convey a sense of coolness, comfort, calm. Don't give the reader a barrage of powerful emotional cues if you don't intend to do so for quite specific purposes such as in promotional literature. In routine corporate documents, keep the tone conservative, understated, and tasteful by using the cooler end of the spectrum—light blues, light greens, light tans, for example. Restrain yourself. Color certainly works, and we all like it, but it can easily become a document designer's trap.

Conclusion

You should consciously make both basic typographical decisions and format decisions when you write. Unfortunately, for too many writers, these issues of physical format are non-issues on the job; only in such "special" documents as job application letters and resumes do many writers consider these issues. Once on the job, they just passively accept whatever format is provided to them. As one writer told us, "It's a 'done deal.' Other people decide my format for me."

Even if most of us do not have full freedom of document format choices, it is a mistake to accept what word processors provide—or what the steno pool and company convention dictate. Don't just follow the same path as your predecessors and colleagues, if their path is not a good one. There is too much at stake. Ineffective format can inhibit readers, but effective use of typography, white space, headings and subheadings, numbering and bulleting, and highlighting techniques can increase your readers' speed, comprehension, and retention of your message. Moreover, these same techniques can improve the subjective impression that your document inevitably creates in the minds of your readers merely because of its physical appearance.

It will not take much time to introduce good use of format cues into your documents, but the payoff can be significant, as Figures 12.9 and 12.10 illustrate. These two figures are job applications and resumes submitted by two successful job applicants, one of them a college junior looking for a summer job before his senior year, the other a hospital employee hoping to land an open position as a supervisor after four years on the job. As the resumes suggest, both applicants were well qualified in substantive terms. Both had done good audience analysis to determine what their readers needed. Both had been selective in choosing what went into their applications and how the applications should be arranged. In addition, however, notice that both writers used many of the same format cues that we have discussed in this chapter. We think that if job applicants can use format cues to enhance the chances of getting a job, they can continue to use many of those same cues on the job. After all, making a good impression and helping readers to read quickly isn't something that only job applicants should consider. These are things you need to consider throughout your career.

NOTES

1. Daniel B. Felker, Frances Pickering, Veda R. Charrow, V. Melissa Holland, Janice C. Redish, *Guidelines for Document Designers*, American Institutes for Research, 1981, 77–78.
2. Felker et al., 79–80.
3. Felker et al., 85–86.
4. James Felici and Ted Nance, *Desktop Publishing Skills: A Primer for Typesetting with Computers and Laser Printers* (Reading, MA: Addison, 1987). Felici and Nance comment, "Although serifs make a great contribution to typographic aesthetics, they are not fundamentally decorative. Serifs are signposts for the eye. By running at counter angles to the predominantly vertical strokes that make up Western letters, they provide a chance for the eye to orient itself, to see the trees despite the forest, so reading is easier" (61–62).
5. Felker et al., 87–88.
6. Felker et al., 81–83. Felker points out that the effect of white space upon readability is difficult to study in isolation; hence there is little direct research evidence on it. However, it is worth noting that James Hartley achieved significantly positive experimental results with texts that featured varied vertical spacing and substantial margins to create short line lengths. His research and his text specifications are described in "Designing Instructional Text," an article in David H. Jonassen's collection, *The Technology of Text: Principles for Structuring, Designing, and Displaying Text* (Englewood Cliffs, NJ: Educational Technology, 1982) 193–213.
7. James Hartley and David H. Jonassen, "The Role of Headings in Printed and Electronic Text," in Jonassen, *The Technology of Text*, 237–63. See also Heidi Swarts, Linda Flower, and John Hayes, "How Headings in Documents Can Mislead Readers," Document Design Project Technical Report Number 9 (Pittsburgh: Carnegie Mellon U, 1980). See also Felker et al., 17–20.
8. For a summary of research on highlighting see D.B. Felker, J.C. Redish, and J. Peterson, "Training Authors of Informative Documents," an article in Thomas M. Duffy and Robert Waller's excellent book, *Designing Usable Texts* (New York: Academic Press, 1985) 43–61. See also Felker et al., 73–76.
9. A. Chute, "Analysis of the Instructional Function of Color and Monochrome Cuing in Media Presentation," *Educational Communication and Technology Journal*, 1979, 251–63. Also see N. Katzman and J. Nyenhuis, "Color Versus Black-and-White Effects on Learning,

April 2, 1990

Scott Edward Dewicki
1929 Plymouth Road, Apt. 4032
Ann Arbor, Michigan 48105
phone: (313) 663-1367

Mr. Jeff Ebersole
Personnel Recruiting
c/o Ford Motor Company
36200 Plymouth Road
Livonia, Michigan 48151

Mr. Ebersole:

At the suggestion of Mr. C. J. Morales, Area Supervisor, Transmission Test Services, I am writing to inquire about summer employment opportunities in product engineering within the Ford Motor Company. As an aerospace engineering student at the University of Michigan in Ann Arbor, I am very interested in the possibilities of using my technical knowledge within automotive engineering related to structural analysis and/or test engineering. Some of my past accomplishments in these areas are:

> Conducting an experiment in modal analysis testing of a Ford F-250 driveshaft at the Ford Livonia, MI TAPME transmission testing site; conceiving and designing an Assured Crew Return Vehicle with a team of classmates participating in Space Systems Design 483; using NASTRAN software to study Finite Element Modeling of structures; wind tunnel testing of aerodynamic shapes; interfacing with Apollo Computer software to optimize airflow around objects and vehicles.
>
> Designing, developing, and upgrading Microsoft Excel and Lotus 1-2-3 spreadsheets to maintain office and faculty fiscal accounts while employed at the University's Nuclear Engineering Department; developing documentation for these spreadsheets; and operating Macintosh, DOS, MS-DOS, and PS/2 computing systems.

Within the enclosed packet of materials, you will find a copy of my resume, transcript, and summer position application which outline my accomplishments and qualifications more specifically. I am very enthusiastic about the possibility of working for the Ford Motor Company and learning from the many fine individuals who have made Ford the most profitable U.S. automaker in recent years. I hope to contribute to its continued success. Please feel free to contact me at any time. My phone number and address are shown above.

Thank you for your consideration.

Respectfully,

Scott Edward Dewicki

Scott Edward Dewicki

FIGURE 12.9 **Résumé and Letter of Application for a Summer Job by a College Senior. Note Use of Format Cues. (He Got the Job.)**

Scott Edward Dewicki

Current Address
1929 Plymouth Road, Apt. 4032
Ann Arbor, MI 48105
(313) 663-1367

Permanent Address
10100 Allen Road
Fowlerville, MI 48836
(517) 223-3141

Objective

Seeking summer employment with a U.S.-based automotive corporation in an area or department related to structural analysis and/or test engineering.

Education

THE UNIVERSITY OF MICHIGAN **Ann Arbor, Michigan**
College of Engineering
Bachelor of Science in Aerospace Engineering, May, 1991

- Conceived, designed, and executed an experiment in modal analysis testing of a Ford F-250 truck driveshaft at the Livonia, MI TAPME transmission testing site.
- Successfully designed a space station crew emergency return vehicle with a team of classmates as part of a NASA-funded course in Space Systems Design.
- Utilized NASTRAN software to study Finite Element Modeling of structures.
- Studied aerodynamic principles and developed airflow surveys of various shapes in University wind tunnels and used Apollo Computer software to optimize airflow characteristics.
- Fluent in Italian. Conversant in Spanish.
- Relevant engineering courses:

Space Systems Design	Structural Mechanics I and II
Aerodynamics II	Technical Writing
Computing Methods	Control Systems Analysis
Aerospace Analytic Math	Aerospace Lab I and II
Engineering Thermodynamics	Introduction to Materials Science

Experience

THE UNIVERSITY OF MICHIGAN **Ann Arbor, Michigan**
Office Assistant, Nuclear Engineering Department, Summer 1989 to Present

- Conceived, designed and maintained customized Microsoft Excel spreadsheets to automate office and faculty fiscal accounting. Provided documentation for these custom spreadsheets.
- Produced technical reports for faculty.
- Operated Macintosh, DOS, MS-DOS, and PS/2 computing systems.

Orientation Leader, Office of Orientation, Summer 1988

- Facilitated topical meetings and discussions regularly.
- Assisted in developing orientation programs for the University.
- Provided detailed recommendations and comments regarding the orientation program to University administration at their request.

Activities

- Licensed private pilot.
- Sexual Assault Prevention and Awareness Center volunteer.
- Naval Reserve Officers' Training Corps.

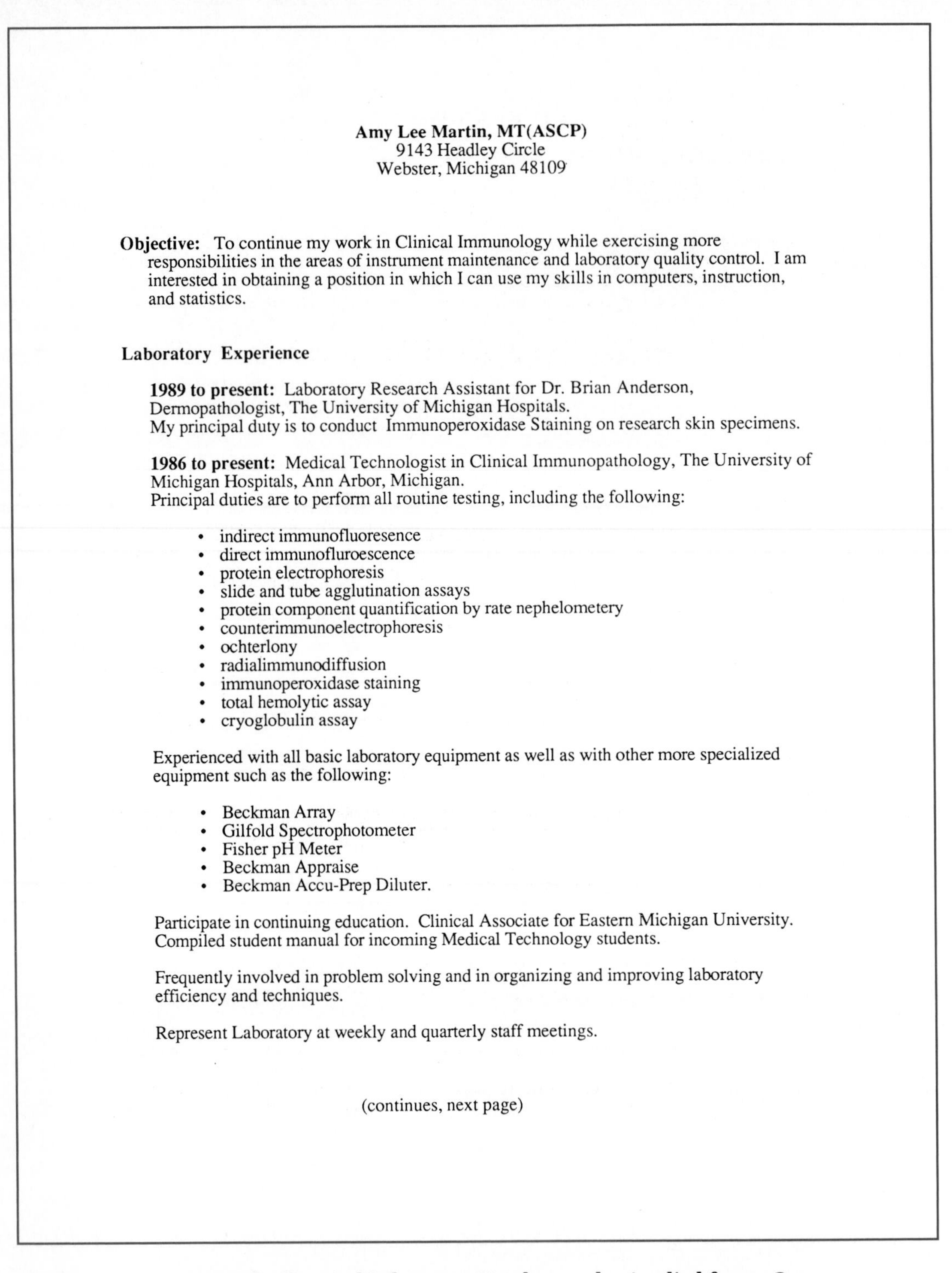

Amy Lee Martin, MT(ASCP)
9143 Headley Circle
Webster, Michigan 48109

Objective: To continue my work in Clinical Immunology while exercising more responsibilities in the areas of instrument maintenance and laboratory quality control. I am interested in obtaining a position in which I can use my skills in computers, instruction, and statistics.

Laboratory Experience

1989 to present: Laboratory Research Assistant for Dr. Brian Anderson, Dermopathologist, The University of Michigan Hospitals.
My principal duty is to conduct Immunoperoxidase Staining on research skin specimens.

1986 to present: Medical Technologist in Clinical Immunopathology, The University of Michigan Hospitals, Ann Arbor, Michigan.
Principal duties are to perform all routine testing, including the following:

- indirect immunofluoresence
- direct immunofluroescence
- protein electrophoresis
- slide and tube agglutination assays
- protein component quantification by rate nephelometery
- counterimmunoelectrophoresis
- ochterlony
- radialimmunodiffusion
- immunoperoxidase staining
- total hemolytic assay
- cryoglobulin assay

Experienced with all basic laboratory equipment as well as with other more specialized equipment such as the following:

- Beckman Array
- Gilfold Spectrophotometer
- Fisher pH Meter
- Beckman Appraise
- Beckman Accu-Prep Diluter.

Participate in continuing education. Clinical Associate for Eastern Michigan University. Compiled student manual for incoming Medical Technology students.

Frequently involved in problem solving and in organizing and improving laboratory efficiency and techniques.

Represent Laboratory at weekly and quarterly staff meetings.

(continues, next page)

FIGURE 12.10 Résumé for Hospital Laboratory Employee who Applied for an Open Supervisory Position. (She Got the Job.)

2

1985-1986: Medical Technologist Intern, The University of Michigan Hospitals. Trained in routine laboratories including the following:

- Blood Bank
- Hematology
- Chemistry
- Microbiology
- Immunology

Gained specialized laboratory experience in Flow Cytometry, Ligand Assay, and Drug Analysis. Performed phlebotomy on inpatients and outpatients.

1984-1986: Laboratory Technologist in Hematology, The University of Michigan Hospitals. Operated and maintained Coulter S Plus 4. Performed routine tests on blood and body fluids. Gained experience on all shifts, weekends and holidays. Shared responsibilities for student worker scheduling.

Other Relevant Work Experience

1982-1984: Laboratory Technician, Webster Research Center, Webster, Michigan. Constructed microscopic heat sensors and performed product quality control analysis.

1980-1981: Laboratory Assistant, Department of Human Genetics, The University of Michigan. Instructed laboratory personnel in computer use. Was in charge of computerizing filing systems for three faculty members. Entered manuscripts for scientific papers into mainframe computer. Assisted faculty members in proofreading.

Education

The University of Michigan, Ann Arbor, Michigan, January 1983-June 1986, Bachelor of Science in Medical Technology, May, 1986.

Albion College, Albion, Michigan, September 1980- May 1982. Pre-engineering major.

Special Skills

Computer skills including Lotus 123 and wordprocessing with several different programs and personal computers (IBM, Mac). Taught Genetics Laboratory personnel to use SPIRES (on MTS mainframe).

Having started my education as a pre-engineering student, I have studied advanced Mathematics and Physics as well as the Medical Technology curriculum. Have exceptionally strong mechanical aptitude (99th percentile on ACT test). Am able to do electrical wiring and am capable of mechanical work such as fabrication of devices (for example, I have built a working generator).

Have also had substantial studies in liberal arts, including six years of German.

Organizations, Awards, and Certification

Member American Society of Clinical Pathologists (Number: 07433272 MMT)
Recipient of Society of Professional Engineers Award,1980
National Honor Society, 1980

Opinion, and Attention," *AV Communication Review* 20 (1972): 16–28.

10. Christopher Velotta, "Safety Labels: What to Put in Them, How to Write Them, and Where to Place Them," *IEEE Transactions on Professional Communication* 30:3 (Sep. 1987): 121–26. Velotta's article discusses use of color and quotes from a proposed American National Standards Institute draft standard, "Specifications for Product Safety Signs and Labels," in which the conventional designations of color are set forth for danger, warning, and caution signs.

Recognizing Additional Issues

CHAPTER 13

Group Writing

Until now, we have been talking about writing as if it were the action of individuals—and of course individuals do write most short technical reports, memoranda, and letters. Admittedly, these documents are often reviewed by supervisors or managers who may provide suggestions and corrections that the individual writer then incorporates before the reports are released. Yet many short, routine documents are largely the products of individuals, even if a review process is normal.

Many longer documents, however, are actually written by groups of professionals, often from different departments. Moreover, these documents can be vitally important to an organization. For example, a formal proposal in response to a Request for Proposals (RFP), a business plan for a corporation, a design report for upper management, or a contract report for a client all require effective group writing if they are to be successful. All of these are complex documents requiring the expertise of professionals from different fields and with different responsibilities. All require considerable time to produce and involve group decision making as well as group writing.

Unfortunately, school experience seldom equips graduates to deal with group writing. Indeed, few students leave school knowing how to produce a group-written report systematically and efficiently, and most leave without having done any group writing at all. Even when they have done some group writing, probably they have done it in an artificial writing context in which students were segregated by discipline, by course, or by classification. Business students in their senior year, for example, might write with other senior business students. Seldom do they write with the engineers, lawyers, foreign language experts, advertising experts, and people of all ages and specialities with whom a professional project group might write in a similar situation or on a similar topic. On our campus, Aerospace Engineering design course students might write with other Aerospace students. Seldom, if ever, do they write with mechanical engineers, electrical engineers, materials scientists, physicists, mathematicians, business people, military officers, and government liaison people, who typically would be involved in many aerospace group projects in the world of professionals.

Not only is group writing important in the professional world; it is much more frequent than most people realize. In the research division of a chemical products corporation, for example, over half of the several hundred professionals with whom we have been associated produce group-written reports as their primary written products. These reports typically have from two to six authors. Similarly, in three separate research projects that we have conducted elsewhere, we have found group-written documents accounting for approximately one-third of all documents. This research has been done within a large city's police department, among a group of military hospital administrators, and among a group of English-speaking Japanese engineers and managers from a number of companies. In each of these very different settings, we have been struck by how often professionals write collaboratively—and by the difficulty of the challenge that group writing presents for them. A particularly interesting example of group writing involved coordinating the written and graphic output of 300 engineers to produce a coherent proposal (Figure 13.1).[1]

Writing groups often consist of persons who are competent individual writers, yet who cannot function effectively as a writing group. When a single person is solely responsible for all aspects of a technical task and producing a technical document about it, the process—though still demanding—nonetheless can be managed efficiently. However, when the hands of several writers and management reviewers are involved, the process itself becomes complicated. All of the difficulties of writing remain, but to them are added all of the difficulties of group dynamics and project management. Often in these difficult situations, groups produce reports like those described to us by a shipyard manager: "The review cycle is endless, and each manager revises the document according to his own interpretation. The result is a collage without stylistic continuity, with no continuity of thought, and with lack of direction to the plan." Our observation is that this comment almost states the rule, not the exception.

Effective group writing is a report writing skill that professionals, supervisors, and managers alike must master. In this chapter, therefore, we explore the typical problems of group writing, explain how group writing is inte-

Group Proposal Writing Project: Martin Marietta, Space Station Division

At a Macintosh Engineering Users Group meeting in Denver, Colorado on February 6, 1987, John Connell summarized the experience which Martin Marietta has recently had with a massive proposal-writing project for a multibillion-dollar space station. Among the points presented were the following:

- Three hundred engineers are creating the text and drawings for the proposal.
- They use 230 Macintosh computers, most of them connected in AppleTalk networks of from 2 to 12 Macs. (They initially tried larger networks, but found that smaller networks were more effective.) The project team also uses 60 LaserWriters.
- Most of these engineers had never used Macintoshes before the beginning of the project.
- So far in the project, each of the 230 Macs averages 74 pages of laser writer text and artwork output per week.
- The total project team output for one month is 68,000 pages. In addition, the project team produces 2500 pages of transparencies per month.
- Of all of the pages produced, approximately 50% are artwork done in MacDraw by the engineers themselves (rather than by graphics specialists).
- In order to ensure that consistent design style is used over the whole document, authors are required to follow a project policy on typography and artwork (fonts, line weights, fill styles, etc.). Also, they are required to have all text and artwork verified by a graphics art group.

Project managers estimate that the savings on the project in terms of reduced manpower and time have so far exceeded the cost of purchasing the Macintosh and LaserWriter equipment for the project by approximately five times.

FIGURE 13.1 Example of Group Writing Project to Produce a Proposal.

grated within a technical project, and present a procedure for managing group writing projects.

❑ 13.1 ❑ Typical Problems with Group-Written Documents

The main problem in group writing is that each individual in the group sees himself or herself as responsible for only a small piece of the whole. The individuals in the group fail to understand the design, scheduling, and production of the entire document and consequently write and schedule their own pieces of the whole according to their own departmental standards and priorities, not according to the standards and priorities of the group project. They produce "their" pieces of the whole, but produce these pieces in such a way that they do not fit well together. In addition, at the last minute pieces or parts of pieces are discovered to be missing—just as the document is due to be assembled and delivered. There is no time for effective integration of the pieces.

Typical problems in group-written documents include all of the following:

- The document addresses different audiences and different purposes between sections.
- Parts of the document overlap in coverage, with the same material being treated more than once and not always in the same way.
- Parts of the document are not consistent with other parts in terms of their structure, content, and design.
- Parts of the document are not consistent with other parts in terms of style, terminology, and method of address to the audience.
- The document is graphically and physically inconsistent from section to section.
- The time available for correction, editing, and production is substantially underestimated, and therefore the final draft appears unfinished.
- Sections of the document do not get written until the last minute (if at all) because no one has had apparent responsibility to write them. This is especially true of preliminaries such as the letter of transmittal, cover, title page, abstract, table of contents, lists of figures and tables, and acknowledgments. It may also be true of appendices and attach-

ments or lists of references. Most important, it may be true of the executive summary—which can be the critical part of the document.

Many of these problems stem from the fact that the responsibilities for parts of group-written documents are usually assigned in terms of subject matter, not in terms of parts of the document. Thus, some parts are overlooked in assignments and found to be missing only on the weekend before the Monday deadline.

❑ 13.2 ❑ Integrating Group Writing Within a Technical Project

Some group-written documents are the result of an entire project, such as when a team is formed solely for the purpose of writing a proposal in response to an RFP. That is, the project group exists to write the proposal. A second-order problem arises when the group-writing project is part of another project—for example, when a final report must be written by a research team or design group. Before we discuss the guidelines for group writing, therefore, we would like to make some suggestions on how to integrate a group-writing project within a technical project.

When the group-writing project is part of a technical project, the group-writing activities must be scheduled and performed as part of the project schedule. Occasionally, the project report can be the responsibility of one person, such as the project director or a technical writer, but even in this situation the planning and writing of the report must be part of the total project scheduling. No matter who does it, the report writing task must be scheduled as an integral part of the project. More accurately, it must be scheduled as a *significant portion* of the project schedule (Figure 13.2). A good rule of thumb is that at least 25% of a technical project schedule should be devoted to writing the final report. The report writing should not be "tacked on" toward the end of the project, as it often is, because the report writing can necessitate additional analysis that has not been scheduled. For example, Figure 13.3 shows a schedule modeled on an actual multi-million-dollar project. The report was completed after the due date, and had to be submitted while some of the investigators were still formulating conclusions. We have found that this schedule too often models the activities of student project groups.

Usually the report writing overlaps other project activities. Only in rather controlled circumstances can the report be scheduled after the prior project tasks have been completed. In other words, writing should overlap the tasks of the project itself (as illustrated in Figure 13.2). This is efficient because of the interrelationship between the analysis and the writing. That is, writing up an analysis of the data often forces a researcher to reinterpret the data or

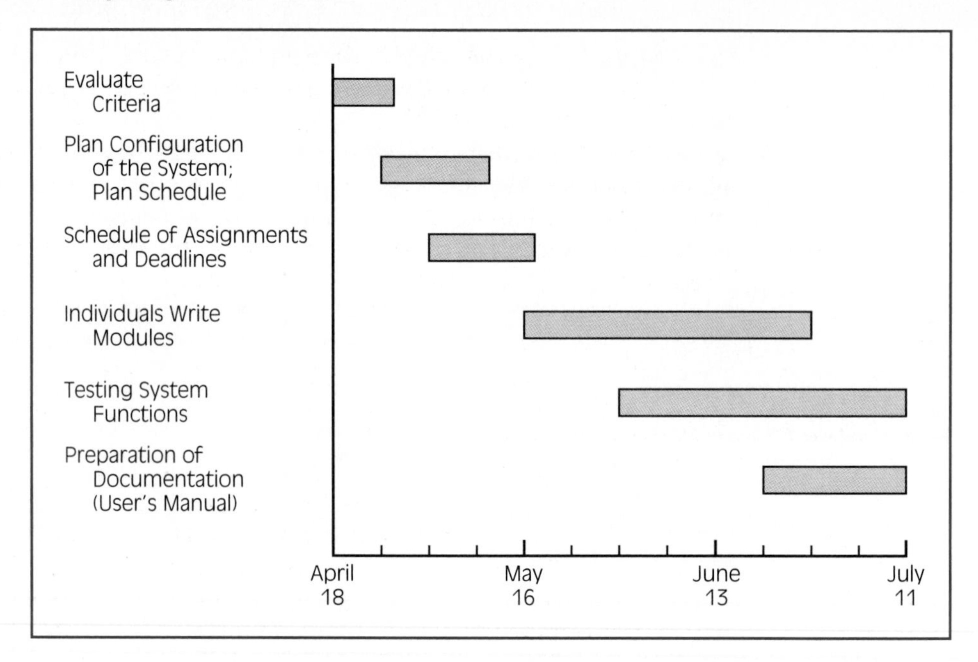

FIGURE 13.2 Schedule for a Three-month Project by a Software Development Group to Develop a Numerical Control System for a Production Department.

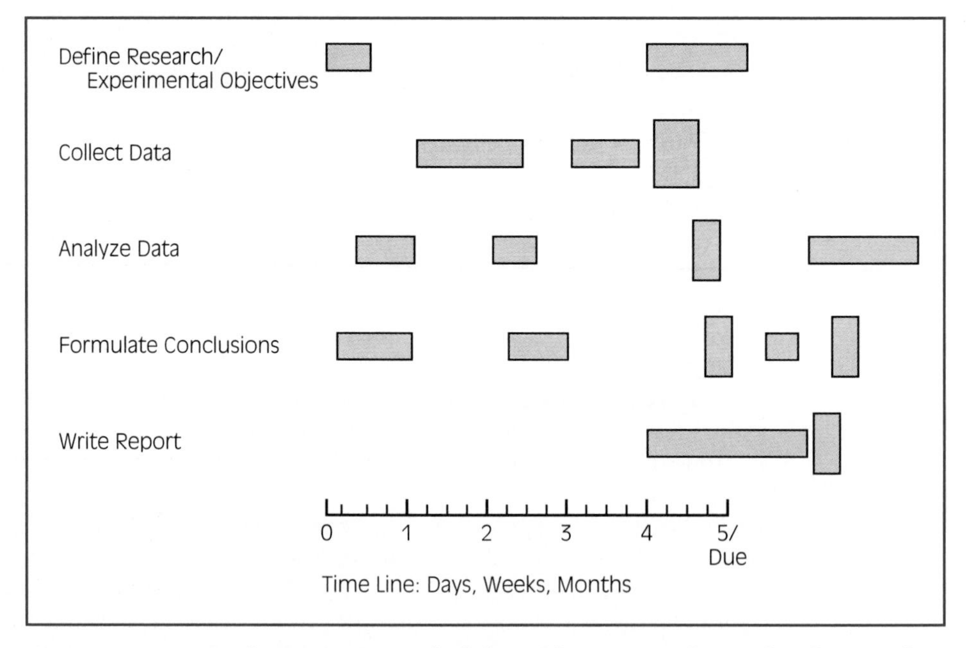

FIGURE 13.3 "Realistic" Project Schedule (with Report Submitted Before Analysis Is Completed).

even to seek new data. Thus, the researchers should start writing up the results before they think they have finished with the data analysis (and formulation of conclusions).[2]

The report writing can overlap all of the other project activities; it does not have to wait until project activities are completed. For example, the objectives of the research or project can be written before a project begins. Sometimes, in fact, the objectives have already been written in the form of a proposal. (This provides most of the introduction to a project report.) Similarly, as the research or experimental objectives are being defined and the methodology designed, they can be written up. (The objectives provide the design basis or criteria sections of a report, and the methodology provides the materials and methodology sections.) The framework for data collection and analysis also can be written up as the data is being collected and analyzed. (This provides the introduction to the analytical section and also appendix material.) Most of the appendices can be prepared during the data recording and analysis stages of the project. Thus, writing the report can occur throughout the project and indeed can be the integrating function for the entire project (Figure 13.4).

To write before completing the technical work is not only efficient; it also helps to clarify and modify the technical work in progress. For precisely this reason, we often suggest to doctoral students that they start writing their

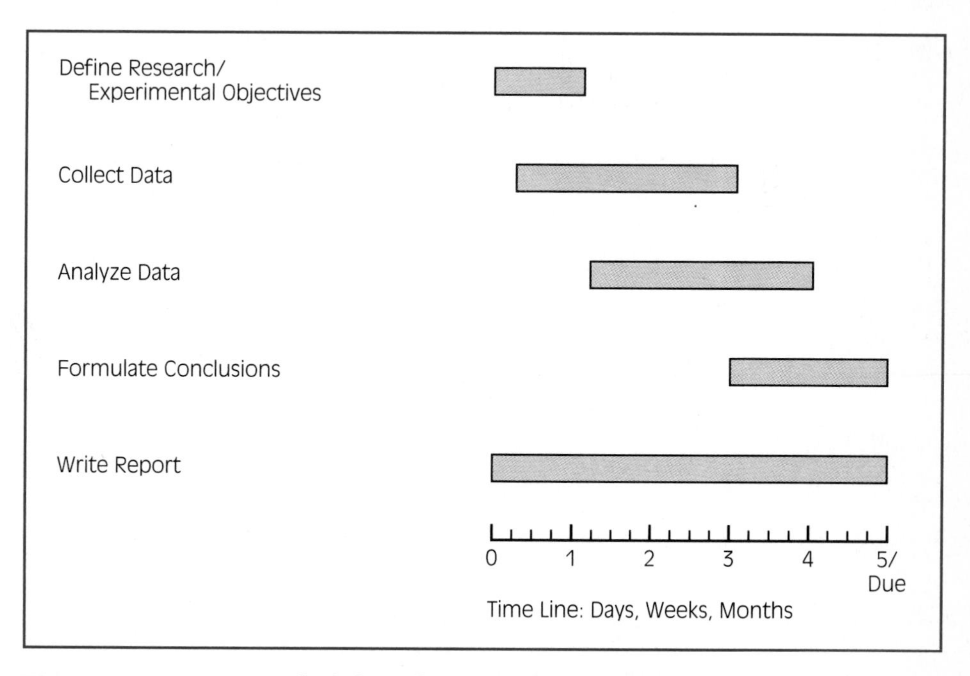

FIGURE 13.4 Project Schedule with Report Writing Serving an Integrative Function.

dissertations before the research is actually done. This forces them to grapple with design and content issues that otherwise they might not think about until too late. Writing the methodology section of a dissertation before doing the research, for example, can actually force a researcher to change his or her method. Similarly, defining the problem, reviewing the literature, or even projecting the results and trying to write about them before they are actually developed can reveal problems in the basic objectives, methods, or tools being used for a project. In the same way, a project team that begins to write its report before the project is completed can discover areas in which more work is needed or in which a different approach might be useful.

Our suggestion is that when a group-written report is part of a technical project group's total activities, the project report be planned and scheduled as an integral part of the project. The group-writing project should be managed as a self-contained process in itself, but it is best integrated if it is not delayed until the end of the project and if at least one-fourth of the total project time is allotted to the group-writing activities.

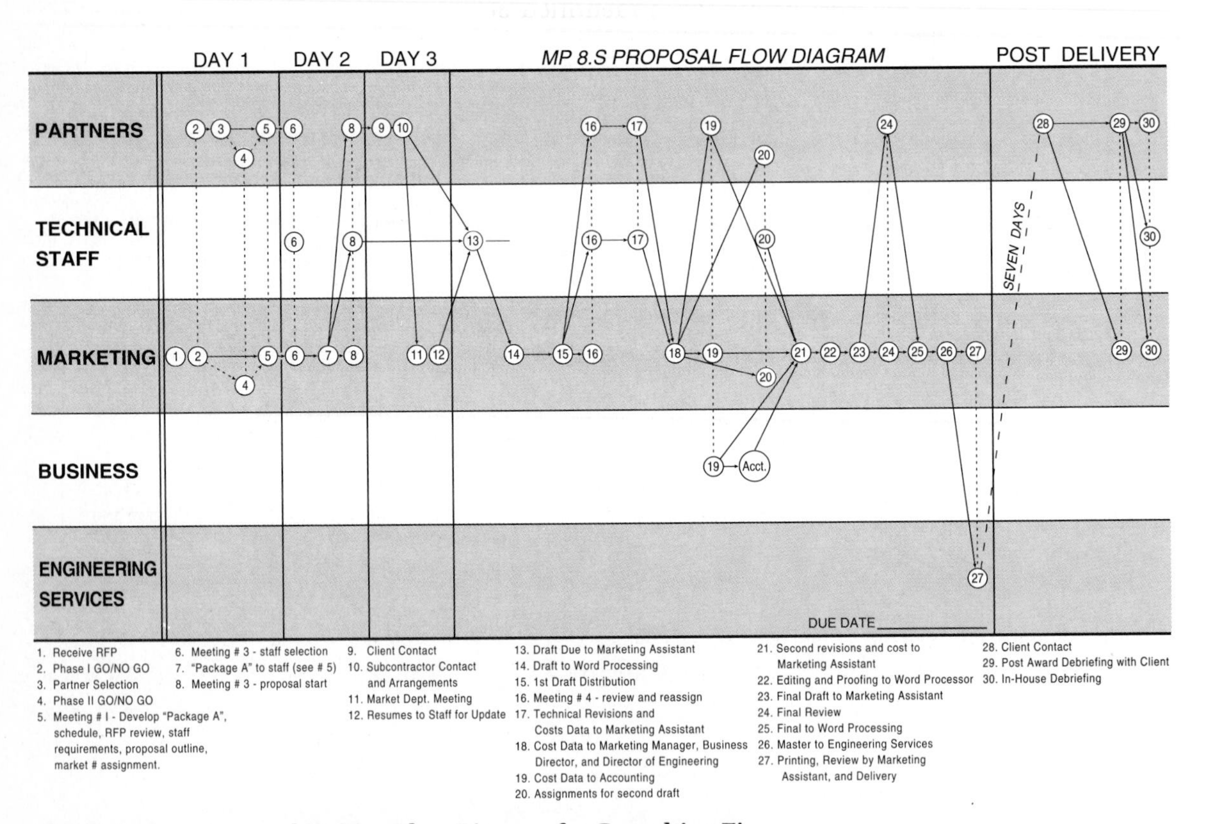

FIGURE 13.5 Proposal Writing Flow Diagram for Consulting Firm.

❑ 13.3 ❑
A Procedure for Producing Group-Written Documents

Depending upon the degree of polish necessary for a final document, the number of stages in a group-writing process (and the consequent number of drafts) varies. A comprehensive procedure might have more than a dozen discrete stages and a minimum of three drafts of the document. We suggest an 8-stage procedure, which can be expanded or compressed according to the scope of a group-written proposal or report. This procedure assumes that a group-writing project should be established as a separate project within the organization, not just treated as an extension of the team members' routine activities.[3]

The group report writing procedure, of course, is embedded in organizational processes that include non-report writing activities. For example, a flow diagram of the proposal process for a consulting firm (Figure 13.5) illustrates how the process of writing the proposal, which is done by the technical staff, is integrated with other activities of managers (i.e., partners), marketing personnel, business planners, and printers (i.e., engineering services).[4] In all, this consulting firm has identified 30 discrete steps in a process, wherein the actual writers of the document (the technical staff) are involved in less than a third of the steps.

Our approach to the problem of group writing is embodied in the guidelines for group writing shown in Figure 13.6. It consists of eight specific stages.

Stage 1. Establish the Project Management and Conceptual Plan

The first stage of the group-writing process is to establish the group writing project. Essentially, this means establishing the task of writing a report as a quasi-independent project, even when it is a stage of a technical project.

Establishing the group-writing project involves the following:

- Designating a project manager responsible for producing the document.
- Establishing schedules and assigning responsibilities.
- Developing a conceptual plan for the document.

When an organization makes a decision to undertake a project, such as developing a proposal that requires written contributions from a number of persons, the first order of business is to form the group and designate a report-writing manager. This manager should be dedicated to the project and should be given all of the authority and resources needed to produce the report.

We stress the need for assigning a project manager with authority and giving that person total responsibility and resources. The manager must be decisive and able to exert control over the project group's activities, sched-

Guidelines for Group Writing

1. **Establish the Project Management and Conceptual Plan**
2. **Plan the Writing and Production of the Document**
3. **Outline Each Section in Detail**
4. **Write the First Draft**
5. **Prepare the Prototype Draft**
6. **Edit the Master Copy (Final Draft)**
7. **Establish a Production Schedule and Follow It**
8. **Follow Up and Debrief the Project**

FIGURE 13.6 Guidelines for Group Writing.

ules, and sometimes conflicting personalities. The role of the report-writing project manager is to establish schedules, assign responsibilities for each part of the report, integrate project activities, and produce the document on schedule. Thus, this person is much more than a coordinator. Although this person often is a staff person who coordinates a group consisting of managers and even upper management, as project manager he or she should be given management responsibility rather than merely coordinating responsibility.[5]

The project manager's initial task is to convene the group and to help it develop a conceptual plan for the document. This consists of analyzing the audiences and clarifying the purpose of the document, identifying the basic sections, and agreeing upon the purposes of each section. Everything discussed in the first several chapters of this book has to be done in order to develop a conceptual plan. On the basis of this plan, the schedules and responsibilities must be assigned and agreed upon. Assignment of responsibilities should cover all sections of the document, including preliminaries, executive summary, letter of transmittal, and appendices and attachments.

We must stress here that merely appointing a project manager and having a group meeting (or several meetings) to discuss the report's structure and schedule are not enough. Getting a group to work together effectively and efficiently is extremely difficult, requiring both excellent planning and superior management. The writing project manager must do his or her homework and must be prepared to make suggestions, reconcile disputes, clarify misunderstandings, and establish agreed-upon directions. He or she must be able to do this when everyone in the room would much prefer to be back in the lab, and when each individual will be quickly satisfied after learning what is expected of "his" or "her" section of the document and when it is due. In effect, the project manager must force the group to work together and to understand not only their own responsibilities but the total responsibility of the group. Seldom will they want to do that. Yet, because the conceptual plan for the document serves as the central guide for planning and producing the document, and because the schedule must be realistic, what happens at these initial stages of a group writing project is critical to the success of the project.

Stage 2. Plan the Writing and Production of the Document

The second stage of the group-writing process involves the entire project team in preliminary planning of the document—the matters discussed in Chapters 4–6 in this book. Although each person has his or her own responsibility, it is important for each person to know what the others are going to do and to know how his or her part relates to theirs.

Preliminary planning consists of meeting three objectives:

- Developing a 3-level topic outline for all sections of the final document, including preliminaries and appendices (as in Figure 13.7).
- Establishing a detailed schedule for completing the remaining stages of the project, including schedules of activities for individuals and subgroups of the entire project team.
- Designing the entire document format so that each person knows how his or her section will be integrated with the others and so that activities such as printing a cover or preparing special graphics can be started.

These three objectives are fully met only if all project members participate in the planning or at least have the results of the planning presented to them both in writing and in face-to-face conversation. Preliminary plans should be summarized in a written schedule.

At the completion of this stage, all project members should have a comprehensive overview of the document design, the specific responsibilities of all project members (themselves included), and a project schedule covering all phases of the production of the document.

PAGE 2

BOOK, VOLUME, PART, ETC / SECTION / SUBSECTION / TOPIC / SUBHEADS

IEFCS FOR BOEING 747/3639
PROGRAM TITLE LOG NO.

SECTION / SUBSECTION / TOPIC / SUBHEADS	VISUALS	ENGINEER	PAGES
SYSTEM DESIGN DESCRIPTION			
GENERAL CAPABILITIES			
PRINCIPAL FEATURES OF THE PROPOSED SYSTEM			
CONFIGURATION OF THE BASIC DUAL FAIL-PASSIVE SYSTEM			
GROWTH TO TRIPLEX FAIL-OPERATIONAL STATUS			
SUMMARY OF THE APFD OPERATING MODES			
SYNCHRONIZATION OF THE APFD COMPUTERS			
ENGAGEMENT IN THE OPERATING MODES			
SYSTEM INTERLOCKS IN ACCORDANCE WITH SCD REQUIREMENTS			
APFD OUTPUTS FOR AIRCRAFT ANNUNCIATOR DISPLAYS			
PITCH AXIS FUNCTIONAL DESCRIPTION			
IMPLEMENTATION OF THE PITCH AXIS FUNCTIONS			
PITCH ATTITUDE CONTROL THROUGH DRIVEN ATTITUDE REFERENCE IMPLEMENTATION			
DYNAMIC PRESSURE SCHEDULING OF ALTITUDE ERROR FOR ALTITUDE HOLD CONTROL			
ALTITUDE CAPTURE TO WITHIN 30 FEET			

FIGURE 13.7 **Form for Preparing 3-level Topic Outline.**

Developing a 3-level topic outline is extremely important. An outline is developed by each person for his or her section of the report, based on the conceptual plan. This outline is discussed by the team as a whole and then revised on the basis of this discussion. A 3-level outline even enables persons to get together to coordinate their sections.

The outline also is used for format purposes. Consistent numbering and a format for heads and subheads are established for the entire document. It is important to establish this framework early on rather than to try to make it consistent at the last moment.

Stage 3. Outline Each Section in Detail

The third stage of the process results in detailed outlines of each section of the document. (Chapters 7 and 8 can be helpful here.) It requires two activities:

- Developing a detailed outline for each section.
- Revising the detailed outlines.

Each member or each subgroup first works independently to develop a detailed outline that fleshes out the topic outline developed in the previous stage.

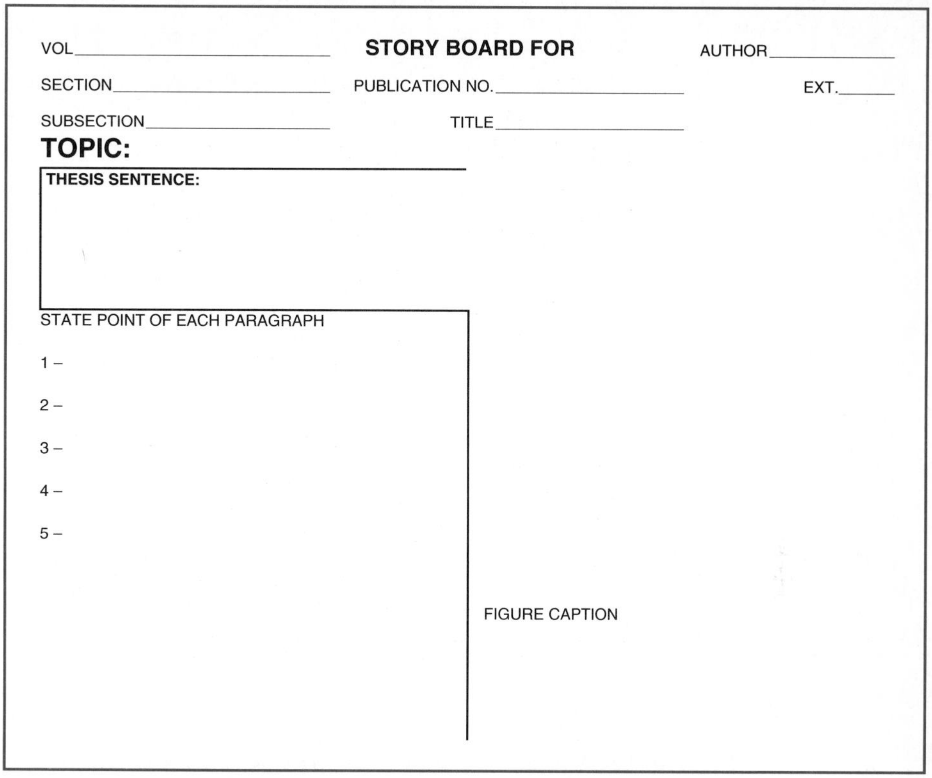

VOL______ **STORY BOARD FOR** AUTHOR______
SECTION______ PUBLICATION NO.______ EXT.______
SUBSECTION______ TITLE______

TOPIC:

THESIS SENTENCE:

STATE POINT OF EACH PARAGRAPH

1 –

2 –

3 –

4 –

5 –

FIGURE CAPTION

FIGURE 13.8 Form for Preparing Detailed Outline for a Section.

This detailed outline includes an inventory of the visuals and tables that are planned.

This detailed outline is prepared in a standard format established for the entire document, as is the case with the "storyboard" approach (Figure 13.8).[6] With a standard format, writers can interact constructively in teams of two or three rather than work in isolation. Important sections of a document usually involve more than one writer, and the detailed outline technique is a means of integrating the writing within a section, just as the group writing procedure is a means of integrating the writing throughout the entire document. With computer conferencing, of course, the approach can be adopted for the entire team. But at this stage it is best for subgroups rather than the whole team to interact.

A "storyboard" amplifies an item of the topic outline—at least one storyboard for each third-level item. The detailed outline suggests the level of detail in the final text (Figure 13.9). It includes sufficient detail to show how the actual prose will look in the document. The detailed outline or storyboard is a means of rethinking and revising initial ideas before the first draft actually is written. This saves considerable agony later on.

After the detailed outlines are prepared, the project team as a whole, or at

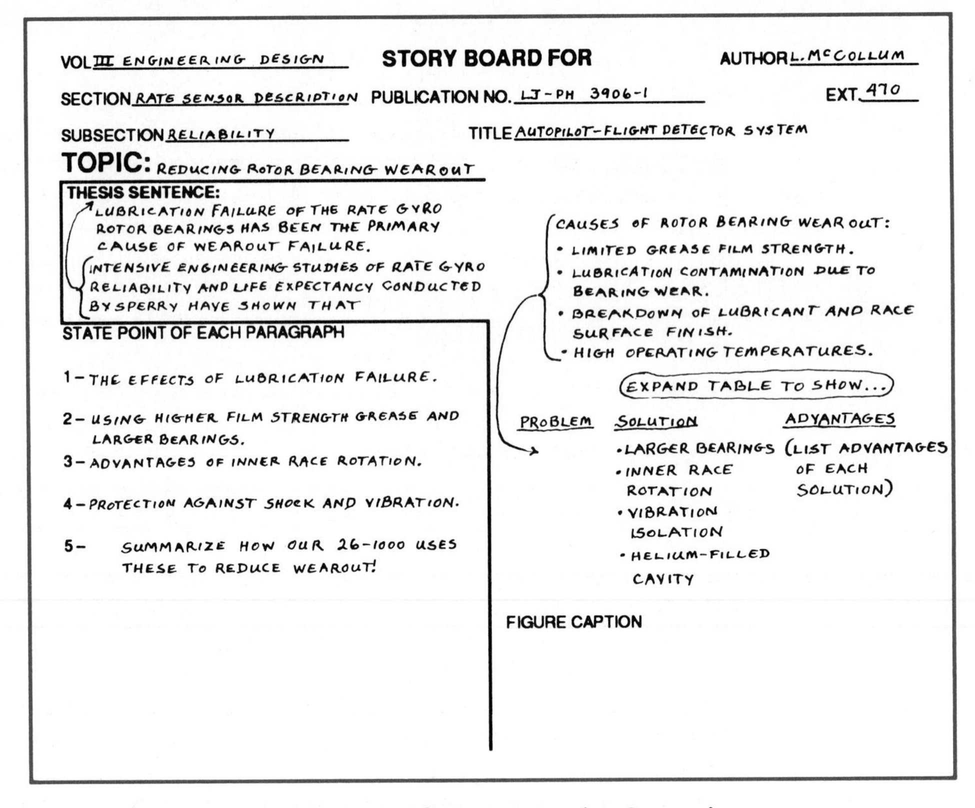

VOL III ENGINEERING DESIGN　　**STORY BOARD FOR**　　AUTHOR L. McCOLLUM

SECTION RATE SENSOR DESCRIPTION　PUBLICATION NO. LJ-PH 3906-1　　EXT. 470

SUBSECTION RELIABILITY　　TITLE AUTOPILOT-FLIGHT DETECTOR SYSTEM

TOPIC: REDUCING ROTOR BEARING WEAROUT

THESIS SENTENCE:
LUBRICATION FAILURE OF THE RATE GYRO ROTOR BEARINGS HAS BEEN THE PRIMARY CAUSE OF WEAROUT FAILURE.
INTENSIVE ENGINEERING STUDIES OF RATE GYRO RELIABILITY AND LIFE EXPECTANCY CONDUCTED BY SPERRY HAVE SHOWN THAT

CAUSES OF ROTOR BEARING WEAR OUT:
- LIMITED GREASE FILM STRENGTH.
- LUBRICATION CONTAMINATION DUE TO BEARING WEAR.
- BREAKDOWN OF LUBRICANT AND RACE SURFACE FINISH.
- HIGH OPERATING TEMPERATURES.

STATE POINT OF EACH PARAGRAPH

1 – THE EFFECTS OF LUBRICATION FAILURE.

2 – USING HIGHER FILM STRENGTH GREASE AND LARGER BEARINGS.

3 – ADVANTAGES OF INNER RACE ROTATION.

4 – PROTECTION AGAINST SHOCK AND VIBRATION.

5 – SUMMARIZE HOW OUR 26-1000 USES THESE TO REDUCE WEAROUT!

(EXPAND TABLE TO SHOW...)

PROBLEM	SOLUTION	ADVANTAGES
	• LARGER BEARINGS • INNER RACE ROTATION • VIBRATION ISOLATION • HELIUM-FILLED CAVITY	(LIST ADVANTAGES OF EACH SOLUTION)

FIGURE CAPTION

FIGURE 13.9 Completed Form with Review Session Suggestions.

least those who are doing any significant part of the writing, meet to discuss the storyboards and to suggest revisions in the detailed outlines. The entire set of storyboards is circulated to the entire group, which then meets to discuss them. This is best accomplished in a meeting with hard copy rather than by computer conferencing.

Discussion of the detailed outlines within the entire group is an important integrating mechanism. The results are that no piece of information is missing from the final document and that all pieces fit together effectively.

An additional benefit of this stage is that it forces all persons to do some actual thinking and outlining by this milestone of the project. (In extreme cases, it enables the project manager to identify weaknesses and perhaps to add or change members of the team before it is too late.)

Stage 4. Write the First Draft

The fourth stage of the process results in a first draft of the text of the document. This involves:

- Preparing the first draft.
- Reviewing the first draft.

Following both the leadership evaluation and suggestions from the group, all members of the group produce a first draft. The objective is to produce a complete first draft of the document so that the project manager can assemble all of the segments as well as the preliminary outlines of the artwork and attachments. There will be some missing pieces, but the draft will reveal what they are and where they fit into the whole. For example, at this point in the process no finished table of contents will exist. Probably there will be no list of figures. And other elements may be missing. Yet, as long as the places for these elements are marked in mock-up form at this stage, the final materials can be added in the next stage.

As the segments are drafted, writers should be given very specific format guidelines to minimize the necessity for revision later in the process. Such matters as margin sets, vertical spacing between sections, type font, type size, depth of indentations, and all other basic formatting details need to be specified. In fact, it is helpful to prepare a written format guide that not only states these details but that shows what the resulting text is supposed to look like. Often it is helpful if the project manager has done this planning before the group meets for the first time. But certainly before Stage 4 of the process, the entire group should know exactly how the final report is expected to look. Much the best way to convey this is by doing a mock-up of several pages and preparing a written set of instructions for text formatting.

Reviewing the first draft provides an evaluation that points up any need for revisions, additions, or deletions. This is a substantive review, not a stylistic review. The result is guidance from the project manager on the recasting, revision, and other changes to be made.

At the first-draft stage, major revisions and significant recasting are relatively easy to incorporate. Later on, they become more difficult. Accordingly, at this stage of the process it is important for the project manager or some person with the ability to evaluate the design of a document as a whole to examine the draft closely to assure that all of the sections fit together, to define what must be done to fix those that do not, and to identify specifically what remains to be done. Again, the manager's authority and personal ability to work with difficult individuals in a stressful situation are essential. At the review stage no member of the team will want to learn that what he or she has done is unacceptable, but often that will be the case. Major rethinking and rewriting may be necessary.

Unlike the previous stages of the process, the first draft should be reviewed by a single person or a small group, by someone who has a comprehensive sense of overview of the whole project and who knows what the final document should be. Sharing the results of these evaluations with the group and soliciting group input can also be useful, but the review must reflect a single and comprehensive view.

Individual writers should have a clearly marked-up copy of their first drafts to work from. They should be given specific and detailed advice about how to revise their sections of the document to assure that the final versions of the parts do fit together. It is not enough for them merely to be told to rewrite or to revise; they need to be told how it should be done.

As the first drafts are being revised, the other sections of the document should be drafted or assembled so that a prototype draft of the document can be prepared.

[Note: Stage 4 often is repeated for complex documents to provide for review and revision of a second draft before proceeding to the prototype draft, Stage 5. This would be the case with a formal design report, for example.]

Stage 5. Prepare the Prototype Draft

The fifth stage involves production of a draft of the complete document. This requires:

- Assembling the prototype draft.
- Reviewing the prototype draft.

A prototype draft is a complete preliminary version of the final product. All elements should be there. The objective is to produce a preliminary final draft that is, except for copyediting, complete. All missing elements should be inserted—cover, title page, executive summary, overviews, table of contents, abstract, letter of transmittal, artwork, list of references, appendices—everything. The assumption is that although some later refinements will be made, the prototype draft is of sufficient quality and completeness that it could—in a pinch—be sent exactly as is.

The first step is to assemble the rewritten sections of the first draft and to supply missing pieces. If Stage 4 has been successful, these sections should all fit together. Then the additional sections are fitted in, along with mechanical devices such as dividers and title pages, if any. All figures and tables also are integrated with the text, both physically and by references.

The second step is to review the prototype draft. Surface copyediting and integrating—making the document mechanically consistent—usually should be done before substantive review by the group. In general, this is best done by one person—perhaps by the project manager but often by an effective copyeditor working closely with the project manager. This person should have a free hand to correct and revise the prototype draft to make it mechanically and stylistically consistent.

Then, the entire project team should review the prototype draft to ensure that it is substantively accurate and that the various sections are effectively integrated with each other. Project members can make stylistic suggestions to clarify the meaning, and at times can suggest even the addition or deletion of material.

When the prototype draft has been reviewed by the entire project team and they agree that the draft is in final form, the team is essentially finished with its work. It is now up to the project manager to produce the final document.

Stage 6. Edit the Master Copy (Final Draft)

The prototype draft is edited and revised according to the suggestions in Stage 5 and then reviewed again. At this point the matters discussed in Chapters 9 through 12 are appropriate considerations. Invariably, inconsistencies reappear, especially if significant revision or recasting has occurred. Usually, graphics need to be revised and integrated to complement the final text. Some of the document format, such as the title page or list of tables and figures, also needs revision.

At times, a fresh reader can be helpful in spotting needed last-minute corrections or making minor stylistic revisions. This person should be new to the project, someone who has not read any of the previous materials, and someone with both the ability and the self-confidence to provide a conscientious review. The external reviewer should be instructed to read carefully for correctness and clarity, not to read merely for overall effect.

Stage 7. Establish a Production Schedule and Follow It

From the beginning of the project, a schedule should be established. To actually produce a bound copy of the document requires that the production activities on the schedule be rigorously adhered to. Artwork, for example, must be finished on time to be integrated with the text. The integration of artwork and text includes such items as putting figure references in the text, paginating the text after the artwork has been integrated, and preparing the list of tables and figures.

Also important is working with the copy center or printer to assure that the final copy is assembled correctly for production. For example, in a document that is intended to be printed on two sides of the paper, it is necessary to insert blank pages at times to assure that pages which should be on the right side (chapter beginnings, for instance) do actually fall on the right. Otherwise the final copy will be inconsistently and incorrectly paginated.

Similarly, if there is artwork in the document, it is necessary to work with the printer to prepare the camera-ready masters in a form that will produce good results. In our experience, most report teams give far too little consideration to these matters, and can end up with a completely flawed product as a result. Just last week, for example, we saw 300 copies of a bound document on which the cover art had turned into a blob of black ink, almost completely obliterating both the design and the print. The master copy looked decent enough, but the photocopy procedure used to reproduce the document had not been up to the task, and the result was expensive and embarassing.

Nobody bothered to check ahead of time, so although the copy center did its best, the result was still $191 worth of unusable copies. That meant a new cover had to be prepared and taken to a printer, and that took more than a week. In the end, the document went out 10 days after it was scheduled to be mailed.

If all aspects of report production and binding are not scheduled, and if nobody bothers to talk to the printer or copy center ahead of time, you can end up with a master copy at the deadline—but a copy that still can't be put together, printed, and bound.

Stage 8. Follow-Up and Debrief the Project

A final activity often ignored by managers of group-written documents is follow-up and debriefing. The objectives are twofold:

- To interact with readers of the document to see if their needs are met.
- To take advantage of the group's experience to identify what needs to be done differently next time.

Follow-up with readers and debriefing are often ignored by an exhausted group and project leader. However, the project should not be regarded as complete until both of these postwriting stages are completed. This should be done as soon after submission of the document as possible, while the experience is still fresh in everyone's mind and before everyone is off on another project.

Follow-up with the audience may alert the project manager to opportunities (or necessities) for further communication or face-to-face meetings. Debriefing provides insights into what should be done differently the next time to improve the quality of the document and the efficiency of the group-writing process.

There is always another report to write and never enough time to write it.

Conclusion

Writing is always difficult; group writing is doubly difficult. Especially this is true if the group members are unable to meet together regularly to plan and to share undertanding of where they are in the process—as is sometimes the case with geographically separated groups. Then, especially, the planning and scheduling aspects of the project management become critical.

To illustrate this point we can draw upon our own experience over the past 2 years during which we have been project members on two geographically separated groups that produced proposals. One group consisted of twelve members representing four locations that were separated by as little as 60 miles and as much as 500. This group met face-to-face three times—for a full day on each occasion.

In the second instance, the project group was made up of representatives from six sites, each of which had a subgroup of up to about fifteen members, and each of which sent two representatives to the group meetings. The sites, located in five states, were as little as 300 miles apart and as much as 2500. The subgroups met regularly at each site, but they never met face-to-face with other subgroups. Instead, the subgroups communicated by electronic mail and by telephone. The representatives of the subgroups met twice over a 3-month period, once in an airport hotel near Chicago and once in Puerto Rico. (For project managers, at least, this group-writing process did have its good moments. Unfortunately we weren't among the managers.)

Both project teams were successful. The first proposal was funded by the state government, and the second became a finalist for an NSF grant after other competitors were eliminated.

In both of these instances, the success of the project group was attributable to strong leadership within the groups and to a disciplined process of planning and drafting the proposals. The dynamics of the two groups were admittedly different: the first group depended heavily upon a single person to redraft and integrate the preliminary drafts of the group members; the second group depended more heavily upon developing a detailed model that individual writers used as a basis for drafting their own segments of the proposal so that they would fit together without extensive rewriting. Yet both groups followed essentially the process we have described, using electronic mail and telephone conferencing more than face-to-face meetings, but still working together effectively.

Computer-based collaborative writing tools are being developed, which will greatly assist the performance of the collaborative process we have described.[7] These tools allow people to share plans electronically, to format a document according to an agreed-upon template, to write notes to each other, to annotate each others' drafts in print or in voice, to keep a log of activities and progress, and to trigger activities automatically when stages of the process are completed. We have even seen a demonstration of the use of compressed video as a collaboration tool between two branch offices of an architectural firm. This tool allowed face-to-face communication as well as sharing of both print and graphic information between two offices separated by more than 60 miles.

These new tools will no doubt make the collaborative process easier in the future than it has been in the past. However, we believe that collaborative writing will still be a process appropriately divided into stages such as we have outlined in this chapter. At least these can provide a starting point for group project planning.

NOTES

1. *INFO-MAC Digest* (on-line), 5:52, February 13, 1987. Although the explanation of the proposal project at Martin Marietta stresses the use of personal computers, you can see that the project team consisted of 300 persons. The article, of course, stresses the output. We are calling your attention to the need and challenge of coordinating the output of 300 persons to achieve coherent reports.
2. This is intrinsic to writing as a problem-solving or design process, as we approach it. The function of writing as an analytical tool is documented in James Paradis, David Dobrin, and Richard Miller, "Writing at Exxon ITD," *Writing in Nonacademic Settings*, ed. Lee Odell and Dixie Goswami (New York: Guilford, 1985), 290–91.
3. Group writing should be a project as defined in management literature; for example, see Melvin Silverman, *The Art of Managing Technical Projects* (Englewood Cliffs, NJ: Prentice, 1987).
4. This flow diagram is from McNamee, Porter and Seeley, a consulting firm in Ann Arbor, Michigan, and is reprinted with permission from Alice Elizabeth Moorhead, *The Rhetorical Design and Function of the Proposal*, doctoral dissertation, U of Michigan, 1984, 134.
5. We are following the suggestion that the project manager be given authority, not just assume the role of project coordinator, as discussed in Silverman, *The Art of Managing Technical Projects*, 33–38. The concept of a document as a product is clearly defined in "Appendix 3: The Technique of Design Review," 313, which is a reprint of a paper by Richard M. Jacobs from the 1979 Product Liability Prevention Conference of the American Society for Quality Control. Our 8-stage procedure with the follow-up and debriefing parallels but does not precisely correspond to that Design Review procedure (309–31). In general, it follows the three-team-meeting procedure outlined by Silverman (167–80):
   - Project-initiation meeting.
   - Design review meeting.
   - Project close-down meeting.
6. We were first introduced to the storyboard approach by Keith B. Roe, who at that time was Manager of Engineering Publications at Sperry Flight Systems. Many companies and consulting firms use this approach, which was developed by James R. Tracy, Head, Writing Services Section, Information Media Department, Hughes Aircraft Company. He discusses it in "The Theory and Lessons of STOP Discourse," *IEEE Transactions on Professional Communication* (June 1983), 68–78. The traditional storyboard technique still is discussed in the literature, as in Robert A. Barakat, "Storyboarding Can Help Your Proposal," IEEE Transactions on Professional Communication (March 1989), 20–25 and Muriel Zimmerman and Hugh Marsh, "Storyboarding an Industrial Proposal: A Case Study of Teaching and Producing Writing," in *Worlds of Writing*, ed. Carolyn B. Matalane (New York: Random House, 1989), 203–21. On-line approaches, however, are being developed, especially in the hypertext environment, but initial programs essentially facilitated editing rather than collaborative interaction for document production. The next generation of programs could well embody the integrative process (discovery and brainstorming) characteristics of the storyboard approach.
7. See Mary D. P. Leland, Robert S. Fish, Robert E. Kraut, "Collaborative Document Production Using Quilt," *Proceedings of the Conference on Computer-Supported Cooperative Work*, The Association for Computing Machinery, 1988, 206–15. A brief overview of the subject is Annette Easton, et al., "Supporting Group Writing with Computer Software," *Bulletin of the Association of Business Communication* (June 1990), 34–37, a special issue devoted to collaborative writing.

CHAPTER 14

Writing in the Multinational Context

Within one generation, the business world has acquired a new dimension: the multinational context. Even as recently as 20 years ago, few business professionals had to concern themselves with audiences in other cultures. Unless they were in an organization directly involved in international trade, most of them wrote letters and reports to native English-speaking readers in organizations within the United States or Canada. The term "multinational corporation" had no currency.

Today, however, most of the Fortune 500 companies are multinational corporations, have subsidiaries in other countries, or routinely do business with companies in other countries. Presidents of major United States corporations, such as Dow Chemical, have climbed a management ladder whose last rungs typically include the positions of Director of the European Division or Vice President for International Marketing. Sales persons, project managers, and engineers in many companies routinely now have assignments abroad—supervising construction of a "turn-key" petrochemical plant in Korea, participating in the design of a textile plant in Yugoslavia, planning alternative agricultural projects in the Gambia, West Africa.

But the multinational context does not just require Americans to go abroad; it requires professionals from other countries to go abroad too. Numerous foreign companies have offices and plants in the United States. For example, more than 100 Japanese companies have facilities in Michigan. In addition, American and foreign companies have jointly owned subsidiaries in the United States. And many American companies are owned by, or are subsidiaries of, foreign companies. No matter what the field—retail sales, banking, manufacturing, service industries, or real estate—in business and industry you must often consider cross-cultural implications even when communicating within the United States.

Writing in this multinational context presents additional challenges in audience analysis and report design. Even when you write in English, as you usually will (especially with non-European cultures), you need to adapt to this context. You should not write a letter or report in a cross-cultural context in the same way that you would in your own organization or to another company with only regional or national operations. The language of the communication may be English; however, if true communication is to occur, you must be sensitive to the underlying cultural expectations of your readers.

Of course, many of the foreign people with whom you will deal professionally have adapted their methods of communication to your expectations. Indeed, we have taught foreign students, managers, engineers, and business professionals from other cultures, many of them non-Western, and they do adapt to American business and communication practices with remarkable success. They learn to write letters and reports according to the Western-based principles taught in this book. They learn, for example, to state the purpose of a document explicitly—when it is intended for American readers. They learn to put the summary "up front" and to get "right to the point" or "down to business," American style. They learn to adopt a persuasive pattern and tone and to make their recommendations direct and explicit. They even learn to address casual business acquaintances by their first names, a uniquely American convention. Many of these features are unthinkable in their own cultures, yet in our experience numerous foreign professionals conscientiously study American-style communication because they realize that to do business with Americans they often will have to do it "American style." The point is that you have to do the same thing if you are to do business in their cultures.

We strongly suggest that you adapt to the cultures of your readers, as they adapt to yours. To establish a sound business and technical relationship requires interactive communication between you and your foreign counterparts. You must learn to be sensitive to the cultural contexts of the communication process, and you must recognize that not all peoples of the world would select or arrange information for letters and reports in the same ways we have described in the earlier chapters.

Our purpose in this chapter is to introduce you to the cross-cultural contexts of technical communication. Although space prohibits going into detail about how to communicate with readers in particular cultures, we hope to sensitize you to the cross-cultural considerations. These can provide a point of departure for your communication and for your further study of the cultural expectations of the particular foreign readers with whom you will deal professionally.

❑ 14.1 ❑ The Cross-Cultural Communication Process

Cross-cultural communication introduces aspects of the communication process that we have not directly discussed.

Your communication is part of a communication process. It responds to prior communication and it leads to subsequent communication, as we explained in Chapter 1. So far, we have interpreted this process primarily as a problem-solving process. In addition, we have assumed that communication usually supports a decision-making process, either by contributing to a decision (as with a recommendation report) or by implementing a decision (as with a purchase order or an engineering change notice). Furthermore, we have assumed that important communication is written or formalized, even if it is preceded by various oral communications such as informal discussions or departmental or project meetings.[1]

These assumptions are based on the role played by written communication in American organizations and on the need to be explicit, precise, and efficient. Written communication in American corporations serves the purpose of codifying decisions and actions; that is, it makes them tangible, observable, concrete. It turns them from "mere talk" into something that can be acted upon. Although every corporation functions upon the basis of informal communication, in American corporations there is also the attitude that "if it's not in writing, it doesn't count."

These assumptions often do not hold in a cross-cultural communication context. You must first concern yourself with the communication process and the role of your written communication in that process. The process itself might be more important than any specific communication act. Furthermore, in a cross-cultural context, oral communication might be considerably more important than written communication, especially as it contributes to decision making. In addition, the relationship between oral and written communication could be extremely important, and quite different from the relationship to which you are accustomed.

In sum, in a cross-cultural context, the function of written communication as we have discussed it can change entirely.

An Extended Example: Initiating Contact with a Japanese Corporation

The difference and the importance of the communication process in a cross-cultural context can be clarified by an example drawn from the experience of a business associate who asked our advice. This example introduces some of the considerations you should make when writing in the multinational context. It involves an initial business contact with a foreign firm. Our associate's American firm—a bank—wanted to do business with the foreign firm, so the communication issue involved establishing the initial contact. Although the example involves a bank, it could apply in any field.

The American subsidiary of a Japanese firm (let's call it Honshu U.S.A., Inc.) had chosen to locate a new marketing facility in an Indiana town (let's call it Mt. Pleasant). Mr. Leafgreen, the Vice President for commercial loans at the Mt. Pleasant branch of the National Bank of Indianapolis, wanted to solicit the Japanese firm's commercial banking business once they had come to town. As a necessary initial step, the main office of Leafgreen's bank initiated contact with the New York office of the Bank of Hiroshima, Ltd., of which Honshu was a client in Japan. However, Mr. Leafgreen's more important step was to initiate direct contact with the General Manager of Honshu U.S.A., Mr. Takashita, who was relocating to Indiana from Ann Arbor, Michigan, to manage the plant. The Vice President had to write an initial contact letter (Figure 14.1), a letter which was sensitive to the cross-cultural context and to the communication process.

His letter is distinctive for its tone and content. It is personalized much more than it is businesslike. It makes no direct reference to any business matter, the covert purpose of the letter; instead it expresses an overt purpose of welcoming Mr. Takashita and his company to the community. A similar letter from one American firm to another probably would have opened with a strong sales pitch summarizing the advantages of the bank's services or product, following with selected technical or financial details to support the pitch. Leafgreen's letter, however, is designed to be part of a communication process that eventually will result in a business relationship. (The process had already been initiated by the contact with the Bank of Hiroshima's New York Office. And on the Japanese side, the process also had been initiated with communication between Honshu and the Hiroshima bank.) Given this particular cross-cultural context, this letter was part of the interpersonal or social stage that precedes the business stage of the communication process with Japanese firms.

Although the addressee, Mr. Takashita, had lived in the United States (and speaks excellent English), the Vice President adopted a culturally sensitive approach. He assumed Mr. Takashita would be familiar with American business practices, but he also assumed that Mr. Takashita would appreciate an approach in keeping with Japanese business protocol. This accounts for the emphasis in the letter on the community and the school system as well as the reference to the weather, all of which would be expected in Japanese culture. (A competing Indiana bank adopted a direct business approach, but the Mt. Pleasant bank eventually received all of the banking business of Honshu and its professional staff. Thus the approach of Leafgreen's letter was successful, other factors being equal.)

The communication process was also important in the follow-up to the

NATIONAL BANK OF INDIANAPOLIS—
MT. PLEASANT
Drawer 543
Mt. Pleasant, Indiana 47862-0687
(313) 662-9900

July 21, 1990

Mr. Iwao Takashita
Honshu-USA, Inc.
Suite 463, Eisenhower Tower
3001 S. State St.
Ann Arbor, MI 48104

Dear Mr. Takashita,

On behalf of the National Bank of Indianapolis—Mt. Pleasant, we would like to take this opportunity to welcome you to Mt. Pleasant and express our gratitude that you have chosen our community for your marketing facility. In this area, we have an enjoyable climate, with four distinct seasons of which each has something to offer.

We think your employees will feel most welcome here in our small town atmosphere. Our local school system is regarded very highly within the State, and we also have a small liberal arts college.

Robert M. Jackson of the National Bank of Indianapolis and I would be pleased to meet with you in Ann Arbor in the near future. We would like to acquaint you with some of the benefits of locating in Mt. Pleasant and to answer any questions you might have about locating in our community, such as about its neighborhoods.

If there is any way we may be of service in facilitating your move to Mt. Pleasant, please feel free to contact us at any time:

Robert M. Jackson
Vice President, International Banking
NATIONAL BANK OF INDIANAPOLIS
111 Circle Drive
Indianapolis, IN 46222
(317) 996-1123

Craig Greenleaf
Vice President, Commercial Loans
NATIONAL BANK OF INDIANAPOLIS
P.O. Drawer 543
Mt. Pleasant, IN 47862
(317) 662-9900

Respectfully yours,

Craig Greenleaf

Craig Greenleaf
Vice President Commercial Loans

CG/jla

cc: Robert M. Jackson

FIGURE 14.1 Initial Business Contact Letter with Japanese Firm.

letter—or lack of follow-up. Mr. Leafgreen did not initiate any follow-up even when over a month went by without a response. If the same thing happened in an American business context, it would cause some apprehension. In fact, the Vice President for International Banking of the bank's head office in Indianapolis pressured Mr. Leafgreen to follow up with either a letter or a phone call. But Mr. Leafgreen wisely waited for Mr. Takashita to make the next move. While he was waiting, Mr. Leafgreen even met Mr. Takashita at a cocktail party for Honshu, sponsored by the Chamber of Commerce, but pointedly he did not mention the letter or do any more than exchange polite social pleasantries.[2] (The officers of the competing bank did try to talk business at the same party.)

Eventually, the Vice President received a response and a visit from Mr. Takashita, informing him that Honshu would open an account with the Mt. Pleasant branch of the National Bank of Indianapolis. At first it was unclear to Mr. Leafgreen whether Mr. Takashita meant a personal account or a corporate account, but later it became clear that Mr. Takashita was giving the Mt. Pleasant bank not only all of the corporate business of Honshu but also the personal banking business of all of its Japanese employees.

A brief summary of an appropriate means of establishing a business relationship with a Japanese firm explains the communication process that dictated the approach that Mr. Leafgreen took with this letter.[3] An initial contact should be preceded by an in-depth analysis of the Japanese firm, its products, its personnel, its financial status, and its reasons for locating in Mt. Pleasant. The analysis should also gather as much information as possible about Mr. Takashita.[4] This background information might be useful in the letter, and certainly will be useful in the first face-to-face meeting. Before the meeting, Mr. Takashita would know a great deal about the National Bank of Indianapolis, the Mt. Pleasant branch, and Mr. Leafgreen, because that is the Japanese way of doing business. Accordingly, an American who hopes to do business with Mr. Takashita should do his or her homework before initiating any contact.

The initial contact by letter had the purpose of getting a meeting with Mr. Takashita. With the Japanese it is important to establish a social relationship before business negotiations: social formalities precede business negotiations. Mr. Leafgreen's initial letter therefore discussed the community, the schools, and the weather, but did not raise any issue of business. Honshu would not have considered locating in Mt. Pleasant if it had not thoroughly investigated the community, its schools, and its other features and found them acceptable. Thus the letter affirms their decision and acknowledges that social values are as important as business values.

The next stage in the communication process was extremely difficult for Mr. Leafgreen. He had to wait for Mr. Takashita to respond. As we mentioned, his manager in Indianapolis was pressuring him to follow up. Mr. Leafgreen, however, had to allow Mr. Takashita to take the initiative in any way he wanted to do so. Mr. Takashita needed time to do further research on the National Bank of Indianapolis and, perhaps, on Mr. Leafgreen. He also needed time for the consensual decision-making process within Honshu to occur. Mr. Leafgreen could not rush things; if he did, he might lose the business opportunity.

Mr. Takashita's eventual response was positive. It led to follow-up meetings and eventual establishment of business relationships. These business relationships would then continue against a background of further community and social relationships among Mr. Leafgreen, his bank, Mr. Takashita, and Honshu. When a business relationship is established with a Japanese firm, social relationships are also established and must be maintained. But these relationships will continue to be kept separate—no talking business when out on the town.

If Mr. Leafgreen had been dealing with an American firm, his approach would have been entirely different. His initial letter would have introduced the National Bank of Indianapolis and all of its business advantages. It would have arranged a follow-up meeting to discuss these advantages in detail. Additional reports, brochures, and documents would probably have been included to explain these advantages. Although social events—cocktail parties and business lunches—might have occurred, these would have complemented or have been secondary to business considerations. Business could and would have been discussed in these interpersonal and social communication situations. Furthermore, Mr. Leafgreen would not have been comfortable until the firm had opened accounts with his bank. He would have been worried that at the last minute they might switch to a competitor. Therefore, he would have followed up and maintained communication—"kept in touch" as Americans would say.

Cross-Cultural Communication Process Considerations

Our point is that communication process considerations are different in a cross-cultural context than they are in the United States, but the process is critical to effective communication. Our own cross-cultural experience has been primarily in doing business with the Japanese and the French, although we have had business transactions with the Polish, West Africans, Dutch, and Germans (and, of course, English-speaking cultures: Canada, England, and Australia). These cross-cultural interactions have sensitized us to process considerations. For example, in many ways doing business with the Japanese is the antithesis of doing business in the United States. This is because with the Japanese the process is primary. The managing director of a documentation firm in Tokyo told us that normally he would spend up to a year establishing personal and social relationships with the appropriate managers of a potential Japanese client before "getting down to business." This contrasts with the American approach, which is direct to the point of being rude in the eyes of Japanese business people. In addition, in the United States any formal arrangements are put in writing, and between corporations the primary mode of communication is written. Nontechnical and business process considerations are relatively unimportant, at least on the surface.[5] Thus, for the most part, throughout this book we could treat report writing as a straightforward rational activity in a decision-making or administrative process.

These cross-cultural communication processes, however, are dependent on the cultures involved. This is illustrated by the American-French process as contrasted to the American-Japanese process. Like Americans, the French tend to be businesslike. However, for Americans, there may be an uncomfortable degree of formality and politeness involved in business communication with the French. In addition, certain social amenities are important. Usually you should not restrict yourself to technical and business considerations when interacting with French businesspersons. Efficiency and directness are not overriding values to the French.[6] For example, if you are a salesperson calling on business clients, you are likely to be able to call on only half as many clients in France as in the United States. That is, allow twice as much time (hours, days, weeks) as you would allow at home. As with the Japanese, long-term business relationships are important to the French, and any business communication is part of a long-term communication process. At the same time, unlike the Japanese, the French tend to keep their social lives and their business relationships separate; in France there usually is little formal socializing among business associates.

In organizational terms, cross-cultural communication is part of a negotiation process. Process considerations affect the purpose (as well as content) of any specific report or letter. You first need to understand the negotiation process appropriate for doing business with persons from a particular culture. You also need to understand the process of maintaining a business relationship with those persons. These processes will determine the relationships between oral and written communication, the interactions between social and business relationships, and the role of written communication in the decision-making process. Our example of doing business with the Japanese illustrates how these relationships, interactions, and roles affect written communication.

Perhaps the most important process considerations involve the roles of oral and written communication and the relationships between them. An important aspect of doing business with the Japanese, for example, is the reliance on oral communication and commitment. When Mr. Takashita told Mr. Leafgreen that Honshu would open accounts with his bank, Mr. Leafgreen did not have to have any further concern. In contrast to American practice, oral communication (and consensual decision making) are the norm within Japanese corporations.[7] The oral commitment is the primary commitment, and any subsequent formal contracts or written agreements are secondary. Our legalistic context is somewhat puzzling to the Japanese. Indeed, they might not be overly concerned with the "clauses" and technicalities of a written agreement; the broad understandings of an oral agreement with a Japanese business organization will create a binding and long-term agreement. The cultural relativity of this practice is illustrated by the contrast with Chinese management practices, where written communication is more important than oral communication (in part the result of central economic planning).[8]

The relationship between oral and written communication therefore is very important in cross-cultural communication with the Japanese, while it is not necessarily important between American firms. In this book we focus on written communication because it is central to the decision-making and administrative processes within American corporations. For us, oral communication and interpersonal communication can be especially important to discovery and investigation, to problem solving, and to planning processes, but usually the result is a written report, memorandum, or letter that affects the organization's activities. Frequently, written reports are not even preceded by oral or interpersonal communication in American corporations, but in a cross-cultural context oral communication can be necessary, important, and even primary.

❑ 14.2 ❑ The Design of Cross-Cultural Written Communication

When you write in a cross-cultural context you should take cultural considerations into account. No matter what the process implications, business and technical correspondence varies across cultures in terms of all of the following:

- format
- salutation and close
- openings
- arrangement
- types of information presented

When you adapt your correspondence, even in English (including memoranda and informal reports between organizations, which usually are in letter format[9]) to conform in some respects to the cultural conventions of your addressee, your communication becomes much more effective. Essentially, this is a matter of audience analysis across cultures.[10]

Cross-Cultural Comparison of Business Letter Organization and Content

The differences among business correspondence in different cultures can be illustrated by a comparison and contrast of some elements of business letters: formal salutation, prefatory paragraph, technical details, cost information, the close, and enclosures (Table 14.1). For comparative purposes, we summarize these elements in American, French, Mexican, and Japanese letters.

TABLE 14.1: Cultural Variations in Technical Business Letters

Technical Business Letter Elements	United States	France	Mexico	Japan
Formal Salutation	?	No	Yes	No
Prefatory Paragraph	No?	No	No?	Yes
Business Paragraph(s)	Yes	Yes	Yes	Yes
Technical and Business Details	Yes	No	Yes	No
Cost Information	Yes	No	No?	No
Polite Close Includes Business Follow-up	Yes	No	No	Yes?
Enclosures	Yes	Yes	No	Yes*

*not referenced

Formal Salutation The formality of the salutation is quite distinctive in some cultures, while the lack of any personal or formal address is as distinctive in others. In Arabic the salutation includes the proper title of the person addressed (and there are a number of different proper titles) as well as a respectful form of address before the title and additional honorific words ("venerable") after some titles. In Latin America, titles are also very important ("Dear Sales Manager Gonzales"). These formal openings contrast with the impersonal opening in French ("messieurs" without names) and Japanese ("gentlemen" without the names). American salutations ("Dear Mr. Jackson") are in between.

Prefatory Paragraph French letters lack a prefatory paragraph before the business paragraphs. (We use the term "prefatory" to distinguish this type of introductory paragraph from an introductory paragraph that introduces the technical or business concerns of the letter, as discussed in Chapter 3.) In contrast, Japanese letters traditionally have a prefatory paragraph with the appropriate (standardized) season greeting, "Although it is still very cold, the ice has started to melt," and an expression of wishes for the prosperity of the addressee's company. American and Mexican letters introduce the business issues in the first paragraph, but also often include familiar or personal phrases.

Technical and Business Details The body of American and Mexican letters includes relevant technical and business details. With Mexican letters these can be more extensive than with American letters because American letters use attachments, whereas Mexican letters tend not to use them. French and

Japanese business letters, however, omit most technical and business details. That is, they present general information without the specific details. An additional difference has to do with arrangement of information. In American letters, the arrangement is **conclusion ⟶ support**. In French letters, the arrangement often is **support ⟶ conclusion**. Mexican letters move from general-to-particular (from conclusion to detailed support and from most-important to least-important). Japanese letters move in the reverse order. Moreover, the Japanese prefer an indirect rather than explicitly persuasive approach tone; therefore, it is not quite appropriate to say that they arrange letters persuasively, if inductively.[11] American persuasion tends to be based upon an almost adversarial approach: "Here is my conclusion, here is why I hold it, here are some objections that might be raised to it, and here is why these objections are not sufficient to overturn the conclusion. QED." By contrast, Japanese persuasion is much more likely to be based upon a cooperative approach to joint problem-solving. "We hope that you agree . . . and we share your concern . . . and we look forward to . . ." If it is necessary to raise a negative point directly, the Japanese writer will soften it—make it almost apologetic—"Quite frankly, we are concerned that . . ." The "quite frankly" is a cue that the Japanese writer is being far more explicit than he or she prefers to be, but that for the sake of getting on with discussion feels it necessary to be explicit. To Western readers, this mild tone seems very gentle indeed. For readers who recognize the cue, however, it is a sign that close attention is needed to restore harmonious relationships.

Cost Information Interesting cultural differences arise with respect to inclusion of details on pricing and costs. French and Japanese business letters do not include the cost details. With the Japanese, this is because quality, service, and business relationships are more important than cost/effectiveness in business decisions. With the French, issues of quality are more important than issues of cost. In addition, costs routinely are standardized in France; that is, negotiations over prices are not the norm—if you prefer a product, you accept the price. Mexican letters include a number of cost-related details, but specific cost information is not included when it is a matter of negotiation, credit, quantity purchase, and so on. In the United States cost information is included because it is of primary importance in most business decisions.

Close The close and signature of business letters vary considerably by culture. American business letters have an informal business close ("We look forward to receiving your order") and a complimentary close ("Sincerely," "Sincerely yours," or perhaps "Very truly yours"). Mexican closes are similar, with perhaps a more personal than business tone. Japanese letters have a personalized close and tend to summarize the theme of the letter rather than to request an action. However, in contrast to their opening, French business letters have a very formal, polite nonbusiness close ("Nous vous prions d'agréer, messieurs, l'expression de nos sentiments les plus dévoués"—literally, "we ask you to accept, gentlemen, the expression of our most devoted regards") but no complimentary close.

Enclosures American, French, and Japanese business letters can include enclosures (the Japanese do not reference them). Mexican business letters usually do not include enclosures.

These variations reflect cultural differences as well as the role of the letter in the communication process. For example, in its arrangement and in its close the Japanese letter reflects the indirectness of Japanese negotiations. The Mexican approach omits enclosures because follow-up communication, invariably in the form of personal contact, is expected. These variations hold true for the technical sales letter as well as the business sales letter.

Cultural Variations of a Technical Business Letter

To illustrate these cultural differences, we show you four versions of the same letter. In Chapter 4, we used a technical sales letter about the "Clutch to disengage turbine starting pump" to illustrate the basic structure of technical reports in English. Using that same letter, we asked experts in French, Mexican, and Japanese business correspondence to rewrite it as it would have been written in those cultures.[12] They didn't just translate the letter. Rather, they took the information in it and recast it entirely, as if they had written the letter themselves in their own cultures and were completely unconstrained about what to put in, what to leave out, how to arrange the letter, and what the tone should be.

This letter is a fairly typical example of a straightforward American technical business letter between two corporations (Figure 14.2). The letter opens with a direct lead into the customer's need. It follows with a summary paragraph that presents the most important information for an American manager. A specific model of clutch is presented as the solution to the customer's need. The cost is specified. The primary reader, Mr. Adams, probably would only skim the rest of the letter, if he read it at all.

The following paragraphs present selected details: a cost breakdown, technical choices (referring to the enclosure), and installation requirements. It closes with an implicit request for an order. The tone throughout is informal. In sum, the letter is direct, has specific cost and technical information, and has an interpersonal tone that Americans establish in their business relationships (note the reference to "Ed Driscoll," not to "Mr. Driscoll" or "Mr. Edward Driscoll").

The French version (Figure 14.3) omits the technical and the cost details. The salutation and opening paragraph are impersonal. The opening introduces the writer's company and generally alludes to the customer's need without specifying it. The second paragraph specifies the clutch but refers the reader to the enclosure for technical details without selecting any to put in the letter. The third paragraph discusses the reliability of the product, a topic not discussed in the American letter (presumably it is assumed).

The fourth paragraph solicits further communication to discuss business details such as cost and delivery. That is, whereas the American letter follows a customer contact and assumes that the sale will be made on the basis of the letter, the French letter assumes a follow-up sales contact rather than an immediate closing of the business deal. The final paragraph is the formal close without the complimentary close ("Sincerely yours") of the American letter. In sum, the French letter is at a high level of generality and is rather impersonal or formal in tone.

The Mexican letter (Figure 14.4) is considerably different from the French letter. It opens with a personalized salutation. The opening paragraph establishes a personal tone (much more so than the American letter, let alone the French letter) and introduces the product with high praise. No reference is made to the customer's needs. The second paragraph (in Spanish the continuation of the first paragraph) stresses the benefits of the product and directly expresses the hope that the customer will adopt the vendor's part in his product.

As you can see from the bracketed comments, the letter would continue with detailed technical and business information—information entirely omitted from the French letter but selectively included in the American letter. Like the French letter, however, the Mexican letter omits pricing information. Also, it assumes a follow-up sales contact before any business is concluded. The complimentary close traditionally is similar to the French, although an Americanized complimentary close could be used. In sum, the Mexican letter has the technical content of the American letter but has a much more personal and solicitous tone. In important characteristics, it is the antithesis of the French letter.

The Japanese letter (Figure 14.5) has the indirectness that characterizes doing business with the Japanese.[13] It opens with the traditional season greeting, which is quite formulaic. Each month has its own characteristics (January is very cold, the middle of winter; March has the hint of spring, early spring; April has the cherry trees in blossom, spring; May has the green trees, the end of spring; June has the rains, the early summer; etc.). The opening paragraph also includes an expression of wishes or a blessing for the customer's prosperity. Finally, it includes a statement that alludes to the established relationship between the two companies. (Recall that in Japan this letter would be part of a communication process that transcends the particular business event that is the subject of the letter.)

Then the letter introduces the business of the letter. The phrase, "by the way," would be typical of the transition in a traditional Japanese business letter. (Others would be: "Well, this time . . ."; "Now, at the end of February. . .") This paragraph is a brief general summary of the business topic that would have been introduced in prior stages of the communication process. No technical or cost details are mentioned or even alluded to in a general way. The next paragraph very briefly (as the next-to-last paragraph in the

W.A. KRAFT CORP.

ENGINE POWER SYSTEMS-GEN SETS-INDUSTRIAL TRANSMISSION EQUIPMENT

45 SIXTH ROAD
P.O. BOX 2189
WOBURN, MA 01888-0389
617/938-9100
TLX 94-0391
FAX 617/933-7812

January 6, 19XX

Mr. Ronald Adams, Manager
Industrial Mfg. Corp.
P.O. Box 263
Centerville, Ohio

Attention: Mr. Andrews, Supervisor, Maintenance Department

Subject: Clutch to disengage turbine starting pump

Gentlemen:

Thank you for the courtesy extended our representative, Ed Driscoll, on his recent visit to Centerville, at which time he discussed with you your requirement of the clutch to be used to disengage the turbine starting pump from the main generator engine.

We understand you wish to mount the clutch on a shaft which would be turning 720 RPM, and that the duty of the turbine pump is 117 HP. Based on this, the torque requirement is 853 pound feet, and a clutch having 16.3 HP per 100 RPM is indicated. Model #CL-310 therefore would seem to fill the bill quite nicely, at a total cost of $369.60.

We refer you to Bulletin #326-B, enclosed. On Page 10 you will find the description, capacities, etc., of this clutch. You will also note, it is available in 2-1/4" and 2-7/16" bores. Accessories are described on Page 11. Two possible spider drive arrangements are suggested, as described in Figs. 2 and 3 on the back cover of the bulletin. The spiders and their dimensions are indicated on Pages 18 and 19.

We are pleased to quote as follows:

1. XA5752 Model CL-310 Clutch in standard bore of 2-1/4" or 2-7/16" $185.60
2. Part #3507 Throwout Yoke 2.25

Diesel and Gas Engines • Generating Sets for Prime and Standby Power • Cogeneration Systems
Clutches • PTO's • Universal Joints • Fluid Couplings • Torque Converters •
Marine Transmissions • Pumps • Variable Speed Drives

ENGINEERED SYSTEMS — SERVICE — SALES — PARTS

PRINTED ON RECYCLED PAPER

FIGURE 14.2 Technical Business Letter in English.

W.A. KRAFT CORP. 45 SIXTH ROAD PO BOX 2189 WOBURN, MA 01888-0389 617/938-9100 FAX 617/933-7812

Mr. Ronald Adams, Manager
Industrial Mfg. Corp.
Mr. Andrews, Maintenance Dept.
January 6, 19XX

3. Part #3039 Hand Lever..$6.70
4. Part #1144-E Operating Shaft 7.95
5. X8152-3-B Spider (maximum bore 2.439/2.441") 167.10

Please note, the X8152-3-B Spider quoted is the through shaft spider described on Page 18, which , you will note, has a hub outside diameter of 4-1/4". On this hub, you would mount the driving sheave. In short, the driving sheave would have to be of such construction that it could be bored out 4-1/4". If this were not possible, then the alternate arrangement (Fig. 2 on back cover) would be applicable, in which case we would recommend Spider Flange A-5932 at $78.65 in lieu of the above $167.10.

Thank you for our opportunity to quote. We look forward to being favored with your order.

Very truly yours,

W.A. KRAFT CORP.

Leo Fitzpatrick

Leo A. Fitzpatrick
Sales Engineer

LAF/jmf
ENCLOSURE

PRINTED ON RECYCLED PAPER

<u>lettre d'information à caractère technico-commercial</u>

à l'attention de la Direction technique

Objet: adoption d'embrayages CL-310 comme équipement de base pour un contrat de pompe à turbine.

Messieurs,

Nous sommes spécialisés dans la fabrication et la commercialisation d'embrayages secs et nous pensons que dans le cadre de vos relations avec la marine française notre matériel peut vous intéresser.

En effet, notre embrayage CL-310 peut être adopté pour votre équipement de Transmission motrice. Nous espèrons que le descriptif ci-joint vous éclairera à ce sujet, dans la perspective d'une consultation technique que nous sommes prêts à vous donner.

Permettez-nous, à ce stade, de souligner la très grande fiabilité de notre produit, ainsi que vous pouvez en rendre compte par les quelques témoignages d'entreprises que nous joignons à la présente lettre.

En outre, nous serions tout à fait disposés à discuter avec vous des conditions (prix, délais) bien adaptées à vos besoins. N'hesitez donc pas à nous solliciter.

Nous vous prions d'agréer, messieurs, l'expression de nos sentiments les plus dévoués.

PAUL DURAND
DIRECTOR DES VENTES
COMMERCIAL

PJ:2

FIGURE 14.3A Technical Business Letter in French.

Attention: Technical Manager

Subject: Adoption of clutch CL-310 as standard equipment for a turbine pump [contract].

Gentlemen,

We specialize in the manufacture and sale of dry clutches. We think that in the context of your business with the French navy our product could be of interest to you.

In sum, our clutch CL-310 can be adopted as standard equipment for your engine transmission. We hope that the enclosed information will explain its use for this purpose, as we are ready to explain in a technical meeting with you.

Permit us, for now, to emphasize the very good reliability of our product, as you can confirm by the comments from our customers that we attach to this letter.

Moreover, we are at your disposal to discuss arrangements (price, delivery) according to your needs. Don't hesitate to ask us about these.

We ask you to accept, gentlemen, the expression of our regards the most devoted.

PAUL DURAND
MANAGER OF COMMERCIAL
SALES

FIGURE 14.3B Technical Business Letter in French (translated).

Estimado Ing. Ramirez:

Aprovecho la presente para enviarle un afectuoso saludo, e informarle de nuestro nuevo y revolucionario producto, el clutch seco de la sene CL-310; el cual ha probado ser un producto bien acogido por los usuarios de bombas de turbina. Creemos que este producto puede ser muy productivo para su uso; ya que según estudios realizados por nosotros, el producto aumentará la vida a su turbina. Esperamos que ustedes usarán el nuevo clutch seco de CLUTCH INTENATA.

[INSERT SPECIFICS ABOUT MARKETABILITY, SERVICES, PRODUCT]*

Si requiere mayor información acerca de las especificaciones técnicas y estudios detallados realizados acerca del clutch, estamos a su disposición para enviarle nuestros folletos y para presentaciones detalladas con cualquiera de nuestros agentes autorizados. [INSERT DETAILS ABOUT METHOD OF CONTACT]*

Sin más por el momento quedamos de usted sus atentos y seguros servidores.

COMPANIA CLUTCH INTENATA

Por________________________
Arg. Luis M. Martínez
Gerente General de Ventas

*Notes to writer

FIGURE 14.4A Technical Business Letter in Spanish (Mexico).

Dear Engineer Ramirez:

I'm using this opportunity to send you an affectionate greeting and to tell you of our new, revolutionary product, the dry clutch CL-310, which has a proven record of acceptance among users of turbine pumps.

We believe that this product will be very good for your application; clearly, according to our actual studies, the product will increase the life of your turbine. We hope that you will use the new dry clutch from CLUTCH INTENATA.

[INSERT SPECIFICS ABOUT MARKETABILITY, SERVICE, PRODUCT]*

If you require more information about the technical specifications and actual detailed studies of the clutch, we are at your disposal to send you our brochures and to give you a detailed presentation from one of our authorized agents. [INSERT DETAILS ABOUT METHOD OF CONTACT]*

This being all for now, we are always ready to serve you.

INTENATA CLUTCH COMPANY

By________________________
Arg. Luis M. Martínez
General Sales Manager

*Notes to writer

FIGURE 14.4B Technical Business Letter in Spanish (Mexico) (translated).

Industrial Mfg 株式会社

ロナルド アダムス部長殿　　　　平成　　年10月6日

W. A. KRAFT 株式会社

販売部長レオ A. フィッパトリック

当社ディーゼルエンジンDH型に関する件

拝啓

秋冷の候、貴社益々ご清栄の段大慶に存じます。

先月、貴社訪問の際は、長時間お邪魔いたしました。ご厚情の程、厚く御礼申し上げます。

さて、その際申し上げました通り、当社では、主発電エンジンからタービン始動エンジンを切り離すクラッチを扱っておりますので、ご採用をご検討くださいますようお願い致しました。重ねてご検討の程お願い致します。販売後の保証は完全でございます。

なお、当社のクラッチについて、お尋ねしたいことがございましたなら、喜んでお答えいたしますので、ご遠慮なくご申し出ください。

先ずは、御礼かたがた状況のご連絡まで申し上げます。

敬具

FIGURE 14.5A Technical Business Letter in Japanese.

Subject: About our diesel engine DH model

Gentlemen:

It is in the midst of Autumn. We hope your company prospers more and more. When we visited your office last month, you spent time with us for many hours, and we express our deepest thanks for your patronage.

By the way, as we mentioned to you at the time, we are going into production with our diesel engine DH model and asked that you consider adopting it. We would like to ask you again to consider this model. We fully guarantee the engine after sale.

Furthermore, if you have any questions about this engine we shall be happy to answer them. So please don't hesitate to ask.

Thank you again very much, and we hasten to inform you of our situation in this letter.

Sincerely,

FIGURE 14.5B Technical Business Letter in Japanese (translated).

French letter) solicits follow-up communication, which undoubtedly will occur within an interpersonal communication situation. The letter closes with a formulaic phrase that closes the communication rather than moving the business deal to another stage (in contrast to the American letter, which closes with a direct reference to a sale or closure of the business deal).

The tone and approach of the Japanese letter reflect the process of doing business in Japan, as previously discussed. Written communication is subordinate and tangential rather than a stage in the communication process. The French and Mexican letters are stages in a communication process, sometimes preceded and usually followed by interpersonal communication stages: they move the process forward. The American letter is a decisive stage in the process, and can be primary.

Other types of technical documents have cultural aspects as well. Manuals and instructions are especially good examples because they can have cultural features that you do not encounter in American documents. For example, in a Japanese manual translated into English the graphics seem to be backward. Americans read graphics left-to-right, top-to-bottom. Japanese—whose books begin "at the back" and go "forward"—apparently read graphics differently than we do. For example, we found that Japanese engineers characteristically called a switch labeled with the terms, left to right, "AUTO" and "MAN," a MAN/AUTO switch, suggesting that they were seeing it right-to-left (Figure 14.6).[14] Additionally, there seem to be differences between organizational patterns and degree of detail in both large and small sections and units of American and Japanese manuals and other documents. These relate to the deductive versus inductive approach, explicit versus implicit logic, and directness and indirectness.[15]

The manual, however, introduces an important qualification in the design of cross-cultural communication. On the one hand, instructions assume a relationship between writer and reader that has cultural implications. On the other hand, depending on the context in which those instructions are used, the instructions have legal and contractual implications as well. Phrasing that might be culturally sensitive might not be legally acceptable. For example, here is a passage from the introduction to a Safety Manual (in English) of a Japanese corporation:

> To avoid unhappy accidents, safety rules are suggested in this SAFETY MANUAL. This SAFETY MANUAL does not include all of the safety rules that might prevent from serious injury to personal and costly damage to the machine.
>
> But at least the persons, who will be concerned with the product, must read and observe this SAFETY MANUAL.
>
> We are very pleased if more safety rules are added to be fit to each plant.
>
> —Safety makes everybody happy.—

As we explain in the next chapter on the legal implications of technical documents, this passage could have serious legal consequences if it appeared in a manual accompanying a product sold in the United States. Yet in Japanese

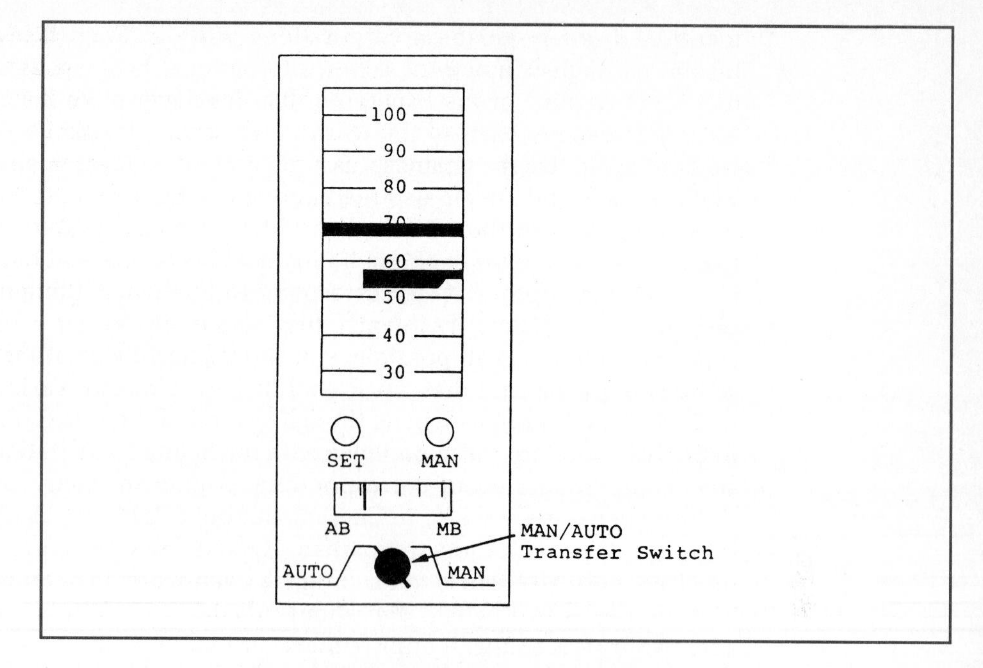

FIGURE 14.6 Graphic from Japanese Instruction Manual.

culture a manual requires this tone of politeness and suggestion; imperative commands are culturally uncomfortable for the Japanese. Likewise, cause-effect relationships are expressed by conditional constructions rather than causal constructions.[16]

The design of cross-cultural communication, therefore, is not quite a simple matter of adapting your content and style to the expectations of readers from another culture. Cross-cultural considerations are only a part of the communication context. Legal, business, and technical information constraints operate as well.

❑ 14.3 ❑ Writing in English Or in . . . ?

A final issue that arises in cross-cultural communication concerns the language in which you write. Do you write in English or in the readers' language? Actually, you have three choices: write in English, write in English and then translate into the foreign language, write in the foreign language. Your choice will depend on the communication situation and the culture with which you are interacting.

You can write in English when interacting with many cultures. As we have

mentioned, many Japanese corporations write in English, even internally. Insofar as English functions as an international business language—rather like a "new Latin" or the Esperanto that has never quite materialized—it is accepted in many parts of the world.[17] For example, the Polish with whom we have dealt, like the Japanese, assume we will conduct business in English. The Germans and Dutch also read and speak English, and most Americans communicate with them in English although some negotiate with the Germans in German. With the French, Spanish, Mexicans, and Italians, however, it is often not appropriate to correspond in English. Although many French people read English fairly fluently, they usually prefer not to use it in written communication. If at all possible, you should negotiate with them in French. (Interestingly, we find that our French friends almost invariably write to us in French, not in English, even though several of them are completely fluent in English.) Our limited experience with the Spanish and Italians is that many are uncomfortable about corresponding in English. Many Latin American businesspersons are fluent in English, but out of courtesy Americans should try to negotiate with them in Spanish. Persons from French-speaking nations in Africa normally refuse to use English even when they know it.

If you write in English, you should design your letters and reports and adapt your style so that it approximates or picks up elements of those of your cross-cultural counterpart (except for other considerations, such as legal, as we have discussed). For example, based on our discussion in the previous section, be formal and polite with the French, and write in generalities; be more personal with the Mexicans, and include details. Even when those cultures adopt American technical and business style in their communication, you should be culturally sensitive, as Mr. Leafgreen was in his letter (Figure 14.1).

When you write in a foreign language, of course, you need to adopt the approach and style of that language; you should not write your document in English and then translate it literally into the target language. In general a literal translation produces an unacceptable document. "Safety makes everybody happy" is a phrase literally translated from Japanese into English; it is neither Japanese nor English; it is what our Japanese friend and colleague, Professor Yoshiaki Shinoda, calls "Japlish" or "Mid-Pacific English," not-quite-English and not-quite-Japanese.

Professor Shinoda, on the faculty of the School of Commerce at Waseda University, Tokyo, and who travels regularly all over the world, puts it this way: "When I get off the plane in a new country, the moment my feet touch the ground, I accept the culture of that country and try to fit myself into it; when I return to Japan, the moment my feet touch the ground, I am once again Japanese." In like manner, when we live and work in a multinational context, accepting the cultures of other people means accepting the ways in which they expect writing to be used and to be presented. We are all citizens of our own cultures, but as professionals in multinational contexts, we are also "citizens" of each others' cultures.

NOTES

1. Herbert A. Simon (*Administrative Behavior*, 3rd ed., New York: Free, 1976, especially Chapter VIII, "Communication") observes that the formal system of communications in an organization is primarily written (157ff). In this book we are concerned with what Simon calls formal communication, as opposed to the informal communication channels in any organization (160ff). The point of this chapter is that in some cultures what Simon considers as informal communication channels are the more important channels; that is, they are the primary communication channels of formal decision-making processes.
2. In Japanese culture it is not appropriate to discuss business during social events, such as dinner or cocktails. This is in direct contrast to American culture, where business is often conducted during social events (e.g., the business lunch; drinks in the bar).
3. A useful introduction to Japanese business practices is Mark Zimmerman's book *How to Do Business with the Japanese* (New York: Random, 1985). Among the aspects we mention, he discusses being prepared (19, 29, 38), socializing (32), being patient (65), oral and written contracts (91ff), long-term relationships (100), and decision making (123–25). Another reference that is useful because it focuses on communication is William V. Ruch, *Corporate Communications: A Comparison of Japanese and American Practices* (Westport, CT: Quorum, 1984). For example, he discusses the importance of oral as opposed to written communication in Japanese business (88).
4. Mr. Leafgreen learned, for example, that Mr. Takashita had been on a panel for a University of Michigan East Asia Business Program on "Negotiating with the Japanese" and had been very well received by the business executives in the audience. During their first meeting he thus was able to compliment Mr. Takashita and therefore to impress Mr. Takashita with the fact that he (Mr. Leafgreen) had done his homework. Whereas an American manager might take this negatively ("buttering up"), a Japanese person will take it positively as a sign you are paying attention to details.
5. Audience analysis, of course, introduces other considerations. Certainly, what is termed "corporate culture" is an important consideration. However, in written communication the tone and content are explicitly technical and business.
6. The French consider the Dutch to be rude and impolite in business negotiations (P. van der Wijst, "Politeness of Requests in Dutch and French," *Abstracts*, 2nd International Eindhoven LSP Conference, Eindhoven University of Technology, Eindhoven, The Netherlands, 3–6 August 1988, 120.). They tend to think the Americans the same. In business dealings they have told us that Americans are "toujours pressés." Interestingly, our American business friends prefer to do business with the Dutch because the Dutch are direct, even blunt.
7. Christina Haas and Jeffrey L. Funk, " 'Shared Information': Some Observations of Communication in Japanese Technical Settings," *Technical Communication* 36:4 (Nov. 1989): 362–67. Also see Ruch, *Corporate Communications*, 88.
8. Herbert W. Hildebrandt, "A Chinese Managerial View of Business Communication," *Management Communication Quarterly* 2:2 (Nov. 1988): 217–34.
9. We illustrate cross-cultural communication with an example of an export sales letter. Our observations also apply to other types of documents in correspondence format, such as a request for information. That is, our discussion applies to working business and technical documents between organizations in different cultures. Other types of technical documents, such as manuals and reports, have cross-cultural implications as well. We introduce these briefly at the end of this section, again to sensitize you to the need to take cultural factors into account so that you can investigate further on your own if and when you are in such a communication situation.
10. Certain cultures such as the Japanese have

started to adopt American business style in their business correspondence in English. Japanese texts on writing business letters in English have adopted models that are similar to those in American business texts. This means that they are teaching the Japanese how to communicate with Americans just as we are suggesting to you how to communicate with the Japanese. Even when you encounter "American-style" correspondence, however, you should take cultural considerations into account in your correspondence (as Mr. Leafgreen did, Figure 14.1).

11. The Japanese preference for being indirect and nonargumentative, a distinct contrast to American business practices, is discussed in Zimmerman and Ruch. A rhetorical analysis well worth reading, especially in conjunction with Zimmerman, is Hiroe Kobayashi, *Rhetorical Patterns in English and Japanese* (Ann Arbor, MI: University Microfilms International, 1984), who discusses the organizational patterns of Japanese documents. If you are interested in Japanese rhetoric, you should become familiar with the concept of "kishotenketsu," discussed in Kobayashi and Mackin (see note 15) and other sources they reference. The traditional Japanese business letter as we discuss it embodies kishotenketsu structure (see note 13).
12. We would like to thank Mr. Michel Gabrielli, M.B.A., Lecturer in Business French, The University of Michigan; Dr. June O. Ferrill, Business Writing Consultant, Houston, Texas; and Professor Yoshiaki Shinoda of the School of Commerce, Waseda University, Tokyo for their assistance. We have drawn on materials prepared by Dr. Ferrill for our comments about business correspondence in Latin America.
13. Professor Shinoda says the pattern of a traditional Japanese business letter follows the ki-sho-ten-ketsu pattern. In his version, the first paragraph would be the ki (introduction) and sho (development) parts, the second and third paragraphs would be the ten (turn) part, and the final paragraph would be the ketsu (conclusion) part. However, there seems to be some disagreement over kishotenketsu (see Kobayashi, 26–33, and Mackin, note 15, 348).
14. Dwight W. Stevenson, "Audience Analysis Across Cultures," *Journal of Technical Writing and Communication* 13:4 (1983): 319–30 (ref. and figure, 328–29).
15. John Mackin, "Surmounting the Barrier Between Japanese and English Technical Documents," *Technical Communication* 36:4 (Nov. 1989): 346–51.
16. Mackin, 350.
17. With certain documents, such as operating instructions for a product that is distributed throughout the world, companies often write only in English. They adopt a neutral, simple English; some use what is called a "controlled language." (Although we do not discuss controlled languages here, you might want to know that it is an approach that some corporations have adopted. For example, "Caterpillar English," is used in service and repair manuals written for Caterpillar equipment used throughout the world. Many other companies have developed similar restricted vocabulary languages in which there is a simple set of allowable nouns and verbs. Other features include the fact that there are no synonyms in controlled vocabulary languages. A brief but relevant introduction to the concept is Kathy L. Strong, "Kodak International Service Language," *Technical Communication* 30:2, (Second Quarter 1983): 20–22. Notice, however, that with some products the instructions come in many languages—English, French, German, Italian, and Spanish. This is especially the case with Common Market products; therefore, American corporations dealing with Common Market economies probably should adopt this approach. With Canada, of course, it is a legal requirement that instructions be in both English and French.

CHAPTER 15

Accepting Legal and Ethical Responsibilities

In this book we have been talking about what is required for effective communication. Throughout, we have assumed that the concept of "effectiveness" must encompass both law and morality. That is, effective technical and professional documents are by definition ethical acts. They must take into account both the legal and ethical responsibilities of writers to readers just as they acknowledge the communication purposes of their authors and the information needs of their readers.

Unfortunately, the clear evidence of numerous lawsuits and newspaper stories suggests that some writers in professional contexts are either unaware of or unresponsive to their legal and ethical responsibilities.

We believe that legal and ethical issues are indeed important parts of the technical writer's consideration and therefore we would now like to address these topics explicitly.

❑ 15.1 ❑ Accepting Legal Responsibilities

When writers do not meet their legal responsibilities, the writer and the writer's company are put in jeopardy of significant loss in legal actions based upon product liability, malpractice, personnel issues, environmental issues, and issues of corporate ownership of intellectual properties. There is not space here to itemize all of those responsibilities, but at least we can identify six basic precepts for you to bear in mind.

There Is No Such Thing as a Confidential Document

First, in your professional life, you might suppose that you can write "off the record" documents, notes, or preliminary versions of documents that don't count as legally significant because they are not ultimately distributed, or because they are designated as "confidential" and are released only under restricted distribution. However, under discovery rules allowed by the *Federal Rules of Civil Procedure*, the plaintiff—the party claiming to be harmed in a legal action—may subpoena the documents of a defendant company, using them in court to prove the harm and to lay a basis for an award for damages. In effect this means that there is no such thing as a confidential document.

Admittedly, a plaintiff must show the relevance of documents before a court will order their release by a defendant company. Yet judges in general agree that "relevance" should be interpreted liberally. This fact is well documented in an article in the *American Journal of Trial Advocacy*,[1] which quotes several judges to the effect that, "This court will not be a party to a suffocation of the truth. The party seeking discovery can better determine the usefulness of an ambivalent document to the presentation of its case and this court will not foreclose that opportunity."[2] The same article quotes another judge, whose view has been cited with approval by still other judges, to the effect that, "In the ordinary case, the judge will order the question to be answered, subject to a [later] finding of irrelevancy."[3]

This concept of broad discovery of documents has important implications for writers of documents in corporate settings. You must always write with the understanding that your words, no matter how privileged you may think them to be, will potentially be made public. Indeed, you must think of your writing as likely to become public in the adversarial context of court, where an opposing attorney will look for every ambiguity and every negative word or sentence to prove his or her client's case. In the warning words of one corporate attorney to his company's technical staff, "You people are today writing reports that *will be used* in product liability suits ten years from now. We don't know what the issues in those cases will be, but we do know that the documents will be used. Be aware that someone will be scrutinizing every word so that he can use the document in court to suggest that [our company] is more cost conscious than safety conscious or that we are indifferent to safety." This statement, by the way, was made in our presence during a consulting visit to one company. The attorney went on to say that, "We are proud of our company and its products. We hope they never cause harm, but, if they do, we can live with our responsibilities as a manufacturer. What we can't live with is losing a case because of sloppy documentation. Just remember, your choice of words is deemed to be the company position. Write accordingly."

A further important implication of the concept of discovery rules is that the term "documents" is not restricted to those that have been typed, signed, and distributed. The term is all-inclusive, encompassing every possible type of record or form of information, as we saw in Chapter 1.[4] Nor are the rights of plaintiffs limited to subpoena of documents. Under the Federal Rules of Evidence, plaintiffs may also obtain permission to inspect property, require physical and mental examinations, or obtain depositions.

Depositions, especially, are important for writers to understand. They are testimony taken under oath and in the presence of both attorneys for use in court. Usually taken in an attorney's office, they are tape recorded and later typed. Frequently, they are several hundred pages in length and involve detailed testimony focussed upon the background, context, and meaning of documents to be used in evidence. For example, you recall the Three Mile Island case. In that case, Mr. Kelly, Mr. Dunn, Mr. Hallman, and Mr. Karrasch testified at length in depositions, which then were used by the Government's attorneys to question them in the Committee's public hearings from which we quoted. In other words, it is not just the documents themselves that will be used as evidence, but what the writer says about those documents under oath.

Writers of technical documents, then, must realize that no document "doesn't count." No document in any of the stages of its development is "off the record." You must write with the recognition that you may have to stand up under oath to defend every writing choice you have made in a document; further, you must write with the recognition that the company, and even you personally, are financially accountable for what you have written. When you write, you become the spokesperson for your organization, and your writing must always begin with your acceptance of that responsibility.

Be forewarned: your documents are never privileged in any of the stages of their history.

You Cannot Put a Time Limit on the Significance of Your Writing

A second point to consider is the time limit of legal significance.

As a student, you have become accustomed to thinking of a document as a temporary thing: today's homework, this week's paper, this term's project report. As a professional, however, you must adopt a longer view. Your reports continue to have legal significance over many years, long after you have forgotten them and perhaps even long after you have left a company for another position.

For example, in a personnel case with which we are familiar, all of the descriptions of rules and procedures in use in an organization over a period of five years were relevant to testimony in a court case. So also were all of the personnel evaluations and records of an individual employee. All of these documents, dozens of them written at different times by different individuals and groups, were entered into evidence and were the basis for testimony in depositions and in court. All of them became exhibits in the case. Many of them were the basis for specific citations in the judge's written opinion in the case, and in a very real sense, the documents—even individual sentences within them—*were* the case.

Similarly, many liability cases hinge upon huge files of documents written years ago, often before current entry-level employees were out of grade school or before they were born. Examples are cases involving a 16-year-old luxury car, a 20-year-old ladder, a 9-year-old industrial crane, and a 50-year-old insulation product.[5]

From the perspective of companies, the way to deal with this long-term use of documents is to implement policies covering document preparation and retention. Under such policies, the appropriate content and design of documents is spelled out, and documents are routinely marked for 5, 10, and 20-year retention, or they are marked to be held indefinitely. At the appropriate times, many companies then review the documents individually, often using both legal staff and professional staff to decide if a document should be discarded or if its retention period should be extended.

In articles addressing corporate policies, one finds advice such as this:

> Records should be kept from the time the product idea first came up. What materials and processes were considered, and why were they discarded? Through how many development and design stages did the product go? Why? Records of corrective action and specification changes are very important pieces of information. All customer complaints and the action taken should be kept, along with service calls and repair information. Finally, advertising labels, tags, booklets, manuals, contracts with customers and vendors, and other product-oriented correspondence should be filed. The company should have a data collection system that will allow for any product to be data-profiled from its birth to the present time, which will show all documents to support the contention that the company has been attempting to build a safe and reliable product.[6]

From the perspective of the individual writer, the way to deal with this long-term use of documents is twofold. First, you must write documents in such a way that they are able to remain useful over an extended lifetime. Second, you should discipline yourself to follow document-retention policies and to keep rigorous records.

Writing documents for long-term use means they must be accurate, complete, and capable of standing upon their own despite inevitable fading of recollection on the part of writer and readers, and despite the disappearance from the company of key personnel whose recollections might fill in gaps about why work was done, how it was done, or what the outcomes were. Disciplining yourself to retain accurate records means knowing and observing the company's policies and keeping your personal files in rigorous order. An article entitled "Product Liability: Some Ounces of Prevention"[7] recommends the following:

> It begins with a log book in which is recorded, chronologically, every event, activity, and conference which has bearing on the design. . . . It is important that every drawing, sketch, and note be signed, dated, and preserved in the design project files as well as being recorded in the project log book. . . . The design engineer should discipline himself to organize his files so that nothing of his early thinking on a design project, or the constructive comments of others relative to it is ever discarded. If it is worth sketching or discussing, it is worth filing.

Naturally, corporate attorneys and managers often view the paper record as both a blessing and a curse. On the one hand, well written records can provide the best defense in legal actions. On the other hand, poorly written records can provide the plaintiff with exactly the ammunition needed to win against the company. It is the latter possibility that unfortunately prompts some corporate attorneys and managers to adopt an ethically and legally questionable response. Fearing sabotage by their own written record, they counsel "talk it, don't write it" or "just report the numbers, don't interpret them; let the supervisor interpret them." (We regret to say that we are quoting one corporate attorney here.[8])

Yet such a position assumes that a written record will be poorly done and that responsible professional work can occur without candid and open writing. We reject these positions. Indeed, we applaud the reaction of a supervisor to the preceding attorney's statement when he said, "He [the attorney] doesn't understand that conclusions are professional judgments supported by evidence." And we applaud the reaction of another who said:

> I have always been proud of this company, its people, and its products—until today, when I heard this lawyer telling us to talk rather than to write, or to write in a way which conceals our professional conclusions. I feel insulted and embarrassed. My God, the company pays me to draw professional conclusions. They aren't just personal opinions; they are careful, objective interpretations of empirical evidence. I am personally prepared to stand on the evidence and to take the consequences; I think the company should be too.

We believe that an individual or company should be able to stand on the written record. Such a stance requires responsible professional work, and it requires both good policies and good writing. But these things can be done, and they are supported by the majority of those who have written about corporate record-keeping. For example, one writer advises, "Document all transactions, activities, and anything else related to your products . . . and keep all writing related to a particular product together in a file. This helps build a defense in the event of a suit. . . . You shouldn't have any doubts in your mind about a product. If you do, get rid of it."[9] Note: the writer didn't say get rid of the writing; he said get rid of the product.

Your Writing Is Your Organization's Promise

A third point to consider is that writing creates understandings between people.

Not all professionals deal directly with clients and customers; however, many do, and most have some input within their organizations to those who do. For example, an individual pathologist in the immunology laboratory of a large hospital, or one of the technologists or technicians on the pathologist's staff, might not have direct contact with patients. Hidden away down in the lab, they probably never actually see patients. Yet reports written by the pathologist and the laboratory staff are directly used by physicians and nurses in health care delivery. This direct interaction with the public establishes understandings for which all professionals are accountable. A carelessly handled laboratory specimen might result in an inaccurate report. That in turn might result in inaccurate diagnosis and inaccurate treatment. And that in turn might result in a malpractice suit against a doctor, or nurse, or hospital.

Similarly, we might read in the product literature about the specific benefits and performances of a product, and on the basis of that decide to buy the product. Having done so, we are then able to hold the company accountable for what it has promised. If the product doesn't perform as promised, we are justified in bringing suit against the company.

Professionals who do deal with the public, and even those whose input to that public is indirect, must understand this notion that they are accountable for their professional work and that promises made are binding. Especially, individuals who deal with customers must understand the notion that warranties may be implied as well as express.[10] That is, a warranty is a promise made by a seller to a buyer, a promise that is part of the basis for a bargain. Some such warranties are "express warranties," which explicitly give assurance of performance and enumerate responsibilities of the manufacturer and rights of the buyer. (The contents of these are rigorously spelled out in recommendations by the Federal Trade Commission.) Others are "implied warranties," understandings created when a seller has a reason to know the end use of goods and the buyer relies upon the seller's skill and judgment in making a buying decision.

Express warranties are normally prepared with great care by qualified people within companies whose responsibility it is to know what must be said. Implied warranties, however, are often created by people at many levels within a company and by people entirely unaware of the legal implications of their words. A telephone call, a business letter, a handwritten note, a passage quoted from one document and inserted in another—all of these can create understandings that a company would be obliged by law to meet.

Again, we can view this responsibility of the seller both from the perspective of the company and from the perspective of the individual writer. For the company, candor, honesty, and accuracy in all communications with clients are essential. For the individual writer, the important thing is to recognize that your written and spoken words to or about clients are never trivial or nonbinding. Even a hastily written, informal note can create a binding position for a company, just as surely as if a team of attorneys had prepared an express warranty and published it in Gothic type. All professionals who deal with the public or who produce materials that others use to deal with the public should behave responsibly, accurately, and openly. Anything less reflects unprofessional behavior and ineffective communication.

You Have a Responsibility to Instruct Adequately

A fourth consideration is appropriate for those who write instructions.

Of course, not all professionals write instructions, but those who do are obliged to write instructions that a jury would define as "adequate" to promote safety. These instructions might cover the construction or assembly of a product, its use, its maintenance, or avoiding hazards associated with it.

A common situation is that professionals might not write these instructions themselves, but do write materials that are used by other people who actually write the instructions. Again the responsibilities can be distributed back through an organization beyond the persons who actually write the final versions of instructions given to users. (Example: A software developer writes a program. The program specifications and instructions written by the programmer are then handed to a technical writer who prepares a user's manual that is sold along with the product. Unfortunately, in many instances whole parts of the specifications and preliminary documentation get incorporated in the final product word-for-word, despite the fact that the programmer never really considered that possibility. In our consulting work, we routinely see this situation, and almost invariably it produces ineffective instructions.)

Professionals involved in writing instructions or contributing to their writing must understand their responsibilities in the eyes of the law. Further, they must understand that the concept of "adequacy" is ultimately a matter for a jury to decide in an individual case. Hence there are no absolutely certain guidelines, and one expert, Patricia Wright, argues that guidelines are inherently limited because they are necessarily overly general.[11] Nonetheless, a jury's expectations of instructions clearly requires the exercise of many skills by writers. In particular, Wright suggests that a writer must employ six kinds of skills to develop adequate instructions.

Task Analysis The writer must perform task analysis to determine precisely what the user must be able to do. Naturally, these actions include the obvious tasks of operating the equipment or performing the maintenance. However, these tasks also involve reading information, finding information, remembering information, choosing alternative paths to "migrate" through information. Moreover, readers seldom read directly through instructions before attempting to follow them. Instead, they dip into the instructions as needed, frequently in random order. Thus, adequate instructions must provide information-handling helps to readers, allowing them to find things easily, to group related things, to see and understand information quickly and easily. The tasks of a reader are not solely performance tasks; there are tasks of perception and cognition as well as tasks of action.

Language Control Writers of instructions must write succinctly, clearly, and in ways that permit one meaning only. To a considerable degree that means following—really following—the familiar advice that research articles and textbooks present: keeping terms simple, accurate, and consistent. Using uncomplicated and direct syntax. Using graphics effectively. Dividing information into manageable chunks. Avoiding sending people off to search for explanations elsewhere in a text. And so on. Simple as it sounds, adequate instructions are well written instructions in which the writer has examined every language choice, has considered alternatives, and has selected the best in terms of the real needs of readers. No level of language choice is exempt.

Control of Graphics and Typography Writers must use cues of structure such as white space, headings, subheadings, numbers, bullets, type font and type size variations, and other format devices to show how the units of a text relate to one another and how they are distinct. These format cues must be complemented by graphics such as ghost drawings, sketches, exploded drawings, photographs, tables, graphs, and schematic drawings. Moreover, writers must understand how graphics and text interact, how they reinforce one another, and how at times they may cause confusion by being combined ineffectively. In short, to write effective instructions, the writer must be a graphic specialist as well as a word specialist.

Interpretation of Research Writers of effective instructions need to be aware of techniques that have been demonstrated to improve reading performance. This includes knowing about the relative legibility of different type sizes and fonts; about the greater effectiveness of "scenario" headings than of topic headings; about the effectiveness of topic sentences in initial positions; and about many of the other specific points that we have attempted to summarize elsewhere in this book. Also, writers need to keep up with what is being learned in new research and know how to find what new information is available.

Pilot Testing Writers of instructions need to know how to test draft instructions. They must know how to use both holistic and analytical evaluation techniques to assure that the design of a document is consistent with their analysis of audience needs. Further, they must know how to design, implement, and interpret performance tests in which documentation is evaluated by being actually used. They must be able to test for missing information; they must be able to verify the accuracy of assumptions of prior knowledge; and they must be able to use testing to determine what readers like or don't like as well as what works and what doesn't work. In short, writing effective instructions requires testing and evaluation as well as planning, design, and implementation.

Managing the Production Process Finally, writers of instructions need to know how to manage the production process that transforms a document idea into a thing that readers can hold in their hands. Paper must be chosen, colors selected, and pages physically designed. With every choice, some trade-offs may have to be made. For example, what the printer or the marketing department thinks "looks good" may not actually be good in terms of information transfer. What the budget will allow may not make instructions effective. And the minor changes requested by colleagues and editors may not enhance the clarity and accessibility of instructions. The writer's task is to balance between these conflicting demands and to choose wisely on behalf of the readers.

To summarize, Wright's advice is that to write effective instructions requires more than just writing. Many of the faults of instructions are traceable to parts of a process that either precede or follow what many writers would regard as "writing." The important point here is that the requirement for "adequacy" in instructions is inherently a subjective thing. A jury decides what constitutes adequacy. As you can see from Wright's analysis, however, the range of factors potentially considered by a jury may be much wider than you first suppose.

Of course, there is much to read about writing adequate instructions. If that kind of writing is your responsibility, it is in your interest to do further research in the area. For example, John A. Conrads, formerly General Service Manager of Deere and Company, describes several factors that Deere and Company have found helpful.[12] These include placing greater emphasis upon graphics; reduction of the number of words used; use of simple language to aid in understanding and to increase the translatability of a text; integration of illustration and text into modules; use of sufficient white space to improve reading and aid understanding; automation of production steps to speed last-minute changes; and provision of a method for rapid updating of all manuals, particularly in regard to safety-related issues. You would do well to examine the entire issue of the journal that contains Conrads' advice. As a special issue of *IEEE Transactions on Professional Communication* devoted to "Legal and Ethical Aspects of Technical Communication," it has advice that goes considerably beyond that which we are able to present briefly here.

Just how difficult it is to write "adequate" instructions may be illustrated by a famous case, *Thomas v. Kaiser Agricultural Chemicals.*[13] In this case, a farmer attempting to fill a fertilizer applicator suffered serious eye injury when liquid nitrogen sprayed into his face. At trial he testified that he had read the warning labels cautioning users to bleed off the air pressure in the tank before refilling it, but he had attempted to fill the applicator without verifying that the air pressure gauge read "zero," although the instructions indicated he should. Also established at trial was the fact that Thomas had used similar equipment for sixteen years and that he had once used an identical piece of equipment. Further, it was established that Thomas had read the instructions. Yet despite all of these points, the court held that the facts did not warrant the conclusion that he had been made to adequately appreciate the danger of unintentionally opening the relief valve. This ruling was subsequently upheld in the Illinois Supreme Court.

Skeptics might say Thomas's case just proves that it is impossible to write adequate instructions; pragmatists might argue that at least we must try.

You Have a Legal Responsibility to Warn

Just as some of you will not write instructions, some of you will not write warnings. However, many of you will write them or contribute to their writing. Therefore, it is important for you to know general guidelines for adequate warnings.

The basic requirements for adequate warnings were suggested by one of our colleagues at Rice University, Professor Linda Driskill. She said, "It is not enough to say 'keep off the grass.'"[14] You must go further to say that there are poisonous snakes in the grass, that the bite of one of them can kill you, and that an alternative route around the grass exists (and show how to use it). Further, she says, you must make the warning in a commanding way at precisely the point at which someone might be expected to need it; and you must make it in such a way that poor readers, non-native English speakers, children, people with poor eyesight, and nighttime visitors can understand it. Finally, you must assure that the warning does not become illegible over time by standing out on the lawn through the damaging effects of rain, wind, and sun.

Although whimsical, Professor Driskill's description of requirements clearly conveys the heavy burden that writers of warnings must bear. A more conventional and more detailed explanation is in Christopher Velotta's article, "Safety Labels: What to Put in Them, How to Write Them, and Where to Place Them."[15] Velotta summarizes the "Specifications for Product Safety Sign and Labels," a draft standard of the American National Standards Institute.[16] That advice, coupled with other information about warnings, suggests the following list of appropriate considerations.

FIGURE 15.1 International Hazard Alert Signal. (*Note:* the signal word should be changed in light of the nature of the hazard. The terms, in decreasing order of seriousness, are "danger," "warning," and "caution.")

Hazard Alert Use a graphic symbol to attract the attention of readers. This symbol, an increasingly familiar international symbol, consists of a triangle with an exclamation point inside (and signal word below). (See Figure 15.1)

Signal Word Announce the nature and severity of a hazard by using appropriate signal words. "Danger" indicates an imminent hazard that, if not avoided, will result in death or serious injury. "Warning" indicates a potentially hazardous situation that, if not avoided, could result in death or serious injury. "Caution" indicates a hazardous situation that, if not avoided, may result in minor or moderate injury. ("Caution" may also be used to indicate other unsafe practices or risks of property damage.) The term "Notice" can be used to warn of potential property damage, but it is inappropriate for any warning of potential human injury.

Color Code If you use color signals, use red to indicate danger, orange to indicate warning, and yellow to indicate caution. Red background requires white lettering; orange and yellow background require black lettering.

Pictograph Signals Use drawings that graphically represent the effects of violating safe practices. These drawings can be created to fit an individual hazard, or they may also be more general, such as those drawings created by the Westinghouse Corporation in its publication, *Danger, Warning, Caution.*[17] (See Figure 15.2)

Nature of the Hazard and Consequences of Ignoring a Warning Along with a graphic showing the nature of the hazard, also use words to clarify the specific nature of the risks that users face in using a product. Explain in words what will happen if they ignore the warning. For example, is the risk one of back strain, eye injury, burns, radioactive exposure, or what?

Avoiding the Hazard Explain what steps users must take to avoid being harmed. What protections are available?

FIGURE 15.2 Graphic Warnings Developed by Westinghouse Electric Corporation in *Danger, Warning, Caution: Product Safety Label Handbook*, 1981.

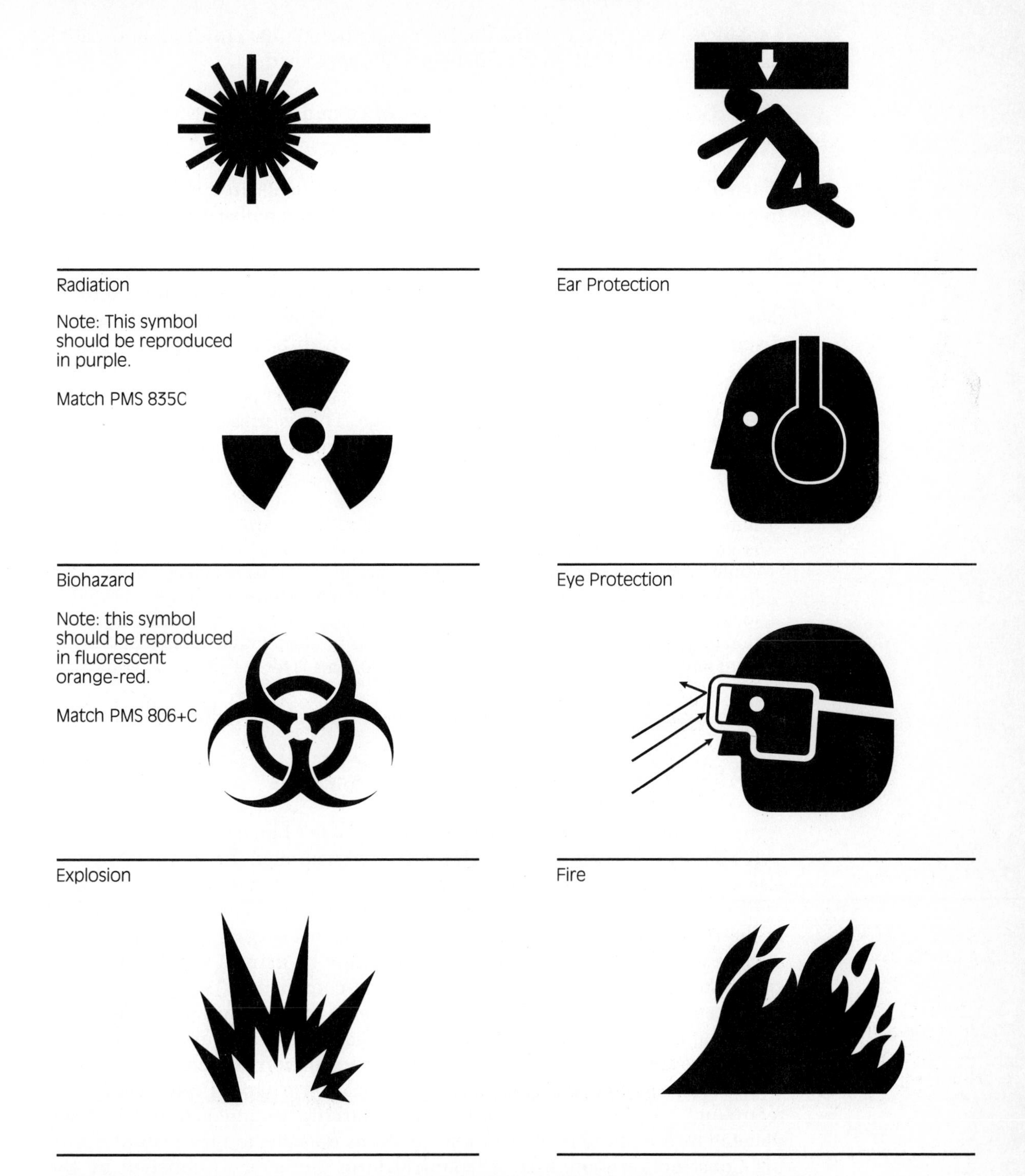

FIGURE 15.2 (continued) Graphic Warnings Developed by Westinghouse Electric Corporation in *Danger, Warning, Caution: Product Safety Label Handbook*, 1981.

Language Keep it short and simple, remembering that children, non-native English speakers, and poor readers must be protected.

Tone Command attention. Use active voice and imperative verbs to command your readers to pay attention.

Placement Place warnings both on the product itself (in durable form) and in any instructions accompanying the product. Remember that the key concept for adequate placement of warnings is that they must appear "at the point of use."

This list of recommendations can be daunting. In fact, it might look so daunting that again you may be tempted to throw up your hands and to dismiss the responsibility to warn as impossibly difficult. Reasonable people ought to be able to understand reasonable things. Right? You can't "idiot proof" the world. Right? Well, in fact the "reasonable person test" is not the one that must be met in adequate warnings. This is presented persuasively in the following statement from a court opinion in the case of *Berkebile* v. *Brantly Helicopter Corporation:*[18]

> It must be emphasized that the test of the necessity of warnings or instructions is not to be governed by the reasonable man standard. In the strict liability context we reject standards based upon what the 'reasonable consumer' could be expected to know, or what the 'reasonable' manufacturer could be expected to 'foresee' about the consumers who use his product. . . . Rather, the sole question here is whether the seller accompanied his product with sufficient instructions and warnings as to make his product safe. This is for the jury to determine.

Let the buyer beware? No, let the manufacturer and the writer beware.

You Have a Legal Responsibility to Be Accurate

A final legal responsibility to consider is the need for accuracy.

It should go without saying that effective technical writing is above all else accurate. Yet there are two aspects of accuracy that we want to stress.

First, there is the accuracy of terminology. As an illustration, an article[19] in the *Journal of Products Liability* points out that several manufacturers of a home insulation product are risking litigation by using a variety of imprecise terms to describe their product. It is "non-combustible" says one; another describes it as "completely non-burning"; still another says it is "fire resistant"; and one talks about "fire control." In fact, all of these terms, when used to describe cellulose and ureaformaldehyde insulation (which will burn) are both inaccurate and dangerous. The accurate term would have been "fire retardant." The choice of that term is not an arbitrary choice. Rather, it is one dictated by the *Standard Definitions of Terms Relating to Fire Tests of Building Construction Materials*, a technical term dictionary published by the

American Society for Testing and Materials. The choice of any other term, while perhaps tempting from a marketing perspective, is legally dangerous.

Not all terms need to be checked in a technical dictionary. Still, in the rush of writing a technical document it is common to be casual about terms, and it is even more frequent to carelessly use several "synonyms" that readers do not know or perceive to be synonymous. The result is dangerous ambiguity.

A second aspect of accuracy is completeness. Accuracy in what is said may be a characteristic of a particular document. However, that same document may fail in its accuracy by what it does *not* say. This particular failure is extremely difficult to spot in normal editing and proofreading procedures, although it might show up glaringly under cross-examination or in deposition. Only the writer can know if he or she has said all that needs to be said. For example, were alternatives considered? You could recommend a choice without mentioning that three alternatives were first considered and rejected. Similarly, you could mention that alternatives were considered but omit the adequate explanation of why they were rejected. In either case, the document would omit important information.

Our position again is that a company or an individual writer should be confident that the record will support the claim of accuracy. There can be no guarantee that your legal defense will be successful, but at least the outcome will not be determined by careless mistakes or by what is left out. The comment of one writer provides an appropriate reminder: "It is a brutal fact that the casual, stupid, or mistaken comment of an employee or agent can result in liability for a product otherwise admirably safe."[20]

To review, then, people who write technical and professional documents need to observe six basic precepts to protect their readers, their companies, and themselves in the legal context. These are as follows:

1. Don't assume confidentiality.
2. Write so that your documents can continue to function effectively for years.
3. Don't promise what your company cannot deliver.
4. Write adequate instructions.
5. Warn your readers of dangers.
6. Be accurate and complete.

As we hope is clear, these precepts go considerably beyond merely meeting specifications. In that respect, they are at once matters of law and matters of ethics. Yet we would now like to turn more explicitly to a discussion of the ethical responsibilities of technical communicators.

❑ 15.2 ❑ Accepting Ethical Responsibilities

Here are two approaches for your consideration. The first is to consider codes of conduct. The second is to look for the ethical implications of our discussion of the communication situation and the nature of your audiences.

Codes of Conduct

Useful codes of conduct for writers are of three types: those relating to the professions, those relating to writers, and those relating to companies.

First, think about the communication aspects of your job in terms of the basic ethical standards applying more broadly to your profession. That is, writing and speaking acts are part of what being a professional means; they are operational aspects of a professional role, not ancillary or unrelated acts. Indeed, precisely because writing and speaking are operational acts—things we actually do in the practice of our professions—they are often the arena in which ethical standards are put to the test. In that light, the standards for writing behavior can be found, at least implicitly, in the codes of behavior that we apply to any of our professions. Law, medicine, engineering, business, government service, education, and a host of other specific professional areas all have codes of behavior that establish high standards and rigorous expectations.[21]

Second, consider codes of conduct for writers. You could argue that few codes of behavior from the professions deal very explicitly with writing and that, therefore, a more specific code of behavior for writers would be appropriate. If so, there is such a code. In fact, there are several, including a 1988 "Code for Communicators" published by the Society for Technical Communication.[22] In part, it states the following:

> . . . Because I recognize that the quality of my services directly affects how well ideas are understood, I am committed to excellence in performance and the highest standards of ethical behavior.
>
> I value the worth of the ideas I am transmitting and the cost of developing and communicating those ideas. I also value the time and effort spent by those who read or see or hear my communication.
>
> I therefore recognize my responsibility to communicate technical information truthfully, clearly, and economically.
>
> My commitment to professional excellence and ethical behavior means that I will
>
> - Use language and visuals with precision.
> - Prefer simple, direct expression of ideas.

- Satisfy the audience's need for information, not my own need for self-expression.
- Hold myself responsible for how well my audience understands my message.
- Respect the work of colleagues, knowing that a communication problem may have more than one solution.
- Strive continually to improve my professional competence.
- Promote a climate that encourages the exercise of professional judgment and that attracts talented individuals to careers in technical communication.

Other similar codes for writers have been published by the Society of Technical Writers and Editors, The International Council for Technical Communication, The International Association of Business Communicators, the American Medical Writers Association, and The Association of Professional Writing Consultants. Most are reproduced in the source cited in note 22.

Third, consider codes of conduct for specific companies. You could argue that these are better sources of information for writers within organizations than the codes of the professions or of professional technical writers simply because they can be more contextually specific.

Again, there are numerous such codes. In fact, three-quarters of the 1500 largest corporations in the United States have such codes of conduct,[23] although it is clear that many of the employees of these companies are unaware of the existence of the codes, uninterested in them, or perhaps even use the codes unethically to set the "outside" limits for behavior.[24]

Without question, codes of each of these three types could be useful as heuristics by which to think about your responsibilities. The act of writing such a code provides a good opportunity to think about ethical issues, and reading it may cause you to think about what you might otherwise ignore. In addition, such codes are useful to introduce newcomers to the expectations of their professions, and they provide measures by which to assess both our own conduct and the conduct of others.

The difficulty is not that such codes are unavailable or difficult to understand. Rather, it is that these codes are treated as unimportant platitudes by many professionals. Or, worse still, they are treated as devices that delineate the extreme boundaries of acceptable behavior, in much the same way that speed limit signs seem to delineate the lowest speed at which motorists will drive. Just as many drivers feel that they can safely "cheat" a few miles per hour above the posted speed, some professionals appear to feel that if there isn't an explicit sanction against it, any behavior is acceptable. Hardly a reassuring ethical stance.

We believe that codes of conduct can be useful to help writers think about ethical issues. At least they can trigger helpful discussion and reflection on ethical issues if we are willing to look for them and take time to pay attention to them.

An Approach Based Upon the Communication Situation

Codes of ethics may provide useful ways to think about what is expected of you as a professional and as a writer. However, we would like to recommend an additional way of thinking about your ethical responsibilities. We see this approach as embedded in our general approach to audience, something that we have been talking about all along, but not explicitly in ethical terms.

Remember our discussion of egocentric organization charts and procedures for identifying, classifying, and characterizing audiences. In that discussion there are six points that have ethical implications. They are as follows:

Your Name Is on It First, we defined your role as a writer as being at the center of the egocentric organization chart. We said you are not an unimportant cog in a vast machine. You are the center, the person responsible for your professional work and for the effectiveness of the communication work resulting from it. You are accountable. In our minds that makes you ethically accountable as well as technically accountable. After all, your name is on the work. Although you speak in a company and for a company, ultimately the work is yours.

If you cannot sign your work with a clear conscience, don't do it. Don't do the work. Don't write the document. And don't sign what you don't believe in. That is an exceedingly tough ethical standard to live by, one that may cause you anguish. Yet if you actually consider the implications of signing your name to something, we believe that the well-being of all of us will be well served.

You Write to Individual Human Beings Second, we emphasized that as a writer—even in an organization—your responsibilities are to individuals. Our system for analysis of audience begins with an identification of individuals who will read what you write. To each of them, you owe respect, honesty, clarity, and completeness in your communication. Even if you cannot see them or identify them individually, when you write you are never dealing with an amorphous mass of readers whose rights, needs, and expectations are so manifold that they cannot be defined or become unimportant. Rather, you are dealing with single human beings to whom you should give the same dignity and the same rights you would expect them to give you. Don't let the size of your audience, or its diversity, or its distance from you blind you to the rights and needs of those individuals.

You Are Responsible for Understanding the Secondary Effects of Your Writing Third, we emphasized that effective communication requires that you address the needs of your primary, immediate, and secondary audiences. You need to give these persons the information and advice they need to act and make decisions. You must also communicate to other people who will be affected by your writing and by the actions and decisions it produces. In a narrow sense that may mean fully dealing with the practical needs of just those people on a distribution list. In a wider ethical context, understanding the consequences of the actions and decisions that your writing might cause will lead you to think well beyond practical effects. Again, that is a tough ethical standard, but one that may merit your consideration many times during your professional life.

You Write Within Communities Fourth, we stressed that communications travel within and between the concentric spheres of an egocentric organization chart. These are communities (in organizational literature called "cultures"), each nested within a larger community. The closest community is your own organizational unit: the radiology lab, the accounting department, the project team, or the programming department. To the members of your own community, certainly, you owe the respect, dignity, rights, duties, and responsibility that go with membership in any community. But your membership in that small community is not your full citizenship. Your citizenship extends to partner groups within the organization; to the organization itself; and beyond to customers, clients, vendors, and regulatory agencies. Beyond all of these, you are part of a profession, an industry, a city, a state, a country, a world. The more narrowly you define the spatial boundaries of your citizenship, the more you risk ignoring the ethical consequences of your writing.

Your Writing Has a Life of Its Own Fifth, in our discussion of audience we stressed the notion that documents, once written, have a life of their own that may extend well beyond the time and place of their first use. A personnel report or a recommendation may be sent out today; it might still be in the file and have effects decades from now. Recognizing that fact and accepting the responsibility it entails may cause you to write much more carefully than if you see your writing as existing in a narrow window of time.

Your Documents Are Intended to Produce Results Finally, in our discussion of audience, we emphasized the concept that addressing an audience is part of a process, a purposeful interaction intended to allow an organization to function. That is, you write to people to give them information and advice; to help them to understand; to help them to decide; to help them to act; and to help them to deal with the consequences of information, decisions, and actions. This purposeful interaction carries with it the implication that your writing is not finished when it rolls out of your laser printer or is deposited in your audiences' mailboxes. It needs to produce results before it is really finished. Follow-up is necessary. Did people get your document? Did they understand it? Did it work? Were the necessary actions and decisions made? Is more information or advice necessary? (Again, think about the Three Mile Island example.) In instrumental terms, the document's value lies in what it produces. In ethical terms, its value cannot be understood in isolation from its consequences.

Conclusion

Meeting the demands of the law and of your conscience will not always be easy in your professional life. You'll live through many arguments, perhaps some legal actions, and not a few bouts of anxiety. We've lived through all of these, and will probably continue to do so. No checklist of legal requirements and no list of ethical precepts will ward them off entirely. The best we can hope is that our discussion here may cause you to think seriously about your writing as both a legally significant and ethical act. If you do that, and if you practice the writing suggestions of the rest of the book, you'll have our respect for your effort and our confidence that your outcomes will be successful.

NOTES

1. Foy R. Devine, "Discovery in Product Liability Cases," *American Journal of Trial Advocacy* 6 (Spring 1983): 241-55.
2. Judge Hemphill, *Duplan Corporation v. Deering Milliken, Inc.*, 397 F. Supp. 1146, 1188 (D.C.S.C. 1974).
3. Judge Kauffman, "Judicial Control Over Discovery," 28 F.R.D. 111, 119 (1961).
4. See John E. Durst, Jr., "Use of Interrogatories in Products Liability Cases in New York," *Trial Lawyer's Quarterly* 15: 2 (1983): 71. Durst lists types of information covered by the term "documents and writings," as it is defined by the State of New York. The list is shown in Chapter 1.
5. In *Hasson v. Ford Motor Company*, 1982, the plaintiff won a $9.2 million award when the brakes on a 16-year-old luxury car failed, seriously injuring a child. The *New York Times* of January 28, 1977 reports the award of $4000 to a plaintiff whose 20-year-old wooden ladder broke, causing injury to the owner. In *California Edison v. Harnischefeger Corporation*, 1981, a plaintiff company was awarded damages when a crane cable broke, dropping a heavy turbine rotor. And in one of the most discussed cases of all, *Basheda v. Johns Manville Corporation*, 1983, plaintiffs were awarded damages as a result of injuries sustained when they installed asbestos insulation in the 1930s and 1940s, years before the carcinogenic nature of asbestos was understood. The Supreme Court of New Jersey held that although the risk was unknown at the time, "the hazard would be deemed knowable if a scientist could have formed that belief by applying research or performing tests that were available at the time."
6. J. Mihalsky and R. Jacobs, "Peeping Through the Keyhole: Why Materials-handling Manufacturers are Scared of You," *Automation* 22:11 (Nov. 1975): 79.
7. R. Weber, "Product Liability: Some Ounces of Prevention," *Automotive Engineering* 92:2 (Sept. 1984): 54–55.
8. For obvious reasons, we must conceal the identity of the company here. However, we were present when these statements were made by the attorney and when the reactions quoted later were presented by those in his audience. We prefer the views of the audience to those of the speaker, and we are happy to point out that the views of this one attorney are at odds with the advice to be found in numerous publications aimed at corporations.
9. Ron Kolgraf, "How to Avoid a Product Liability Suit," *Industrial Distribution*, Feb. 1978, 39.
10. For a good discussion of warranties, see John McGill, "Warranties: What Are They and How They Can be Used in Risk Control," *Journal of Products Liability* 2 (1978): 105–15. Also see Michael Weinberber, "The Law of Implied Warranty: What Hath *Martin* Wrought?" *Trial Lawyer's Quarterly*, 13: 2 (1980): 21–34.

11. Patricia Wright, "The Instructions Clearly State. . . . Can't People Read?" *Applied Ergonomics*, Sept. 1981, 131.
12. John A. Conrads, " 'Illustruction': Increasing Safety and Reducing Liability Exposure," *IEEE Transactions on Professional Communication*, 30: 3 (Sep. 1987): 134.
13. Peter P. Hill, Jr., "*Caveat Venditor*: Failure to Heed Instructions Is Not a Defense to Illinois Products Liability Actions—Thomas v. Kaiser Agricultural Chemicals," *DePaul Law Review* 30 (Winter 1981): 477–98.
14. Linda Driskill, in an unpublished lecture entitled "Trends in Liability Affecting Technical Writers." Presented at the University of Michigan, 1980.
15. Christopher Velotta, "Safety Labels: What to Put in Them, How to Write Them, and Where to Place Them," *IEEE Transactions on Professional Communication* 30: 3 (Sep. 1987): 121–26.
16. "Specifications for Product Safety Sign and Label," ANSI Draft Z535.4, August 1984.
17. *Danger, Warning, Caution: Product Safety Label Handbook*, Westinghouse Electric Corporation, 1981. As the book points out, other information relative to labeling is required by the Department of Transportation, the Environmental Protection Agency, the Occupational Safety and Health Administration, the Nuclear Regulatory Commission, and the Consumer Product Safety Commission, as well as by other state and local regulatory bodies.
18. *Berkebile v. Brantly Helicopter Corporation*, 337, A2d 893, quoted by Richard E. Shandell, "Failure to Warn—A Drug Manufacturer's Liability," *Trial Lawyer's Quarterly* 14:1 (1982): 5.
19. David Orloff, "Potential Fire Hazards and the Role of Product Literature," *Journal of Products Liability* 3 (1979): 29–41.
20. Lamar Tooze, "How to Help Your Members Avoid (and Defend) Product Liability Suits," *Association Management*, Jan. 1978, 40.
21. See Rena A. Gorlin, ed., *Codes of Professional Responsibility* (Washington: Bureau of National Affairs, 1986). This book presents twenty-two codes of behavior from a spectrum of different professions.
22. These codes are presented in R. John Brockmann and Fern Rook, eds., *Technical Communication and Ethics* (Washington, DC: Society for Technical Communication, 1989): 93–97.
23. This fact was pointed out by Mark Clayton in a 1986 article in the *Christian Science Monitor* and subsequently quoted in Brockmann and Rook, *Technical Communication and Ethics*, 91.
24. Andred Abbott, "Professional Ethics," *American Journal of Sociology* 88 (Mar. 1983): 855–85. Cited in Arthur E. Walzer, "Professional Ethics, Codes of Conduct, and the Society for Technical Communication," *Technical Communication and Ethics*, 104–05.

CHAPTER 16

Using Electronic Communication Tools

Scattered throughout this book are enough references to computer technology for you to have recognized that we are computer users ourselves and strong proponents of the use of computers for writing. In a word, we are hooked on computers, can't imagine how we ever managed to write without them, and hope you feel the same way. Yet we need to talk a little bit more explicitly here about electronic communication tools, for indeed this technology presents opportunities, challenges, and problems for writers. Accordingly, in this chapter we would like to discuss five aspects of electronic communication:

- Using the Computer to Create Documents
- Accessing Information On-Line
- Publishing Documents With "Desktop" Publishing Tools
- Using E-Mail
- Learning New Tools: Hypertext and Multimedia Documents

In the process, we would like to emphasize both the opportunities and the problems that they present, all of which you will face during your life as a professional, and some of which you are probably facing even now.

❑ 16.1 ❑
Using the Computer to Create Documents

We start with three assumptions.

First, we assume that you are a computer user, that you have done a reasonable amount of writing and editing on a computer, and that by choice or by necessity you will continue to write that way for much of your professional life no matter what we say here.

If we are wrong about this basic assumption, our very strong advice is that you get out and learn to use a computer for purposes of writing. Also, if you don't know how to type well, learn that too. Computer illiterates and hunt-and-peck typists are significantly handicapped in the workplace already; we just can't imagine how they will get along in the workplace of the future. We have seen many instances of these handicaps among mid-level and older professionals, but they didn't have a chance to learn about computers in school, as you probably did. They should learn. You must learn.

Second, we assume that you like the experience of writing on a computer. Most people do. Both the popular folklore and the early literature on computer-assisted writing are full of the testimonials of true believers. "It has tripled my output." "It has cut my writing time in half." "I could never spell correctly, but with a computer I never have to worry about spelling errors." "My typing speed has doubled since I started using a computer." And so on. Overstated claims certainly, but evidence that most people who write on a computer do like it.

Third, we assume that you have not read much of the literature on computer-assisted writing and that you are therefore largely unaware of what research is telling us about it now that most of us have had access to computers for writing for a little more than ten years. It is on this latter point that we would like to focus briefly here.

What does the research literature tell us?

On the positive side, despite the glowing testimonials of users, the list of points generally agreed upon among researchers is surprisingly short. In fact, there are just two main points that everyone seems to agree upon.[1] First, students who write at the computer compose longer texts with fewer mechanical errors than when they compose by hand. Second, students enjoy writing more on a computer than when they write with pencil and paper.

Other positive findings have been less well established, although they may include the following points: Students writing at computers appear less affected than pencil writers by an inability to get started; they are more willing to engage in peer editing of their texts; and they are more able to distance themselves psychologically from what they have written than are students who compose on paper.[2] When they use computer planning aides, computer-using students seem to outperform pencil planners by coming up with more to write about, although the quality of their actual writing doesn't seem to differ from that of pencil writers.[3] Students writing at computers seem to be willing to experiment more than pencil writers, although the amount of experience required to reach this stage is substantial.[4] And writers who use computer editing tools such as spelling checkers outperform human editors in detecting surface-level errors such as typos and misspellings, but only human editors can detect substantive errors.[5]

On the negative side, the research suggests that, although current writing tools do offer helpful features, their strengths are related to what computers can do, not what writers need done. In particular, current writing tools are strong at dealing with the bottom-level concerns of writers, such as spelling, typography, layout, and word choice. Unfortunately, they are weak with the more important, top-level concerns such as planning and large-scale structure—the very concerns that we have emphasized throughout this text. Although computer writers do appear to revise more than their pencil-writing colleagues, they tend to limit their revisions to tinkering with surface details. Even though computer writers could move text around easily and change it significantly in the process of planning, writing, and editing, they don't often do that. They may like to write at the computer, but the products of computer writers are not actually better than those produced by pencil writers.[6] In fact, contrary to the popular wisdom, research suggests that there is no apparent relationship between productivity and computer writing.[7] Finally, the promise of true writer's aid software, which exploits the knowledge gained from research on comprehension and combines it with Artificial Intelligence techniques for natural language processing, is still a gleam in the eye of software designers.[8]

All of these negative points might lead you to think that we would be dubious about computer writing. In fact, quite the opposite.

Writing at a computer *is* easier and more fun than writing with a pencil. It does encourage people to write more. It does seem to help some writers to get started and to experiment more frequently. And it does produce better surface editing and a more handsome product than writing by hand. All of these are real benefits, even if they fall short of the theoretical benefits that we would like to achieve. More importantly, we believe that if you are aware of the limitations and seductions of computer writing, you can work to improve the way you do it. It isn't the computer that is at fault; it is the writer.

Start with the recognition that planning is known to produce better writing, whether it is done on paper or on a computer.[9] Writing an outline does pay off, and it is extremely easy to do on a computer, as we have indicated. But you have to do the prewriting planning and the outlining for it to work. You shouldn't just start writing, as researchers have discovered to be the composing habit of many writers who use computers.

In one significant study of experienced writers, researchers found that members of the study group fell into two general categories: those who just started to write in order to find out what they had to say, and those who carefully planned out their writing by creating outlines and by thinking about what they had to say before they wrote.[10] Both strategies can work, but for the first sort of writers, the computer was not a very satisfactory tool. They were uncomfortable with the fact that they could not see much of their text as they wrote. They did little significant revision, instead limiting their editing to surface-level tinkering. They felt compelled to save text that perhaps they would have thrown away had it been in messy handwriting. They didn't like the fact that they couldn't mark up the text with arrows and circles and handwritten notes. Conversely, the writers who developed good plans before beginning to write tended to like the computer because they had solved most of their problems of content and structure before they began to write. Thus they could focus on implementing the plan and dealing with the lower-level issues in their texts.

The difference here is not between computer writers and pencil writers; it is between the writing strategies of individuals. (We admit that we assume that professionals in an organizational context often know their purpose in writing—what they have to say—before they start to write.)

We strongly recommend the planned approach as the better approach for you to take. It is the approach that we have presented throughout this book. Plan the document before you write it. Deal with the large structural and content issues before you deal with the surface-level issues. Think and work both hierarchically and recursively. Edit contextually before you edit for surface details.

Granted, you may have the urge to just begin to write as a way of thinking about your topic, but recognize that you should throw away the "thinking drafts" before you really get down to business. When you write with a computer you need to plan first, then to print out the text frequently so that you can read over what you have written, looking at the whole, not just at the parts. You may need to make major revisions in content and structure as well as in surface details. And certainly you need to put the document through numerous drafts until you really are satisfied with it. You shouldn't stop with an early draft just because it looks good as it comes out of your laser printer.

For writers whose habit it is to work from a plan, we suggest that the computer is a natural planning tool for you. Use it as a device for outlining. But don't neglect the same advice that we gave to the writers who use a first draft as a way of thinking about their topics. Print frequently, read what you have written, edit recursively, and revise on both structural and surface levels until the draft is polished on all levels.

A further point about editing on a computer. One study found three types of editing strategies employed by computer writers.[11] There were linear editors, who finished the draft quickly, went back to the beginning, and edited linearly through the text. There were intermittent editors, who wrote fairly long passages of text and then occasionally stopped to revise episodically. And finally, there were the recursive editors, who wrote shorter units of text than either of the two other types of revisers. Unlike the other two types of revisers, the recursive editors tended to focus their revisions at the point at which text was being generated. That is, they read what they had just written, then they experimented as they added new text, often playing around with three or four different versions of a new text element before going on. They scrolled back and reread very frequently, but they generally didn't go back more than a few lines.

Although the linear editing style characterized both weak writers and strong writers, the linear editing approach was generally characteristic of the weaker writers. When they had poor plans, these writers wrote disconnected, poorly thought-out texts. Then they failed to correct the substantive problems with their linear editing. Admittedly some linear editors worked from strong plans and thus were able to produce good texts even though they edited linearly. Recursive editors, on the other hand, tended to be the stronger writers. To quote the study, "They evidenced a highly embedded form of revising, typically making changes at a variety of levels . . . during revision episodes."[12] Again, there were also poor writers among the recursive editors, but in general writers who had good plans and who experimented on a number of levels as they wrote tended to be more successful than those who either raced through the first draft or worked through it intermittently.

A final observation about editing. Even surface-level editing is something for you to do, not something for the computer to do for you. Text analysis programs can count your words and alert you to overly long sentences. However, just remember that you could reverse the order of the words in a short sentence, and the computer would still think it looked fine. Similarly, you can use a spelling checker, but don't expect it to know the difference between "their" and "there." Also remember that the computer can alert you to the presence of a passive verb, but it can't begin to tell you whether the passive verb is appropriate or not.

To summarize, then, the good thing about writing at a computer is that it is easy. However, we can also say that the bad thing about writing at a computer is that it is easy. If you choose to use a computer's power to assist you at all stages of the writing process, it can be a wonderful tool. Yet, if you let it, the computer can also cause you to concentrate on surface details and to confuse neatly typed text with well-written text. Our advice is merely to use the computer to enhance all stages of your writing process; don't limit your use of the computer's power to those lower-level things it does best, working with surface-level details.

❑ 16.2 ❑ Accessing Information On-Line

Part of the process of communication is accessing information. We used to do that by going to the library or to the file drawer. Now our electronic tools are making it possible for us to access information by getting it "on-line." In companies, for example, it is now common for computer records of all reports and documents published in-house to be available on-line. Sometimes, these records only contain abstracts, and thus require the second step of actually obtaining a "hard copy" of the document in the traditional way. For example, the library system of our university has an on-line catalog that we can access from any terminal or workstation on campus. This allows us to find any of the material in our collection, yet we still must go to the library shelves themselves to obtain the actual documents. Increasingly, however, electronic records consist of full-text files, which allow users to display the documents themselves on the screen. And as computers become more powerful and more available, we can expect this trend to continue.

Records available to us "on-line" are certainly not limited to those available within our own organizations. Because of computer networking and commercially available database information tools, we literally have the information resources of the world at our fingertips if we know how to use on-line information searching tools, or at least know that they are available for professional librarians to use for us.[13]

To illustrate, suppose you are interested in finding books published on an esoteric subject unlikely to be represented in your local library. If you went to the library to browse the catalog, and didn't find books on your topic, you still wouldn't know whether such books existed. Yet if you accessed File 145 of Dialog Information Services, one of the basic on-line information services, you would be able to browse electronically in *Books in Print*. You could search for books on your topic without knowing their titles or the names of their authors (although you could search in these traditional ways too). Instead, you could search under "keywords." Using a list of terms that might provide clues to your topic, you could combine and recombine terms until you narrowed down on your very specific topic. Having done so, you could pull up a list of every trade book published on your topic in the United States. Still within Dialog, you could examine any of its 250 files, which include all disciplines and encompass such diverse information as dissertations, laws and regulations, and business information.

Or consider another example. You are interested in specifications governing the design of a piece of equipment, let's say headlights used in automobiles. You have a fresh idea for a new design, but you need to know what specifications such a design would have to meet.

You could go to the library and consult the various specifications published by such organizations as the American National Standards Institute, the International Standards Organization, or the International Electrotechnical Commission. Yet, you could just as easily again use Dialog, File 113, to find voluntary and federal specifications and standards applicable to your topic of interest.

Having found the applicable standards, you might then want to search through patent information to see whether your idea for an improved headlight design has been patented by someone else. If so, you could look again in Dialog, Files 23–25, to find such information.

You could also search through numerous databases that index literature, such as that published in 30,000 or more of the world's scientific and technical journals. If you happened to read Japanese, you could search for articles published only in Japanese. Or you could specify English journals only. Or you could limit your search to publications that have appeared within the past six months. Or you could specify that you are interested only in articles written on your topic in Russian, after 1989. There is almost no limit to your ability to find books, journals, standards, patents, technical reports, conference proceedings, vendor catalogs, or software if you know how to use on-line information searching tools, or at least if you know that they are available to library personnel whom you can ask for help in your search.

To be an expert in this type of searching, you must have specialized training. Yet you can be an amateur and still work effectively with professional librarians to use information searching tools. You will have to plan carefully, for as you can see you could easily be deluged with information unless you narrowed your search sufficiently to get at those few documents you really need. Such planning can be done by carefully defining the types of information you want and by specifying the words and phrases that might be keys to that information. Then you must do an exploratory search to see if your search strategy is a good one. (A trained librarian can help you with this search.) If your search plan is a good one, you can get instant results. If not, you can redefine, modify, enlarge, or narrow your search on-line. You can see the abstracts of documents of interest on the screen. You can then either print the screen yourself or order the abstracts to be printed "off-line," at night when the rates are low, and they will be mailed to you the next day. You can also take advantage of the on-line mail service that will allow on-line ordering of full-text of most materials found in the search. You can even store your search so that you will be able to consult the database in the future to see what has been added since your last search.

For writers, there are two important implications of these search tools:

First, if you don't use this powerful resource, you are not taking advantage of the electronic tools available. Knowing about it and still not using it is even worse than not knowing at all, but we often find in our consulting work that professionals don't know about such tools. For example, we helped one company find information in a short session of on-line searching, information that the project leader had no idea could be found on-line. His project had been going on for some months when he happened to mention his need for particular information about vibration in automobile steering gears. A 15-minute on-line search found the information he needed. We doubt he would have found it otherwise.

Second, if such information sources are to be the repositories for information that you produce—and they will—you must design the information so that it can be found by people interested in it. This means that titles, abstracts, and keywords are important retrieval tools. If you do not plan and write these well, the information that you produce can disappear into electronic "files," never to be seen again.

Since the first commercially available database came on-line in 1969, the information access of professionals has undergone a revolution, and that revolution continues with such developments as full-text storage of documents. Learning to access these tools and learning adequately to provide information to these tools is, in our view, part of the challenge and opportunity of being a communicator in an electronic age.

❑ 16.3 ❑ Publishing Documents with "Desktop" Publishing Tools

Desktop publishing is a recent technology that depends upon the availability of hardware and software first released in 1985. In simple terms these tools allow you to do the following:

- Produce a document design that looks as if it had been typeset, either creating your own design or "plugging into" prepared formats for many basic documents such as newsletters, flyers, proposals, and other types of standard business communications.
- Combine text and graphic materials that have been created in separate word-processing and graphic application programs or by optical scanning of photographs, documents, or drawings.
- Manipulate text and pictures either collectively or separately on the screen, either editing, resizing, trimming, repositioning, or adding and deleting both text and pictures.
- Print your document either as a camera-ready master copy for printing purposes or as multiple copies suitable for distribution.

Already, this new technology has had a huge impact upon corporate communication by allowing professionals in all fields to do many of the things that typesetters and graphic artists have traditionally done. At their own desks, professionals can prepare camera-ready copy for reports, contracts, forms, newsletters, flyers, brochures, articles, warranty cards, manuals, specifications, and a host of other corporate documents. Less than $10,000 will buy a very powerful desktop publishing setup consisting of computer, software, laser printer, and optical scanner. Yet with these tools you can produce documents that are elaborately designed and beautifully printed, and that allow you to combine text and graphics without depending upon the help of typesetters, graphic designers, or even secretaries.

These new tools do not replace the tools that traditional publishers have provided; however, they do bring three important advantages that professionals in all fields have been quick to recognize and that we would encourage you to exploit as well.

First, the new tools can reduce the time required to produce polished documents by allowing you to eliminate the step of sending them out for typesetting or printing. You can run any of your word-processing files through desktop publishing software in a matter of minutes—or at most a few hours—to provide a polished level of document appearance that you have not been able to get previously without "going outside."

Second, desktop publishing tools can reduce the cost of document production both by eliminating typesetting and by permitting small-volume publication, which print shops cannot provide inexpensively. At a typical print shop, the first 1000 copies are expensive; after that, the cost of additional copies falls rapidly. However, on a laser printer in your own office you can produce a first copy that is no more expensive than any other, and all copies are cheap. In practical terms, then, you can afford to produce a "published" document today as a single copy. Until 1985, you really had to need a large number of copies before you could justify the time and cost associated with sending a document out for printing.

Third, and from our point of view most important, desktop publishing tools can give you direct control of your document design, permitting a full range of choices from very simple "template" formats to very elaborate, customized designs for special products. Moreover, you can exert this direct control without having to depend upon the messy and time-consuming traditional tools of scissors, glue stick, tape, "white-out," and ruler. On your own screen, you can "paste up" documents combining text and pictures, you can play with the design until you are really satisfied that it best presents your message and addresses your audiences, and you can do these things in a way that is clean, accurate, and fast.

Of course, taking advantage of desktop publishing requires spending some money on hardware and software. It requires time to learn. And most importantly, it requires learning about such unfamiliar things as layout principles and printing concepts and terms such as "kerning" or "leading" (LED-ING), terms that traditionally meant something to only a small community of typesetting professionals. Yet if you have a good working knowledge of desktop publishing, the technology opens up a wide new range of possibilities for producing technical and professional documents in-house. These tools may have opened up a wide new range of necessities as well as possibilities. That is, technology defines possibilities, but it also defines expectations. Thus, just as the word-processing revolution taught readers to expect a good deal more than the standard typewriter could provide a decade ago, now desktop publishing is teaching readers to expect more in document design than traditional word processors provide.

These tools may become necessary for you to learn in another sense. They are quickly becoming embedded features of office software. Word-processing programs of 5 years ago could provide only one font and permit only single-column text that looked much like typewritten text. Now most of these programs permit desktop publishing features such as multiple fonts and type sizes, and they permit multicolumn text as well as a variety of graphic devices such as shaded boxes, arrows, and elaborate line designs. Other programs make it possible for you to design 35 mm slides or transparency masters for use in oral presentations. Similarly, spreadsheet programs of 5 years ago could provide only cells into which numbers could be put for purposes of automatic calculation; now those same programs allow graphic representation of numerical data in a wide range of graphic forms such as line graphs, pie diagrams, and bar charts. In other words, desktop publishing tools used to be separate tools; now they are rapidly becoming embedded features of standard office programs. You can ignore these tools, but they are there to be used, and your readers increasingly will expect you to use them. Even if there were no advantages in terms of cost, time, and quality, that may make learning these skills necessary for you.

In some of our own classes, we have required students to learn to use desktop publishing tools. For example, in a software documentation course (where students need to be able to create complex page designs), we required our students to pick 3 pages at random from popular magazines and then to duplicate the design of those pages using basic desktop publishing tools. We found that in an average of 7 hours all of the students in the class, all of whom already knew how to use word-processing tools, could reproduce reasonably complex page designs that included artwork and scanned photographs.

A few hours in the computer lab will suffice to get you started. If you are really going to exploit the power of desktop publishing, however, you will need more information of a specialized sort. In fact, let us sound a cautionary note here: a limited knowledge of these tools can quickly give evidence of the old adage that it takes a computer to really mess things up. To provide a starting point, we recommend a good bibliography of materials on the topic in *Desktop Publishing: A Critical Bibliography*, by R. John Brockmann.[14] This bibliography lists books, magazines, and videotapes on the topic, and in each category it identifies the "best" sources of information available.

We believe that desktop publishing is an important part of the quiet revolution in corporate communication that has been made possible by our electronic tools. Whereas writing and publishing were once separate acts, now they are rapidly becoming combined in the communication responsibilities of professionals. Writing isn't just writing any more; it is publishing as well.

❑ 16.4 ❑ Using E-Mail

For years, "hackers" have been exchanging notes and permitting each other access to files on mainframe computers. Within the past decade, however, electronic mail, "E-Mail" as it is usually called, has become the daily tool of professionals and ordinary people in all parts of our society. Through local area networks, colleagues electronically exchange information within the workplace. For example, since approximately 1980, most of the faculty and administration on our campus have become users of our electronic mail system and often can be reached more quickly by that means than by telephone, visit, or internal mail. Some faculty and administrators still don't use E-mail, but most do, and the business of operating a large university has become an electronic business within the decade. A message can be sent to an individual or a group, and when the recipients sign on to their computers, they are notified that a message is waiting for them. They can then display the full text of the message, read it, and respond directly in a matter of seconds. Students on our campus are users too, and many students send and receive messages from us daily, having learned that a good way to get quick "turn-around" with faculty is to approach them electronically. They also talk back and forth to each other that way.

Nor are we limited to local exchange of information. Through gateways to external networks, we send and receive electronic messages to colleagues and business associates across the country and around the world. We even use E-mail to keep in touch with family members scattered around the country.

Again, this is a quiet revolution, one with many obviously positive features and one that creates new types of problems and new types of "documents" for communicators.

E-mail combines features of telephone calls, "post-it" notes, face-to-face oral communication, and CB-radio chatter. Messages are almost always short. For example, among the eight messages waiting for us this morning, the average length was six lines. Second, they are point-blank, action oriented, asking for information and decisions or providing them with no letterlike politeness up front, no background information such as one might find in memoranda or reports, and certainly no up-front summary of their contents. Third, rapidly written, they tend to be surprisingly careless in language choices and are frequently riddled with errors and telegraphic sentences. No one, it seems, really edits electronic mail. Fourth, they are often humorous and indiscreet. People will say things in E-mail that they would never write in a letter or memo.

For busy professionals, E-mail messages are a wonderful tool for taking care of business rapidly. Pull up the message, read it, respond, and move on to the next. Yet, as businesses shift from paper-based and conversation-based communication to E-mail-based communication, significant problems are appearing. For example, where is the record of that communication? Many users keep no electronic files at all, some keep disorganized files of messages, and almost none of them prints and files hard copy records of E-mail sent and received. This means that important decisions, actions, and information are communicated essentially without record, or at least with chaotic and incomplete records.

We are also seeing problems stemming from the assumption that a message sent is a message received. Not all users of E-mail sign on regularly, yet people sending messages to them assume that if a message has been sent, it has been received, read, understood, and acted upon. However, many users have come to dread signing on to their computers because they know what a pile of work waits for them. For example, in one study eighteen members of the staff of a research laboratory were interviewed about their use of E-mail.[15] The interviews revealed that they received between 12 and 50 messages per day per person, and they sent 1 to 10 messages per day. Among senior personnel, the number increased significantly: one senior administrator received 75 messages per day and sent 30. This same study found that some employees saved as few as 10 messages. Others saved everything and had files with as many as 1350 messages. Some cleared out files regularly; others not at all. Some organized the messages into orderly files; others kept a jumble of messages. Some recipients (two of the eighteen) limited their reading of messages to once a day during a specified time. Six read their messages two to three times per day. And the rest read E-mail as soon as it arrived, treating it in the same way they might treat an interrupting telephone call.

In response to this barrage of E-mail, some of those interviewed felt overwhelmed with as few as 20 messages per day. Others were able to keep on top of as many as 75 messages per day, and still others felt "on the edge" in their ability to keep up.

This study shows that we are still just learning how to use E-mail technology despite the fact that it has been around for several years and is widely used already. Users have not yet learned how to view the new communication mode: Is it oral communication? Or written communication? Do you treat it like a telephone call, or like a memorandum? Does it need to be kept, or can it be discarded? Does it need to be organized, or can it be left in a chronological clutter? Similarly, the study shows that although the current tools allow us to send and receive messages, they have not yet given us effective ways to deal with the basic problems of work management, time management, or archiving of old messages. Indeed, the central conclusion of the study is that designers of E-mail systems have a good deal of work yet to do.

From our perspective, E-mail may be a new communication mode, but it still carries with it the audience and purpose considerations with which we have dealt in this book. Granted, its document content, structure, and design differ from those of letters, memoranda, and reports. It is a communication system in which the organization chart has been completely replaced by the network of wires and electronic signals traveling within and between organizations. Yet, we can only emphasize that effective use of this new tool has given users and system designers alike a new challenge. Whatever else you may say about E-mail, it is here to stay, and we have to learn to use it.

❑ 16.5 ❑
Learning New Tools: Hypertext and Multimedia Documents

Written documents, although hierarchically structured, are capable of being read linearly, from top to bottom. Further, they contain words and static pictures of a variety of types. Yet two of the most exciting new developments on the horizon for writers, hypertext and multimedia documents, challenge even these basic characteristics of writing.

"Hypertext" is a term probably unfamiliar to many readers. It refers to a type of document structure that is already being used (especially in on-line computer documentation) and that will unquestionably be used more widely in the future. It is not a new idea, but until recently it was a concept that our tools couldn't handle.[16] Now it is possible, and it is being used.

The concept of hypertext can best be understood if you think of a stack of cards. Each card contains some module of information in any of the forms that can be presented upon a computer: words, photographs, drawings, computer animations, or sound. The "cards" in the stack can be of any size—a single sentence or visual image; or something as long as a chapter in a book, or the book itself. Like cards, the modules of information can be shuffled and viewed in any order, but unlike cards—which are tangible—the modules of information are not visible unless they are called for. In other words, the deck of cards exists electronically, always out of view, becoming evident to the user only as he or she searches for additional information, either by following predetermined "links" between cards, or by browsing.

On our campus, in the lobby of Student Union, you will find a computer setup with a hypertext "tour" of the university. Suppose you are hungry when you arrive, and you decide to look for a place to eat. You could go to the information desk and ask, but you could also call up a file called "restaurants," which contains a list of most restaurants in the city. In it you find restaurants arranged by nationality: Greek food, Japanese food, German food, and so on. Because you are only interested in, let's say, German food, you have no need to see any of the other types. So you choose a card relating to German restaurants. There you find a list of all of the German restaurants in the city, along with general descriptions and such things as telephone numbers and addresses. But you also find that the restaurants are grouped by price ranges, from inexpensive to expensive. Again, you can choose. If your wallet is fat, you might select only the expensive German restaurants and ignore all the others. If so, you would again get a list, but this time you would get more information about them, information such as descriptions of the decor or details about their location. Seeing one that looks inviting, you can call up a card that lists its full menu and prices as well as a map showing you how to get from where you are now to the restaurant.

In your tour, then, you have seen perhaps three or four "cards" of information that have allowed you to select a place to eat for the evening. Meanwhile, among the cards you have not needed to examine are cards relating to every aspect of life at the university: colleges, departments, faculty, students, parking lots, dormitories, recreational facilities, and so on. A huge library of cards exists, but you have had no need to see it, so you have not been bothered with it.

Contrast that with a guidebook that the university might provide. Again, it would have modules of information on the aspects of university life previously mentioned. Yet because a guidebook would be tangible, you would have to handle all of the information and to search through all of it to find that one German restaurant. Having done so, you certainly could not expect to find such detail as the full menu of the restaurant or the map showing you how to get there. If such details were included in a guidebook, it would have to be absurdly large. In addition, such a directory would be expensive to produce and impossible to keep current. For hypertext documents, however, size is no problem, and they can be updated quickly and inexpensively.

Hypertext, then, is a powerful information tool, perhaps the most powerful ever devised. Yet using it requires approaching writing in a distinctive and new way. It requires writing modularly so that text elements are self-sufficient and comprehensible. It requires establishing "links" that allow users to connect modules of different types in predictable ways. And it requires establishing cues to allow for purposeful searching even when predetermined links have not been established. From the user's perspective, such tools are spectacular aids to information searching. From the writer's perspective, however, these tools pose important new design considerations.[17]

Perhaps the most exciting challenge and opportunity of hypertext tools is that they allow the creation of multimedia documents as well as of static word-and-picture documents. That is, computers can present static words and pictures, but they can also produce sound, animation, and video. Given that capability, there is no reason that a hypertext "document" cannot consist of all of these modes of communication in combination.

In the case of pictures, personal computers are already well advanced. They can show color photographs as clearly as 35mm slides projected on a screen. Similarly, their capability for handling sound is impressive, and, as any computer game player knows, the animation possible on computers today is breathtaking.

In the case of video, the technology still has some distance to go, but already it is possible to put video windows onto the screen of a personal computer for as little as $400 worth of hardware and software. Shortly, it should be possible to insert video just as easily as we now insert photographs and drawings into printed text. With this capability, you will be able to create new types of information—words, photographs, animations, drawings, voices, and video clips—all of them presented with great effectiveness on an inexpensive computer. The photographs and video especially will be impressive because of the high resolution of a computer screen, roughly three times that of a TV set.

If you have not seen hypertext and multimedia presentations, such technology may sound exotic. But such technology is already here and being used by some writers. All we must do now is learn to use it. That is an enormous challenge, certainly, but it is also an exciting opportunity.

Conclusion

Our electronic tools have caused, and are still causing, an astonishingly rapid revolution in communication. In our own experience, that personal computer revolution has so far lasted just a little over 10 years. During that time we have gone from experimenting with writing on a primitive personal computer with 4 K of memory to everyday use of a personal computer with 140 megabytes of memory. In our first workshop on personal computers, we remember the instructor saying that "within 10 years, you will be writing on a PC with the power of a mainframe and price tag under $1300." We couldn't believe him at the time, and his price estimate has turned out to be optimistic, but he was almost right.

During that 10 years, we have had nine different machines with memories ranging from 16 K to 140 megabytes, and we have used countless software packages. The programs we are now using require more than two-hundred times the memory capability of our first machine just to load the program. And today's software is infinitely more powerful and flexible and easier to use than the cumbersome programs with which we began. Ten years ago, we had to print our documents on a pathetic dot-matrix printer; now we can send them to laser printers, page printers, or photo-typesetters both inside and outside our institution. Ten years ago, a library search meant a trip to the library; now it means a session at the keyboard. Ten years ago, if we wanted a picture included in a publication, we either duplicated it on a copy machine and pasted it in, or we created and pasted in a hand-drawn original; now we create the artwork on our screens or we scan it with optical scanners, inserting and manipulating it electronically. Ten years ago, we sent messages by campus mail or U.S. mail; now we send them by E-mail and FAX.

This revolution in hardware and software has been exciting, helpful, and sometimes frustrating for us, for our students, and for professionals in every part of our society. Yet despite all of the change that this electronic revolution has brought, it has not changed the essential act of writing. Good writing is still good thinking. Audience and purpose considerations are still the best guides for writers, and we believe that the document design process that we have presented in this book can still apply. Our advice is that you should use the new tools, but don't lose sight of the fact that, as powerful as they are, these tools still put the burden of every writing choice on you. They've given you new opportunities, but they still present you with the fundamental challenges writers have always faced.

NOTES

1. G. E. Hawisher, "Research in Computers and Writing: Findings and Implications." Paper presented at the Annual Meeting of the American Educational Research Association, New Orleans, Louisiana, April 1988. Hawisher's synopsis has been cited with approval and supported in further research by Michael M. Williamson and Penny Pence in "Word Processing and Student Writers," a chapter in *Computer Writing Environments: Theory, Research, and Design*, ed. Bruce K. Britton and Shawn M. Glynn (Hillsdale, NJ: Lawrence Erlbaum, 1989).
2. Shawn M. Glynn, Denise R. Oaks, Linda F. Mattocks, and Bruce K. Britton, "Computer Environments for Managing Writers' Think-

ing Processes," *Computer Writing Environments: Theory, Research, and Design*, ed. Britton et al., 12.

3. H. Burns, *Stimulating Rhetorical Invention in English Composition Through Computer-Assisted Instruction, Dissertation Abstracts International*, 40, 3734A, University Microfilms, 1979.
4. Williamson and Pence, 120.
5. James Hartley, "The Role of Colleagues and Text-Editing Programs in Improving Text," *IEEE Transactions in Professional Communication* 27 (1984): 42–44.
6. C. Haas, *Computers and the Writing Process: A Comparative Protocol Study*, Technical Report 34, Carnegie-Mellon U, Communications Design Center, 1987. Similar findings were reported by R. Kurth in "Using Word Processing to Enhance Revision Strategies During Student Writing Activities," *Educational Technology*, Jan. 1987, 13–19.
7. R. T. Kellogg, "Writing Method and Productivity of Science and Engineering Faculty," *Research in Higher Education* 25 (1986): 147–63.
8. David E. Kieras, "An Advanced Computerized Aid for the Writing of Comprehensible Technical Documents," *Computer Writing Environments: Theory, Research, and Design*, 165.
9. John B. Smith and Marcy Lansman, "A Cognitive Basis for a Computer Writing Environment," *Computer Writing Environments: Theory, Research, and Design*, 25–26.
10. L. Birdwell-Bowes, P. Johnson, and S. Brehe, "Composing and Computers: Case Studies of Experienced Writers," *Writing in Real Time: Modeling Production Processes*, ed. A. Matsuhashi (London: Longman,1987) 81–107.
11. Williamson and Pence, 115–17.
12. Williamson and Pence, 117.
13. We are indebted for our own understanding and use of these tools to our colleague, Maurita P. Holland, Head, Engineering Libraries, the University of Michigan. In our discussion here we draw heavily upon her lecture, "Using Computer Technology to Find Technical Information," delivered in the Engineering Summer Conference at the University of Michigan, July 20, 1988. For readers interested in further information about on-line searching, we strongly recommend her videotaped course, "Technical Information Search and Acquisition Techniques," available through the Association for Media-Based Continuing Education for Engineers, 225 North Avenue NW, Atlanta, GA 30332. The course, consisting of eight videotapes accompanied by a *Sourceguide*, both explains and illustrates on-line search techniques for professionals.
14. R. John Brockmann, "Desktop Publishing: A Critical Bibliography," *IEEE Transactions on Professional Communication* 31: 1 (Mar. 1988): 30–36.
15. Wendy E. Mackay, "More Than Just a Communication System: Diversity in the Use of Electronic Mail," *Proceedings of the Conference on Computer-Supported Cooperative Work*, September 26–28, 1988, Portland, Oregon, 344–53.
16. "The History of Hypertext," *Simply Stated: The Newsletter of the Document Design Center*, American Institutes for Research 79 (Oct. 1988): 1–4. The history of hypertext can be traced back to 1945, when Vannevar Bush proposed a system for managing information being developed for the government by as many as 6000 scientists. Later, in the 1960s, Ted Nelson coined the word "hypertext" to describe systems that he was developing for storing modules or files of information that, although created separately, could be accessed through a variety of paths by users who might wish to view modules in different orders and to pursue a topic in different degrees of detail.
17. For an interesting discussion of the challenges of modular writing, see Janet H. Walker, "The Role of Modularity in Document Authoring Systems," *Proceedings of the ACM Conference on Document Processing Systems*, Santa Fe, New Mexico, December 5-9, 1988, 2–8.

Acknowledgments

We are grateful for the permission of many professionals, organizations, and publications to use their letters and reports, sections of reports, and figures in this book.

The professionals include:

M. A. Bowen, Manager, Chrysler Corporation
Ruthan Brodsky, School Administrator
Susan Carruthers, Mechanical Engineer
Sandra L. Conrad, Computer Graphic Artist
Michele Craven, Electrical Engineer
Michael DeBoer, Sergeant, Department of Police
Lisa S. Ede, Ph.D., Professor of English, Oregon State University
Scott Edward Dewicki, Aerospace Engineer
June O. Ferrill, Ph.D., Business Writing Consultant
R. J. Fisher, Jr., President, Central Concrete Products
Michel Gabrielli, Lecturer in Business French, University of Michigan
Donald H. Gray, Ph.D., Professor of Civil Engineering, University of Michigan
Gerald S. Holyzsko, Chemical Engineer
Kelly Katt, Assistant City Manager/Economic Development Coordinator
Howard Klee, Jr., Ph.D., Project Director, AMOCO Corporation
Craig Leafgreen, Bank Chief Financial Officer and Senior Vice President
Andrea A. Lunsford, Ph.D., Professor of English and Vice-Chair for Rhetoric and Composition, Ohio State University
Mark B. Luther, R. A., Architect
Amy Lee Martin, Medical Technologist
Alice E. Moorhead, D.A., Director of Writing, English Department, Hamline University

Peter D. O'Connell, District Judge
Ben L. Ross, Test Engineer, Chrysler Corporation
Charles R. Schwenk, Ph.D., Professor of Management, Indiana University
Yoshiaki Shinoda, Professor of English, School of Commerce, Waseda University
Stanley Vitton, Ph.D., Assistant Professor of Civil Engineering, University of Alabama
Eugene Wegner, former Manager, Dow Chemical Company

The organizations include:

American National Standards Institute, for figures from *Scientific and Technical Reports*, ANSI Standard Z39.18-1987, copyright 1987, American National Standards Institute, New York, NY. (Permission for Figures 5.10 and 5.11.) Copies of this standard as well as *Preparation of Scientific Papers for Written and Oral Presentation*, ANSI Standard Z39.16-1979, and *Writing Abstracts*, ANSI Standard Z39.14-1979, may be purchased from the American National Standards Institute at 1430 Broadway, New York, N.Y. 10018.

Community Systems Foundation (CSF Ltd.). (Permission for Figures 4.13, 5.9, 5.12, 5.13, and 5.14.)

McNamee, Porter, and Seeley, Consulting Engineers. (Permission for Figures 3.6, 8.3, 11.9, and 13.5.)

National Shipbuilding Research Program and the University of Michigan Transportation Research Institute (UMTRI), for figures from *Writing Shipyard Reports*, NSRP 0290, UMTRI Marine Systems Division, Ann Arbor, MI, 1988, a product of Project DTMA 91-82-C-20022, funded by the U.S. Department of Transportation Maritime Administration and supported by the Society of Naval Architects and Marine Engineers Ship Production Committee Education and Training Panel (SP-9). (Permission for Figures 2.11, 4.2, 4.6, 4.12, and 5.8.)

NEC Technologies, Inc., for selections from *COMPREHENSIVE USER MANUAL, A Complete Reference for the ProSpeed 286*, NEC Home Electronics, Wood Dale, IL, 1989, First Edition. (Permission for Figures 8.1, 8.7, and 8.8.)

Random Games & Toys, for *Crossword Bingo*, Garrett J. Donner and Michael S. Steer, Random Games & Toys, Ann Arbor, MI, 1986. (Permission for Figure 8.6.)

Tec-Ed Technical Communication and Graphics Services, Inc. (Permission for Figures 8.1, 8.7, and 8.8.)

W. A. Kraft Corporation (Permission for Figures 4.3 and 14.2.)

Westinghouse Electric Corporation, for figures from *Danger, Warning, Caution: Product Safety Label Handbook*, Westinghouse Electric Corporation, 1981. (Permission for Figure 15.2.)

The publications include:

Michael Macdonald-Ross, "Graphics in Texts," *Review of Research in Education*, ed. L. S. Shulman, 5: 1977, Itasca, Ill.: Peacock, 49–85, American Educational Research Association, Washington, D.C. (Permission for Figure 11.12.)

J.C. Mathes, "Three Mile Island: The Management Communication Role," *Engineering Management International*, 3 (1986), 261–268, Elsevier Science Publishers, Physical Sciences and Engineering Division, Amsterdam, The Netherlands. (Permission for Figures 2.2 and 2.5.)

Charles R. Schwenk, "Conflict in Organizational Decision Making: An Exploratory Study of Its Effects in For-Profit and Not-For-Profit Organizations," *Management Science*, 36:4 (1990), 436–448, The Institute of Management Sciences, Providence, RI. (Permission for Figure 3.9.)

Dwight W. Stevenson, "Audience Analysis Across Cultures," *Journal of Technical Writing and Communication*, 13:4 (1983), 319–330, Baywood Publishing Company, Farmingdale, N.Y. (Permission for Figure 14.6.)

"The Attitude Gap," *Research News*, The University of Michigan, Division of Research Development and Administration, November–December 1987, 15–20, Regents of The University of Michigan, Ann Arbor, MI. (Permission for Figure 3.8.)

Index

M

N

O

P

R

S

T

U

V

W